SAMPLING AND MONITORING IN CROP PROTECTION

The Theoretical Basis for Developing Practical Decision Guides

M.R. BINNS
Eastern Cereal and Oilseed Research Centre
Agriculture and Agri-Food Canada
Ottawa, Canada

J.P. NYROP
Department of Entomology
Cornell University
New York State Agricultural Experiment Station, USA

and

W. VAN DER WERF
Laboratory of Theoretical Production Ecology
Department of Plant Sciences
Wageningen University
The Netherlands

CABI *Publishing*

CABI *Publishing* is a division of CAB *International*

CABI Publishing
CAB International
Wallingford
Oxon OX10 8DE
UK

Tel: +44 (0)1491 832111
Fax: +44 (0)1491 833508
E-mail: cabi@cabi.org
Web site: http://www.cabi.org

CABI Publishing
10 E 40th Street
Suite 3203
New York, NY 10016
USA

Tel: +1 212 481 7018
Fax: +1 212 686 7993
E-mail: cabi-nao@cabi.org

A catalogue record for this book is available from the British Library, London, UK

Library of Congress Cataloging-in-Publication Data

Binns, M. R. (Michael R.)
Sampling and monitoring in crop protection : the theoretical basis for developing practical decision guides / M.R. Binns, J.P. Nyrop, and W. van der Werf.
p. cm.
Includes bibliographical references (p.).
ISBN 0–85199–347–8 (alk. paper)
1. Agricultural pests--Control--Statistical methods. 2. Plants, Protection of. 3. Sampling (Statistics) I. Nyrop, J. P. (Jan P.) II. Werf, W. van der (Wopke) III. Title.

SB950.B46 2000
632′.6--dc21

99–053498

ISBN 0 85199 347 8

Typeset in Goudy by Columns Design Ltd, Reading.
Printed and bound in the UK by Biddles Ltd, Guildford and King's Lynn.

Contents

Preface		vii
Acknowledgements		xi
1	**Basic Concepts of Decision-making in Pest Management**	**1**
1.1	Introduction	1
1.2	Agriculture, pest management and the role of information	1
1.3	Injury, damage, loss and threshold concepts	3
1.4	The effect of uncertainty on the decision process: probabilities of management action	6
1.5	The sampling effort: the OC and ASN functions	8
1.6	On fuzziness and satisfaction	9
1.7	Monitoring a system through time	11
1.8	The formal design of methods for sampling and monitoring: Why and how?	12
1.9	On learning by doing and farmer empowerment	13
1.10	Summary	14
2	**Basic Concepts of Sampling for Pest Management**	**17**
2.1	Introduction	17
2.2	Randomness	17
2.3	Bias, precision and accuracy	19
2.4	Dealing with bias	22
2.5	Criteria for a trustworthy sample	23
2.6	Spatially explicit data versus data lists	24
2.7	Simulation of the sampling process	25
2.8	The variance of sample means	27
2.9	The distribution of sample means – the Central Limit Theorem	29
2.10	What does this mean for decision-making?	33
2.11	Calculating a probability of decision (OC) function	34
2.12	Summary	39

3	**Classifying Pest Density**	**41**
3.1	Introduction	41
3.2	Classification versus estimation	41
	3.2.1 Estimation	42
	3.2.1.1 Choosing the sample size, n, to achieve specified precision	44
	3.2.2 Classification	45
3.3	Reducing sampling costs: classification by sampling in batches	46
	3.3.1 Two batches	46
	3.3.2 More than two batches: batch sequential sampling plans	52
3.4	Variance–mean relationships	53
3.5	Summary	60
4	**Distributions**	**61**
4.1	Introduction	61
4.2	Spatial pattern, frequency distribution and probability distribution	61
4.3	Random patterns and the Poisson probability distribution	63
4.4	Fitting a distribution to data using the maximum likelihood principle	66
4.5	Testing the goodness of fit of a probability distribution	67
4.6	Aggregated spatial patterns and the negative binomial distribution	71
	4.6.1 Estimating the negative binomial k from a variance–mean relationship	77
4.7	The effect of sample size on fitting distributions	78
4.8	The mechanics of fitting distributions to data	80
4.9	The binomial distribution	82
4.10	The beta-binomial distribution	83
4.11	Relationships among the distributions	89
4.12	On the choice of a probability distribution model	91
4.13	Summary	91
Appendix	Computer generation of aggregated spatial patterns	94
5	**Sequential Sampling for Classification**	**97**
5.1	Introduction	97
5.2	From batch to sequential sampling	97
5.3	Iwao's sequential procedure	98
5.4	Wald's sequential probability ratio test (SPRT)	108
5.5	Converging lines	116
5.6	Green's stop boundary for estimation	120
5.7	Which stop boundary?	122
5.8	Summary	123
Appendix	Calculating aggregation parameters for use in the simulations	127
6	**Enhancing and Evaluating the Usefulness of Sampling Plans**	**131**
6.1	Introduction	131
6.2	The cyclical process of learning, improvement and design	132

	6.2.1 Proposing or modifying a sampling method (phase 1)	133
	6.2.2 Determining and evaluating the performance of the proposed method (phase 2)	133
	6.2.3 Implementing the decision guide (phase 3)	133
6.3	Design ingredients of sampling plans that affect usefulness	134
	6.3.1 The size of the management unit	134
	6.3.2 The definition of the sample unit	135
	6.3.3 When and what to sample	136
	6.3.4 The sample path in the crop	137
	6.3.5 Procedures for recording data	137
6.4	Evaluation by users, field tests and simulation	138
	6.4.1 Field evaluation	139
	6.4.2 Simulation, sensitivity and robustness	140
	6.4.2.1 Critical density	140
	6.4.2.2 Sensitivity with respect to sampling distribution	140
	6.4.2.3 Size and representativeness of sample	141
6.5	The economic value of sample information	141
	6.5.1 Best control strategy with no sampling	142
	6.5.2 Calculating the value of sample information	143
	6.5.3 The value of sample information in practice	144
6.6	Summary	153
7	**Binomial Counts**	**155**
7.1	Introduction	155
7.2	What is binomial sampling?	156
7.3	Presence–absence binomial sampling with the Poisson distribution	156
7.4	Tally numbers other than zero	158
	7.4.1 The tally number and precision	162
7.5	Binomial sampling based on the negative binomial distribution	165
7.6	Incorporating a variance–mean relationship	170
7.7	Binomial sampling using an empirical model	173
7.8	Estimation	177
7.9	Summary	178
Appendix	Variance and bias of binomial count estimates of the mean	180
8	**Multiple Sources of Variation**	**183**
8.1	Introduction	183
8.2	Nested or multistage sampling	184
8.3	Analysis of variance and two-stage sampling	185
	8.3.1 TPL and the application of ANOVA in decision-making	187
	8.3.2 Use of the ANOVA to test for correlation within *psu*	187
	8.3.3 Use of the ANOVA to estimate optimum values for n_s	188
8.4	Patchy environments	191
8.5	Predictable patches	192
8.6	Unpredictable patches – VIS	192
	8.6.1 A VIS look-up chart	198

8.6.2 OC and ASN functions for VIS plans 201
8.7 Summary 202
Appendix The variance of the sample mean in two-stage sampling 204

9 Resampling to Evaluate the Properties of Sampling Plans 205
9.1 Introduction 205
9.2 Principles of resampling 206
9.3 Typical properties of OC and ASN functions estimated by resampling 208
9.4 Collecting data for resampling: the number and size of the basic samples 212
9.5 Sequential sampling 216
9.6 Binomial sampling 217
9.7 A comparison between resampling and model-based methods 221
9.8 When is resampling a better method to evaluate sampling plans? 223
9.9 Summary 224
Appendix Confidence intervals for negative binomial *k* estimated by TPL 226

10 Sampling over Time to Classify or Estimate a Population Growth Curve 227
10.1 Introduction 227
10.2 Time-sequential classification of a population trajectory 228
10.3 Time-sequential estimation of a population growth curve 236
10.3.1 Bayesian and regression approaches 236
10.3.2 Implementation of the regression approach 238
10.4 A comparison between the methods and other considerations 242
10.5 Summary 243

11 Monitoring Pest Populations through Time 245
11.1 Introduction 245
11.2 Terminology of monitoring 246
11.3 Tripartite classification sampling plans 246
11.4 Principles for quantitative evaluation of monitoring protocols 249
11.4.1 Direct calculation based on PC and ASN functions and the trajectory 250
11.4.2 Calculation based on simulation and the trajectory 251
11.4.3 Principles for designing monitoring protocols 252
11.5 Population trajectories for use in evaluating monitoring protocols 253
11.6 Evaluation of monitoring protocols based on cascaded tripartite classification 255
11.7 Further considerations 263
11.8 Summary 264

12 Epilogue 267

Glossary 273

Index 281

Preface

The Purpose of the Book

Efficient field sampling is a cornerstone of pest management: having some idea of the status of pests provides growers and consultants with the necessary basis to choose between management options. Many books have been written about pest management, and some include extensive discussion of sampling methods (see, e.g. Kogan and Herzog, 1980; Pedigo and Buntin, 1994). Most of these books are collections of chapters written by individual authors and, although they may be of high quality and cover a good deal of ground, they lack the unified structure which is required to make them suitable for learning or teaching. This book aims to provide a unified approach to sampling for decision-making in pest management, that can be used as an introduction to the subject. Moreover, the book is accompanied by a set of modern computer-design tools that allow users to design and evaluate their own sampling programs.

The Intended Audience

The book is aimed primarily at graduate or final-year undergraduate students who are specializing in pest management. Those working in pest management, including extension specialists, consultants and researchers, should find the book helpful as a reference, as well as a tool that they can use in their professional endeavours.

Scope

This book covers the statistical concepts of sampling in agricultural pest management. These concepts can be summarized quite simply as: how best to get sample data from the field and how to use the data in decision-making. The focus is on collecting samples and analysing the data, with a view to choosing which of two or more pest management actions is most appropriate to use. Except where specifically noted, we always assume that one of these actions is to do nothing. Others may include introducing or re-introducing natural enemies, applying a pesticide, or adopting a wait-and-see approach by scheduling another sample to be taken after an interval of time. Wider problems of sampling for estimation in ecological or modelling studies are not covered, except where necessary for the main purpose.

Prerequisites

We have tried to keep mathematical and statistical requirements as minimal as we could. We have introduced only as much statistical theory as is necessary to understand why certain procedures are recommended for sampling biological organisms, and also to give the reader tools to infer (from sample data) properties of biological populations which are relevant for pest management. The text should be understandable to anyone with introductory college mathematics: the ability to follow and understand simple developments of mathematical formulae is required in some parts of the book, but not (we hope) essential to understanding the concepts. The ability to work with computer software is probably necessary, even if only to understand how results might be obtained. Of course, we expect a general understanding of the context in which sampling for pest management is done, in particular a basic knowledge of agricultural pests and how they interact with agricultural crops.

The Text

The book starts with two introductory chapters on general principles of decision-making, decision guides and sampling for pest management. Chapter 3 explains the difference between sampling for classification and estimation, and develops a rationale for sequential sampling by comparing the performance of sampling plans that use a fixed sample size and those that use batches, whereby sampling is terminated as soon as enough sampling evidence is collected to make a decision. Chapter 4 presents four probability distributions, one of which is relatively new to crop protection, that can characterize frequency distributions resulting from sampling, and which play a central role in the simulations of sampling performance in subsequent chapters. Chapter 5 presents three alternative designs for developing sequential sampling plans for classification, one of which is completely new to crop protection. Chapter 6 discusses how the techniques described in previous chapters fit into practical pest management, and lists criteria and a quantitative methodology for evaluating the usefulness of sampling information in crop protection. Chapter 7 looks at sampling plans that are based on the notion that pests are not enumerated on sample units (as in Chapter 5), but that the sample units are classified as either *positive* or *negative* with respect to some classification criterion concerning pest density. This is called binomial count sampling. Chapter 8 explores variability in pest counts arising from the aggregated distribution of pests and a nested structure of sample collection. A procedure known as variable intensity sampling is described, which guarantees that the field is adequately covered by presetting the number of locations at which sample units are taken, but which takes an adaptive number of sample units at each location such that, when all sample units have been collected, a reliable decision may be reached. Chapter 9 explains (bootstrap) resampling as a method of evaluating pest management decision guides, when none of the probability distributions discussed in Chapter 4 (or any other published in the literature) adequately characterizes the sample data. Chapter 10 looks at methods for classifying or estimating population growth curves, or trajectories, by sampling over time. The first is an extension of a sequential classification plan (introduced in Chapter 5) to the time domain, and the second uses sample data collected over time continually to re-estimate a growth curve model. Chapter 11 deals with monitoring populations over time: how and when to take the next sample, based on current sample data. This is especially relevant for pests with multiple generations per season that need to be kept in check for an extended period of time. In these two chapters (10 and 11) there is a jump to a higher level of complexity, because time comes in as a factor that affects sampling. An epilogue rounds out the entire book, explaining how we think the principles described can be used to improve the everyday practice of growing crops and managing pests. A glossary is added at the end as a *vade mecum* for those who do not feel on friendly terms with the statistical notation and ideas used in the book.

Electronic Book Chapters

Throughout the book we have included worked examples (exhibits) to illustrate key concepts and applications. These are a vital part of the book. The calculations involved have been programmed in the mathematical software program MathCad™, and are available as electronic chapters on the Internet (www.nysaes.cornell.edu/ent/faculty/nyrop/cpdm). The electronic chapters are interactive mathematical and simulation tools that allow you to check our examples, and explore your own sampling problems and ideas. We have found that there are so many avenues which are worth exploring with each worksheet that we have decided to leave it to you, the reader, to follow where your fancy dictates. The sort of thing that can be done with these electronic book chapters includes the following:

- investigating the effect on decision-making of decreasing or increasing the sample size
- comparing the properties of sequential and fixed sample size plans
- investigating the difference in accuracy and effort between different types of sequential plans
- comparing binomial and full count sampling plans
- generating aggregated distributions and comparing different sampling strategies on them
- investigating control by sampling over time

amongst many other topics.

Not much expertise is required to use the MathCad™ computer tools. Beyond fearlessness in the face of something new, all that is required is the ability to change numbers (input parameters) on a screen and a little patience to wait, while simulations are running, for results to appear on your computer screen. A MathCad™ Explorer that will allow you to use our electronic tools, is provided free of charge on the Internet (http://www.mathsoft.com). Although we have tried to make the software as error-free as possible, we accept no responsibility for any errors or misunderstandings arising from its use.

Captatio benevoluntatis

We started the enterprise of producing this book and software because we believed that there is a need for an accessible introduction to sampling and decision-making in pest management, as well as a need for easy to use interactive tools to see 'living math' as related to sampling and decision-making on the screen of the computer. We hope that it will provide insights and useful applications to you and the people that you work with.

References

Kogan, M. and Herzog, D.C. (1980) *Sampling Methods in Soybean Entomology*. Springer-Verlag, New York.

Pedigo, L.P. and Buntin, G.D. (1994) *Handbook of Sampling Methods for Arthropods in Agriculture*. CRC Press, Boca Raton, Florida, 714 pp.

Acknowledgements

We would like to thank all the institutions and agencies that have given their trust and support to the three of us, during our collaborative adventures in sampling. First of all, we want to thank our home institutions, department chairs and colleagues for making our mutual visits and collaborations feasible. Thus, M.R. Binns thanks the Eastern Cereal and Oilseed Research Centre of Agriculture and Agri-Food Canada for their hospitality, J.P. Nyrop thanks the New York State Agricultural Experiment Station at Cornell University and W. van der Werf thanks the Department of Theoretical Production Ecology at Wageningen University and its chairpersons for the stimulus and opportunity that they provided.

Several institutions provided financial support for our work together. These include the United States Department of Agriculture, which funded a small intensive workshop in The Netherlands in 1989 (with J.P.N. and W.vdW.) on pest sampling, the Rockefeller Foundation, which provided a Fulbright Fellowship to W.vdW. for spending a sabbatical with J.P.N. in 1992, the North Atlantic Treaty Organization, which provided travel money to the three of us during 1993–1995, and the Wageningen University (WU), which offered a Visiting Scientist Grant to J.P.N. in 1994. The C.T. de Wit Research School for Production Ecology (part of WU) specifically provided Visiting Scientist Grants to both M.R.B. and J.P.N. in 1996 and 1997 for the purpose of writing this book. The job was bigger than we thought, and more money was needed to complete it. The Dutch Technology Foundation STW and Wageningen University funded additional stays of W.vdW. with J.P.N. and M.R.B. to work on finalizing the book and software during 1997 to 1999. All of this financial support was instrumental in getting the job done, because it required face-to-face discussion, brainstorming and mutual encouragement. Several colleagues provided useful comments on our writing. We would like to thank especially Mike Hardman for his encouragement and suggestions for improvement and clarification of the text. Last, but not least, we thank our spouses for their loving support during the whole of this project.

Basic Concepts of Decision-making in Pest Management 1

1.1 Introduction

Knowledge and information related to all aspects of the crop–pest system are the foundations on which modern pest management decisions are based. In this introductory chapter we categorize these foundations, and discuss their relevance to pest management. We show how pest injury, crop damage and crop loss are related to concepts of acceptable thresholds for pest density in a crop. Sampling pest numbers in the crop is a direct way of comparing the actual density in a field with the acceptable threshold, and thus recommending a management decision. Sampling thus plays a central role in pest management. We introduce principles and criteria whereby different sampling strategies can be compared with each other. We discuss the formal process for choosing test sampling plans and comparing them with each other before recommending a plan for practical use.

1.2 Agriculture, Pest Management and the Role of Information

When humans settled in communities and started farming agricultural crops, many thousands of years ago, they introduced a new element into the existing state of nature: an organized attempt to fill land areas with selected plants, and to exclude unwanted plants and animals. The attempts were successful, supporting larger communities and paving the way for civilization as we know it today. Nowadays, farming has become highly productive in many parts of the world, with large areas devoted to monoculture, or some other kind of organized planting scheme. It is not surprising that organisms other than humans try to cash in on the bounty laid out before them. In particular, what we humans call *pests* try to share the results of our labour, whether they be arthropods, vertebrates, other plants or pathogens.

Present-day growers are well equipped to protect their crops from excessive damage, and they can draw upon a large body of knowledge and management options, both preventative and curative. We have extensive knowledge about the ecology of pests and their natural enemies in cropping systems. We have ever-growing knowledge about the genetics of plant resistance, and powerful techniques for building resistance into plant material. We have a long experience in the use of cropping plans and cultural practices for deterring pests. There is a wide choice of effective chemical pesticides – which should, however, be used judiciously to minimize deleterious side-effects. There is the challenge, therefore, of combining the available management options in the most appropriate way into the practical business of day-to-day farming. How can we identify strategies that integrate the available techniques in such a manner that crop losses are prevented, or reduced to acceptable levels, with minimal harmful side-effects?

Questions on how to design systems for integrated pest management can be formulated at different temporal and spatial scales. The management decisions that growers make for their crops in any field and at any time during the season are embedded in the larger framework of what their plans are for their entire farm operations over a time frame of 1, 2 or many years. Furthermore, what each grower decides at the farm level is done in the context of the larger plans and desires of farming communities and of society as a whole. Policy decisions taken by any level of government have a profound impact on the behaviour of individual growers and of grower groups. This book focuses attention at the field level: the crop protection decisions that growers should make in their fields at any moment during a growing season, given the local situation and the expected (and/or possible) future course of events in these fields and on the market. Such a decision depends upon the growers' objectives and on the peculiarities of the production system.

The knowledge and information affecting decisions about a production system at the field level come from a variety of sources, such as scientific research, economic and social studies, field sampling, community discussions and personal opinions. It is useful to categorize this knowledge and information:

1. *Process knowledge* – knowledge about the dynamic behaviour of the component processes in the system, and how they interact, in relation to the natural environment. This is generally valid biological knowledge about how the system 'works'. It includes: how rapidly pest populations grow (or decline) and when certain development stages occur in the season; how pests affect plant growth and crop production; how natural enemies affect pest populations; how abiotic factors drive population processes; and how pest organisms are distributed in the crop. Such knowledge is required to make sensible assessments of how pests affect crop production, and may also be used to devise system structures for avoiding pest problems.
2. *Socio-economic information and management objectives* – knowledge about the socio-economic framework in which a pest manager makes pest control decisions; personal and public perceptions as well as normative judgements relating to the crop and its management. This includes information on the value of the crop now, and as it might be at harvest, costs of any management action taken to realign the system, how individual growers deal with uncertainty about a pest's impact on final yield, and the acceptability of certain management actions (e.g. pesticide sprays).
3. *Knowledge of state* – knowledge about the current state of the system in the field. Information on the current state of the system allows a pest manager to specify which aspects of the general knowledge framework are relevant at a certain time

and place. An essential component is some measure of the abundance of pests and/or natural enemies. Other important components include the state and growth stage of the crop, and the current and future weather. This information is collected specifically to help decide on a course of action. The action might be: using a pesticide or biological control agent; using a cultural technique; or possibly even doing nothing. Without information on the current state of the system, a grower must resort to some kind of automated decision process, such as calendar spraying. In certain instances, a regional estimate or forecast provides enough information on the current state of the system for decision-making, especially for very mobile pests (e.g. potato blight), but in general a measure of abundance for each field is required. We shall refer to the process of collecting data to estimate a measure of abundance in the field as *sampling*. From this point onwards, we shall often refer to the field, orchard or area for which a decision sample is taken as a *management unit*. The objective of this book is to describe how best to collect and use data from a management unit for the specific purpose of deciding on a course of action against a pest.

1.3 Injury, Damage, Loss and Threshold Concepts

The presence of pests in a crop may result in physiological *injury* to the crop (e.g. change in leaf area or photosynthesis rate). Injury may result in *damage* to the crop (i.e. a reduction in the amount or quality of harvestable product). The final effect on financial revenue from the crop is called *loss*. Injury is expressed in crop physiological terms, damage in units that express the amount or quality of the product, and loss in monetary units. It is often convenient to combine injury due to pests and damage consequent on injury, referring simply to damage caused by pests; we shall follow this course.

Pest abundance is generally estimated in terms of average number (or density) of pests contained in some specified unit of the crop, such as a leaf or stem of a plant. Anticipating Chapter 2, we refer to this unit as a sample unit. By 'pests' we mean the number of aphids, borers, pustules, weeds and so on per sample unit, not the number of pest species. Suppose that we have estimated the amount of damage (reduction of harvestable yield in kg) caused by a range of pest densities in the absence of any control measures. Suppose also that we know the monetary value of each kg of harvestable crop. We can then determine the level of pest density at which the reduction in revenue from the crop, due to pest injury, equals the cost of controlling the pest. This level of pest abundance is called the *Economic Injury Level* (*EIL*). It is calculated from the equation

$$C = EIL \times K \times D \times V \tag{1.1}$$

where C is the cost of control (\$ ha^{-1}), *EIL* is the pest density at which control is precisely worth the effort, *K* is the proportion of the pest population killed by control action, *D* is the proportion of crop damaged per unit pest density and *V* is the value of the harvested product (\$ ha^{-1}).

This simple derivation of the *EIL* is based on the assumption that the relationship between pest density and damage is linear (Fig. 1.1). Much the same derivation can be made for curvilinear relationships. The total costs due to the presence of a pest are the sum of the costs of control and of crop loss. Much has been written about the concept of economic injury level (Higley and Pedigo, 1996). A few points to keep in mind are as follows:

1. The concept of the *EIL* is focused on an economic evaluation. Other criteria that a grower may find important when deciding on an intervention are ignored. Such criteria include:

- the time available to the grower to implement a control treatment
- the possibility of combining interventions against multiple pests into one management action (lower costs)
- a desire to minimize pesticide use
- a desire to conserve natural enemies

The *EIL* concept is therefore narrow in scope. Growers are likely to consider additional criteria beyond the *EIL*.
2. The *EIL* deals with preventable damage. If damage is no longer preventable, because the injury is irreversible or the plant cannot compensate for damage already done, or if there are no curative treatments (chemical, cultural or biological), the *EIL* paradigm is irrelevant.
3. Using a threshold density to decide on a pest management action implies that there is variability in pest density among management units (otherwise, there would be no need to look), and that this variability is relevant to management.

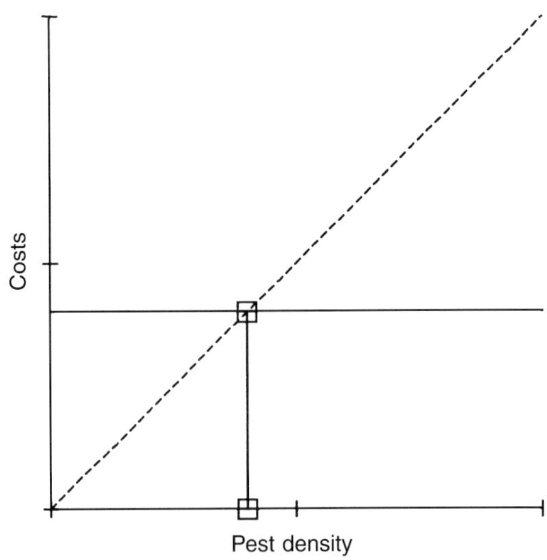

Fig. 1.1. A pictorial description of the derivation of *EIL*. Preventable crop loss (- - -), defined as KDV × pest density, intersects the cost of control (——) at the *EIL* (□).

This is not the case for many diseases, whose control follows a preventative rather than a curative approach. For instance, the control of *Phytophthora infestans* (late blight) in potatoes and *Venturia inequalis* (scab) in apples is geared towards preventing infection. Warning systems are therefore primarily based on weather conditions, and do not take into account actual infection levels within the crop.

4. The *EIL* paradigm is a theoretical framework, that assumes perfect knowledge about the quantities involved. In reality, we do not have that knowledge, and decisions have to be made under uncertainty. Uncertainty affects thresholds.

5. The *EIL* concept assumes that there is a clearly defined instant in time when the desirability of a control action should be evaluated. In practice, this can be too restrictive. Many pest problems evolve over time, and decisions may be required at different times for the same pest. The *EIL* equation (Equation 1.1) is not equipped to deal with such uncertainty.

6. The *EIL* concept may cause an undesirable locking-in to a reactive mode of pest control, which fosters pesticide dependence and is inappropriate for agriculture in the present age. Crop production systems should be designed and developed to exploit the ability of natural systems to prevent serious pest problems. Because such mechanisms may occasionally fail, an efficient technology is needed to make informed decisions about the necessity of corrective action. This book is about the design of such technology.

Despite these important shortcomings, the *EIL* concept provides a useful guideline in decision-making for crop protection. For the remainder of this chapter (and for the rest of the book) we shall assume that some kind of threshold, or decision criterion, is appropriate for management actions. In determining such thresholds, management objectives other than profit may have been taken into account.

In the literature, a useful distinction is made between the *EIL*, the break-even density at which monetary costs and benefits of control are equal, and a lower pest density (the *economic threshold*, *ET*), which is actually proposed as the decision criterion for the initiation of corrective intervention. *ET* is lower than *EIL*, because between the time when pest density is estimated and the time when a control treatment can be applied, pest numbers may increase beyond the break-even *EIL* value, or even change to such a degree that the proposed corrective action may not be ideal. In this book, we often use the term *critical density* (*cd*) for the density to be used as a decision criterion. We use this more neutral term, unburdened by historical connotations, for several reasons:

1. When uncertainties in process knowledge are taken into account, it may be necessary to use lower thresholds than suggested by either *EIL* or *ET* reasoning to avoid small probabilities of large damage.

2. Even if the evaluation is in purely economic terms and there is perfect knowledge about processes and the value of the crop, it may be desirable to choose *cd* unequal to *ET*. One reason is that it is impossible to know the pest density in the field precisely. Estimation errors have to be accounted for. If erroneous decisions not to intervene are costly, because of excessive pest damage, it may be best to have *cd* less than *ET*, both to reduce the risk of excessive events occurring and to minimize the expected total costs. We say more about this in Chapter 6.

3. Many more criteria can be used to define *cd*. A more complete and ecologically sound evaluation of the pros and cons may be made than is the case for a purely economic threshold.
4. There may be multiple critical densities. For example, if there are three decision choices, two critical densities are required: (i) high density (greater than cd_2) – intervene immediately; (ii) intermediate density (between cd_1 and cd_2) – check again after a week; (iii) low density (less than cd_1) – no action necessary.

Use of the term *critical density* emphasizes that this density is used as a 'critical' cut-off point, when making a decision. It is no more and no less than that. Nothing is implied about the background of its quantification.

1.4 The Effect of Uncertainty on the Decision Process: Probabilities of Management Action

Let us assume that we have perfect knowledge about our crop production system, allowing us to quantify the economic threshold precisely. Suppose that $ET = 5$ pests per sample unit. Then, what should our decision be? The answer is obvious (Fig. 1.2a): if pest density is 5 or below, the decision should be not to intervene (to save time and costs and avoid side-effects), and if it is above 5, the decision should be to intervene (to avoid losses greater than the costs of control). However, the underlying assumptions are not realistic:

1. We never have perfect knowledge about the economic threshold, because our process knowledge is imperfect; there may be uncertainty about prices, and there may be ambivalence about management objectives.
2. We are never able to know the pest density in the field with absolute certainty unless we go out and count the whole population.

Let us first consider fuzziness in socio-economic information and management objectives. How does it affect the decision? There are different approaches to answering this question and we come back to it in Chapters 6 and 12, where we discuss performance indicators. Fuzzy socio-economic information includes – for instance – a variable price that is unknown at the moment of decision-making. It also includes uncertainty about the damage for a certain level of pest injury; for instance, in relation to future weather. A high price would shift the *EIL* down, and – conversely – a low price would shift it up. The *EIL* for a crop in one field will not be the same as the *EIL* for another field, due to differences in productivity or the pest density–damage relationship. One might argue that because the 'real' *EIL* is unknown, gradually sloping curves as in Fig. 1.2b would be acceptable. The range of densities over which these curves change from one to zero (or 0 to 1) would then reflect the expected range of uncertainty about crop loss. However, if we can characterize the expected distribution of crop loss for each pest density, then we could in theory still calculate a precise *cd*, below which – averaged over all prices and damages, and weighted for the likelihood of crop loss – it is best never to intervene, and above which it is best always to intervene, as in Fig. 1.2a. We shall explore this in Chapter 6.

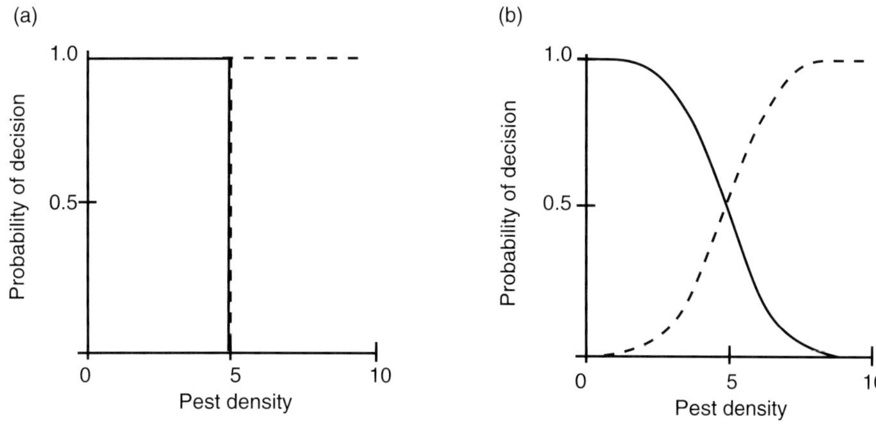

Fig. 1.2. Probability of decision curves: - - -, intervention; ——, no intervention. (a) Assuming quantifiable uncertainty about the pest–loss relationship leading to a well defined threshold, and perfect knowledge about pest density; (b), as (a), but with information about pest density obtained from sample data.

The strategy described in the previous paragraph is best 'on average', but it will certainly be incorrect in specific instances. It is therefore equally defendable to argue that over a limited range of pest densities it is unimportant which decision is taken, because the advantages of either decision are more or less in balance with the disadvantages. For instance, intervening has the advantage that the pest issue may be 'put to bed' for some time; it has been taken care of and the grower has peace of mind. On the other hand, not intervening saves time, money and resources. The step curves (Fig. 1.2a) are for those people who are prepared to integrate all uncertainty in the calculation of the critical density. The smooth curves (Fig. 1.2b) are for those people who think that it is not of importance to balance all advantages and disadvantages of a certain decision in detail, and are therefore prepared to accept either decision within a certain area of indifference for pest density. In both cases, there is a critical density, but only in the first case do the curves step at precisely this value. In the other case *cd* is near the halfway point of either curve.

In establishing the critical density, one may be less interested in minimizing the average total costs than in minimizing the chance of particularly nasty events; for example, (almost) total crop failure. This strategy might be called *risk-sensitive*, while the strategy that minimizes average total costs might be called *risk-neutral*. Note that the pure *EIL* strategy is risk-neutral. Depending upon the approach that is taken, different values for *cd* can be determined. Risk-sensitive thresholds are lower than risk-neutral thresholds. Thresholds that minimize average total costs are generally lower than the *EIL*, although this result is not intuitively obvious (Rossing *et al.*, 1994a,b).

The difficulty of obtaining precise knowledge of the state of the crop–pest system will always result in smooth probability of decision curves, as in Fig. 1.2b. Obtaining precise knowledge of state is costly in practice, because it requires extensive sampling. We must be content with a level of information that is sufficient to provide a basis for making a reasonable decision. With less than perfect information, we may find ourselves inadvertently doing the wrong thing (Fig. 1.3): at densities above *cd* there will be some instances in which the decision is *not* to control, and vice versa at densities below *cd*. Hence, the probability of decision curve, as a function of density, will not 'step' at *cd*, but will change in a smooth fashion from 1 at low densities to 0 at high densities (or from 0 to 1), with the halfway point near *cd*. The fewer sample units we take, the flatter these functions will be. To make them close to a step function, we must collect very many sample units. Because this is costly, and because the evaluation of multi-dimensional pest management decision problems may be fuzzy anyway, probability of decision functions for practical sampling plans tend to be smooth.

1.5 The Sampling Effort: the OC and ASN Functions

The information collected in a field, orchard or other management unit, and used to help assess the state of the system, is called a sample. Each sample consists of data recorded on a number of sample units, the smallest entities for which information on pest abundance is independently assessed. We shall show in later chapters that the number of sample units to be collected for a particular field is not necessarily known in advance. The rules for collecting sample units often determine the number of units collected: the number of sample units actually collected depends on the average number of pests per sample unit and on the *cd*. Steeper probability of decision curves usually require more sampling effort. This is a cost that must be considered. Each sampling plan therefore has not only its own probability of decision curves, but also an average number of sample units curve. This is illustrated by an example based on a sampling plan developed in Chapter 5 (Fig. 1.4).

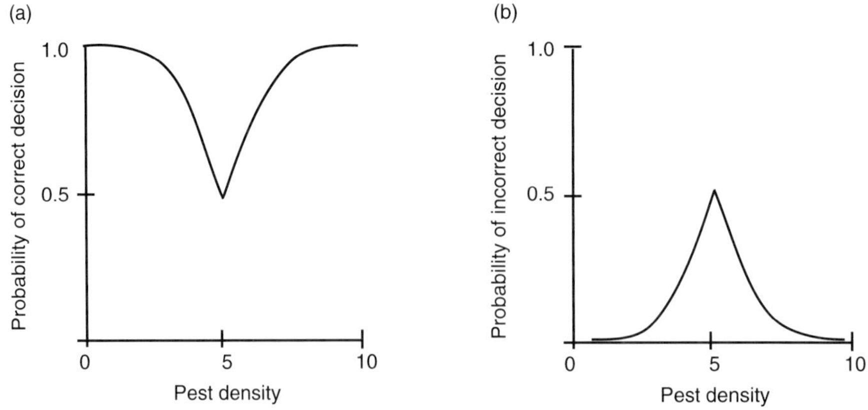

Fig. 1.3. Alternative presentations for Fig. 1.2b. (a) Probability of correct decision. (b) Probability of incorrect decision.

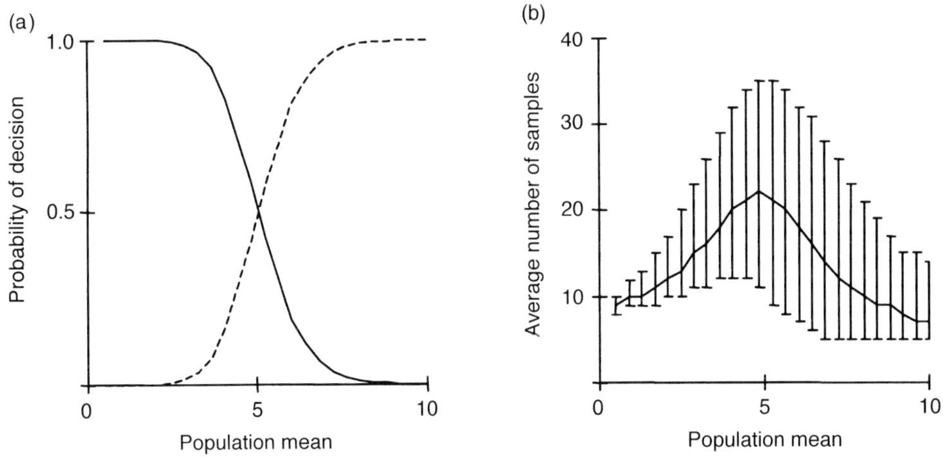

Fig. 1.4. Probability of decision (a) and average number of sample units (b) curves, estimated by simulation, for a sampling plan based on one of the methods of Chapter 5; minimum (5) and maximum (50) numbers of sample units were specified in the plan. The OC function is the probability of deciding not to implement a management action (——), the probability of deciding to implement a management action is the complement of the OC function (- - -). The ASN function is the continuous curve in b; the intervals in b represent the 10th and 90th percentiles of the distribution of numbers of sample units in the simulations. The 10th percentile, for example, is defined as the number below which lies 10% of the distribution.

At this point, we need to accept that sampling for making decisions is not a new concept, and that certain technical terms have been used for many years now in, for example, industrial quality control. Whether we like it or not, the use of already accepted technical terms is necessary to avoid confusion when we want to use results obtained outside agricultural pest management or when we want to understand people working in other areas of decision-making.

The term 'operating characteristic function', often shortened to OC function, is widely used to refer to one of what we have called 'probability of decision curves'. Specifically, the OC function is equal to the 'probability of not implementing a management action against the pest'; for example, the curve in Fig. 1.4a, which starts at 1 for low pest density and decreases to 0 for high pest density is an OC function. Similarly, the term 'average sample number function', often shortened to ASN function, is widely used to refer to the 'average number of sample units curve'; for example, the curve in Fig. 1.4b is an ASN function. Henceforth, we use the terms OC function and ASN function.

1.6 On Fuzziness and Satisfaction

The concept of OC functions can be presented as a simple table (Table 1.1). What we require of a sampling plan is to give an 'intervene' decision if the density is very

Table 1.1. Preference for chosen management action as a function of pest density.

	Actual pest density				
Decision	Very low	Low	Around *cd*	High	Very high
No intervention	Always	Most often	Sometimes	Seldom	Never
Intervention	Never	Seldom	Sometimes	Most often	Always

high, and a 'don't intervene' decision if the density is very low. At densities closer to the critical pest density (however defined), we are more willing to accept deviations from this preferred pattern. And if pest density is close to the threshold, we might not be too much concerned whatever the decision is. Pest densities close to the threshold form a kind of indifference area, where either decision is acceptable, because there are so many uncertainties involved. Note that this is not a 'couldn't care less' attitude, but a realistic assessment of all the available information, including that provided by the sample on the state of the system.

Growers may compensate for a decision in several ways. For instance, if a decision not to intervene has been made, but the risk of pest loss remains possible, a grower may decide to be more watchful in the immediate future. Growers may also be willing to refrain from using pesticides if they expect to be able to combine the intervention later on with another management action that will have to be made. The point is that models which are used to describe sampling processes and pest–damage relationships and uncertainties therein can never capture all of the decision elements that matter to a producer. Hence, some fuzziness in the OC function is quite acceptable.

As we noted above, some crops are more valuable than others, some growers are prepared to take greater risks than others and, if the action to be taken is the application of a chemical, there may be side-effects. These factors may have been considered in specifying *cd*, but they should be considered also for the shape of the OC function. It is reasonable that the OC function for a valuable crop would be steeper than that for a less valuable one, in order to reduce the probability of a big loss. Growers who are less prepared to take risks might prefer a steeper OC function. In effect, where the actual value of *cd* is regarded as critically important, the steepness of the OC function also becomes critical.

The most helpful type of sampling plan is one which takes into account all relevant process, state and socio-economic knowledge, so that the user gets the best decision at the end. The cost of arriving at a decision should be commensurate with the value of the recommendation (discussed in detail in Chapter 6). Many types of sampling plan are discussed in this book, and each plan has enough flexibility to produce almost any shape of OC function (steepness and *cd*). It is in tailoring sample theory to practical needs that skill, as well as wide knowledge and understanding, is required.

1.7 Monitoring a System through Time

So far, we have implicitly considered the situation in which sample information is collected just once, a management decision is made and implemented, and the process ends. For pests that pose a continual threat throughout a lengthy period during the season, such a procedure is inadequate. Such pests must be regularly checked by taking samples on several occasions. We call this *monitoring*. We define monitoring as an overall strategy which includes setting up a schedule of potential sampling occasions, with the added feature that the results of each sample specify either: (i) a positive management action or (ii) when, in the schedule of potential sampling occasions, to resample. We further define a *monitoring protocol* as a procedure that directs how sampling resources are to be allocated. This is best explained using an example.

Plant feeding mites can be deleterious to crops in agricultural systems. These organisms can pose problems over an entire growing season, and the fluctuations in population numbers may not be predictable over more than a relatively short time horizon. As a result, it is necessary to check a population repeatedly to be sure that intervention is not needed. Also, when a control action has been taken, further checking may be required. Each time the population is checked, we propose a sampling plan that can give more than two recommendations:

1. Intervention now.
2. No intervention now, but resample soon.
3. No intervention now, but wait some time before resampling.

Details of how this can be done will be described in later chapters. The main difference between sampling where there are only two possible recommendations and where there are three is in the OC function and the use of a critical density. There are now two critical densities (cd_1, cd_2) dividing the range of pest densities into three parts: less than cd_1, between cd_1 and cd_2, and greater than cd_2. The rationale for this set of management options for pest management is that: (i) some pest populations are already high and require treatment immediately (greater than cd_2); (ii) some are not that high, but risk being too high within a short period of time (between cd_1 and cd_2); (iii) some are low and can be expected to remain low for some time (less than cd_1). Category (iii) may be further subdivided, so that the recommended time for resampling can be after one of several time intervals.

An OC function for each sample occasion would have three curves as shown in Fig. 1.5, and to prevent confusion we shall refer to sets of OC functions such as these as *probability of decision functions* (*PD functions*). The two basic principles behind the procedure are: (i) that predictive models for the pest are not reliable, except in the short term, and (ii) that managing a crop does not require estimates of pest density *per se*, but only inasmuch as they aid decision-making. We believe that what we here call monitoring is done in some form or other by a great number of people. We shall describe methods of optimizing the procedure. The procedure itself is analogous to scheduling inspections of kitchens in public restaurants for health safety: restaurants with good records of safety are inspected less frequently.

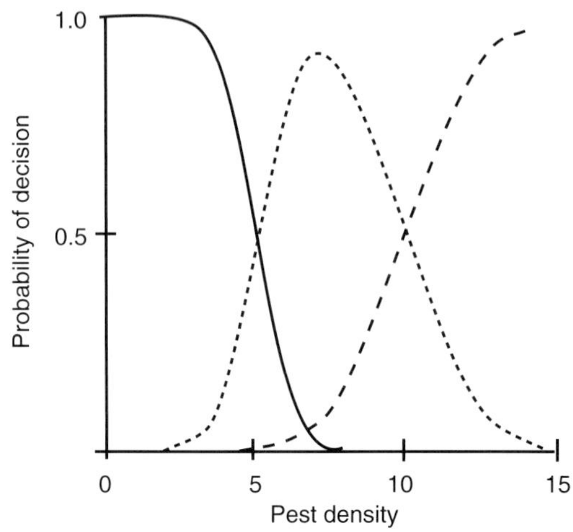

Fig. 1.5. Probability of decision functions for a sampling plan with three alternative decisions and two critical densities, 5 and 10. ——, Density classified as smaller than 5; - - -, density classified as greater than 10; ----, density classified as lying between 5 and 10.

Assessing the cost of monitoring over a season is not done as in one-time sampling with a single decision. In one-time sampling, the OC and ASN functions describe the performance of the sampling plan for each mean pest density (Fig. 1.4). For a monitoring protocol, similar performance measures can be calculated, but these measures now refer to population numbers over the entire monitoring period, called a *population trajectory*. Thus, for each possible population trajectory, there is a single probability of intervening, corresponding to one point on a probability of decision curve. Corresponding to the ASN, there are sample number data, such as the total number of sample units collected during the entire monitoring period, and the number of times the population was actually sampled. There are other considerations and other performance criteria, which will be discussed in later chapters.

1.8 The Formal Design of Methods for Sampling and Monitoring: Why and How?

This book emphasizes formal methods for designing sampling and monitoring plans. In practice, protocols developed via this method are often modified by users to fit their particular needs, limitations and preferences. As a result, a natural question is whether formal analyses and design of sampling plans and monitoring protocols are necessary or useful. Our unequivocal answer to this question is 'yes'. Even when plans or protocols are modified from those proposed, it is important that the foundation of the modifications be sound. Practitioners may take a plan developed for another purpose and modify it simply by altering the sample size, or start with the sample size they are prepared to use and go from there. Such a pragmatic approach may be successful, but if no attempt is made to investigate its properties and properties of alternative schemes, users may easily miss out on a superior scheme. Basing the design of plans and protocols on formal methods ensures that explicit basic data and assumptions are used (instead of implicit or potentially incorrect ones) and that the purpose and restrictions of the plans have been well thought through. As a result, the plans are verifiable and their operation is understood. Moreover, when well-tailored to specific needs they potentially perform better than any *ad hoc* plan.

The design process that we outline here, and which will be explored in greater detail in subsequent chapters, entails the construction of 'models' that relate pest densities to decision recommendations. Pest density is always an unknown quantity but, depending on the parameters of the sampling plans, these models define the characteristics of calculable performance criteria such as the OC and ASN functions. During the design process, different types of sampling plan can be tried out and, for each type, its parameters can be varied to explore the effect on the performance criteria. Based on the weight that is given to the different performance criteria, an 'optimal' plan may be selected. Such a plan may not be strictly optimal in a technical sense, because there are usually several criteria. Some criteria are directly conflicting (e.g. sample size and precision), so compromises must be made. Moreover, in practice, the parameter space of sampling plans can be only partly explored, and some useful possibilities may be overlooked. Nevertheless, the explorative approach using models immensely widens the selection from which choices can be made, and a better end-product is more likely.

Another advantage of the formal approach is that whenever any model is constructed, whether it be a mathematical model or a conceptual model, the process of constructing the model and explicitly describing its components engenders a level of understanding that otherwise would not be possible. The increased insight may suggest yet better solutions.

1.9 On Learning by Doing and Farmer Empowerment

It may be impossible to develop acceptable agricultural production systems for the future that rely solely on natural internal mechanisms of pest control. We believe that, in agricultural production systems which exploit ecological processes to keep pest populations under control, efficient sampling and monitoring tools are required to keep the pulse of the system and to schedule interventions – of whatever nature – where and when necessary. We do not advocate that sampling plans be designed and extended to practitioners as prescriptions. The final word is always with the user. Moreover, users are likely to adapt plans, originally based on formal methods, to their own needs and restrictions. In this way, sampling plans are taken up in a stream of learning by doing. Therefore, the two approaches of design, by 'engineering' and by 'learning by doing', are complementary. To emphasize the responsibility of the user, we prefer to use the term 'decision guides' rather than 'decision rules'. Labour is expensive in highly productive agricultural systems. It is therefore imperative that sampling and monitoring methods are effective (i.e. they produce good results), but require a minimum of input. This book provides tools for designing sampling and monitoring plans that fulfil this criterion.

1.10 Summary

1. Growers make decisions at several spatial and temporal scales.
2. Three constituent types of information play a large role in decision-making:

- knowledge about the dynamics of the production system
- socio-economic information and expectations
- knowledge about the actual state of the system

3. Pests reduce production and cause loss of revenue, depending on their density. The calculation of the threshold level of pest density above which some kind of intervention is necessary requires an integration of all types of information, possibly accounting for uncertainties in knowledge. There are several potentially useful threshold concepts: *EIL*, *ET* and modifications thereof.
4. Sampling looks at the state of the system. The two basic performance indicators of sampling plans are:

- the operating characteristic (OC) function – that is, the probability of a decision to do nothing about the pest, as a function of pest density
- the average sample number (ASN) function, also a function of pest density

5. Sampling uncertainty along with information uncertainty determine ideal shapes of OC and ASN functions.
6. A decision about the timing of future samples (if taken at all) may be needed, in addition to a decision on what to do about the pest directly following sampling. Monitoring is sampling through time.
7. Properties of sampling and monitoring strategies can be calculated and then optimized. The general structure of this optimization and design procedure encompasses setting up potential plans, calculating their performance using statistical or simulation methods, and choosing the preferred plan, based on explicit performance characteristics such as sample size and average expected loss.
8. The formal design process described in this book leads to procedures that may be adapted to the specific requirements of individual users. They then become part of the evolving production system.

References and Suggested Reading

Brown, G.B. (1997) Simple models of natural enemy action and economic thresholds. *American Entomologist* 43, 117–123.

Dent, D.R. (ed.) (1995) *Integrated Pest Management*. Chapman & Hall, London, 356 pp.

Dent, D.R. and Walton, M.P. (eds) (1995) *Methods in Ecological and Agricultural Entomology*. CAB International, Wallingford, 387 pp.

Higley, L.G. and Pedigo, L.P. (eds) (1996) *Economic Thresholds for Integrated Pest Management*. University of Nebraska Press, Lincoln, Nebraska, 327 pp.

Norton, G.A. and Mumford, J.D. (eds) (1993) *Decision Tools for Pest Management*. CAB International, Wallingford, 279 pp.

Pedigo, L.P. and Buntin, G.D. (eds) (1994) *Handbook of Sampling Methods for Arthropods in Agriculture*. CRC press, Boca Raton, Florida, 714 pp.

Rossing, W.A.H., Daamen, R.A. and Hendrix, E.M.T. (1994a) Framework to support decisions on chemical pest control under uncertainty, applied to aphids and brown rust in winter wheat. *Crop Protection* 15, 25–34.

Rossing, W.A.H., Daamen, R.A. and Jansen, M.J.W. (1994b) Uncertainty analysis applied to supervised control of aphids and brown rust in winter wheat. Part 1. Quantification of uncertainty in cost–benefit calculations. *Agricultural Systems* 44, 419–448.

Basic Concepts of Sampling for Pest Management 2

2.1 Introduction

In this chapter we start with the concept of randomness, and show how a lack of randomness in sampling can lead to bias. Bias, precision and accuracy are defined and described, and ways of avoiding bias when collecting sample data are presented. Four attributes that determine the basic trustworthiness of sampling plans for pest management are described: representativeness, reliability, relevance and practicality. The variances of sample data and of sample means are defined. The precision of estimated population parameters is commonly characterized by the variance. Using data on the distribution of Colorado potato beetle in potatoes as an example, we demonstrate the reduction of variance with larger sample size. Using the same data set, we demonstrate that sample means for large sample sizes are described by the normal distribution. Using this result, we show how to compute a probability of decision function (operating characteristic (OC) function) for a sampling plan that classifies density using a sample mean. Simulation is introduced as a tool for characterizing the properties of sampling plans.

2.2 Randomness

We introduce the concepts of randomness and random sampling by describing two simple examples. Very often, bias is a consequence of non-random sampling. These examples also allow us to illustrate some technical terms we need to use later on.

EXAMPLE 1: RANDOM NUMBERS 1–100. The human mind is generally incapable of producing truly random samples, even from a defined population. We asked people in the Department of Theoretical Production Ecology at Wageningen University to write down, independently of each other, five numbers that they were to choose randomly from the population of numbers 1–100. The results are given in Table 2.1. As might have been expected from such an aware group of people, the numbers were not too bad as random numbers, but even here there is

Table 2.1. The tabulation of 185 numbers (five numbers from each of 37 people), according to value. For example, the first row shows the total numbers received in the range from 1 to 10: two 1's, two 2's, three 3's, three 4's, six 5's, one 6, five 7's, three 8's, seven 9's and one 10.

	Units										
Range	1	2	3	4	5	6	7	8	9	0	Total
1–10	2	2	3	3	6	1	5	3	7	1	33
11–20	4	2	2	3	0	0	3	1	3	2	20
21–30	1	2	5	0	0	4	1	0	2	0	15
31–40	1	2	2	1	3	2	4	2	2	1	20
41–50	1	0	2	2	8	1	2	0	0	2	18
51–60	3	0	1	4	0	0	1	0	2	0	11
61–70	2	1	1	5	1	1	5	2	3	0	21
71–80	3	1	1	2	3	0	2	5	0	0	17
81–90	2	0	2	2	1	1	1	4	1	1	15
91–100	0	1	2	2	2	0	0	4	3	1	15
Total	19	11	21	24	24	10	24	21	23	8	185

a marked preference for numbers less than 10, and an aversion to numbers ending in 0 or 6.

An inability to produce a random sample, even from so well defined a population as the numbers 1–100, has important practical ramifications when sampling for pest management. When we sample a population, we usually assume that sample units are drawn randomly, so that each sample unit has an equally likely chance of being selected. In this example, the sample units are the numbers 1–100. When sample units are not selected randomly, the information collected may have a slant or tendency that does not truly represent the population. As a result, estimates obtained by sampling may not be accurate; the estimates might be systematically greater or less than the true population parameter. In this example, the average of all chosen numbers was 45.3, whereas the average of all numbers from 1 to 100 is 50.5.

EXAMPLE 2: SAMPLING MARBLES. Imagine a sack that contains many marbles. The total collection of marbles is defined as the *population* of all marbles. Some of the marbles are red and the remainder are green. Now suppose that we wish to estimate the proportion of red marbles in the sack without having to check all the marbles. To do this, we decide to draw 20 marbles from the sack and we note the colour of each. We might then infer that the proportion of red marbles in the sack is equal to the number of red marbles drawn divided by 20. In order for our estimate of the proportion of red marbles in the sack to be accurate, we must assume that marbles were drawn from the sack randomly.

Sampling is said to be random if each item or set of items in the population is equally likely to be drawn. For the most part, we would be willing to accept that sampling the marbles in the sack is random, but there may be circumstances which would cause this assumption to be questioned. For example, it might be that the red marbles were placed in the sack first, the green marbles were then added, but the marbles were not mixed before sampling. It might also be that the red marbles are heavier than the green ones, which would make it more likely for the green marbles to be on top in the sack and hence more likely to be selected. If we repeated the sampling experiment several times under these conditions, our samples would systematically contain more green marbles than if samples had been truly random.

A central difference between these two scenarios is that the *expected value* of the sample proportion of red marbles is different. The expected value can be defined heuristically as the long-term average of the sample means (i.e. proportions). Let us imagine the experiment being repeated with random sampling (after proper mixing) and with non-random sampling (without mixing). After a great many repetitions all possible sets of 20 marbles would have occurred, some more often than others. The relative frequency with which any one set of 20 marbles occurs would be approximately equal to the probability of it occurring on any one occasion. When sampling is random these probabilities are equal to each other, but when sampling is non-random the probabilities of samples with high numbers of green marbles would be relatively greater. Thus, with random sampling, the expected value of the sample proportion of red marbles would be equal to the true proportion of red marbles in the whole sack, whereas, with non-random sampling, the expected value would be less. The actual value with non-random sampling would depend on properties of the marbles themselves, of the sack, and of the person (or machine) drawing the marbles: it might be only fractionally less than the value under random sampling, or it might be much less. Therefore, if our goal was to estimate the actual proportion of red marbles in the sack, but were unaware of any weight difference between the red and green marbles or inadequate mixing, our estimates would be biased downwards to some unknown degree.

At this point it is useful to distinguish between the sample unit itself and the property (or properties) being investigated. In the marble example, the sample units are marbles and the property being investigated is their colour. Even when marbles are randomly selected (no *selection bias*), if the person doing the sampling cannot readily distinguish between the colours red and green, the data collected (colour) may be incorrect. This would lead to a different type of bias, *enumeration bias*.

2.3 Bias, Precision and Accuracy

When samples are taken from a population to estimate a parameter such as the mean number of weevils per plant in a lucerne field, the estimate will almost certainly not be equal to the true parameter. If the same population is sampled many times there will be several different estimates of the mean, and these estimates will be scattered about the true mean in some pattern. The central value and shape of this pattern depend on the bias and precision of the sample estimates. *Bias* is any systematic deviation of the sample estimate from the true parameter. In the marble sampling model, we described how non-random sampling can lead to systematic error or bias. This systematic error is distinct from random error because it does not balance out on average. Bias is defined as the size of the difference between the expectation of a sample estimate and the population parameter being estimated. If m is an estimate of the population mean μ, then bias is defined as $E(m) - \mu$, where $E(\ )$ denotes expectation. Bias is usually impossible to measure in practice, because we rarely know what μ is. Certain types of estimators such as ratio estimators are inherently biased because of their mathematical formulation, but in sampling for pest management decision-making, a more important potential source of bias is the way in which data are collected or analysed.

Precision refers to how close to its own expectation we can expect one single estimate to be. If m is an estimate of the population mean μ, then precision is quantified by $E(m - E(m))^2$. With 'E( )' occurring twice, this may look daunting, but it can be put into words:

- $(m - E(m))^2$ is the squared difference between an estimate and its long-term average
- $E(m - E(m))^2$ is the long-term average of these squared differences

$E(m - E(m))^2$ is actually more a measure of imprecision than precision, because it increases as estimates get more erratic. Note that if the difference, $(m - E(m))$, were not squared, the result would be zero.

Accuracy denotes the closeness of population estimates to the true population parameter. Accuracy therefore incorporates both bias and precision. For example, an estimate may have high precision, but low accuracy because of high bias. The relationships among bias, precision and accuracy can be visualized by considering again the 'random' numbers of Example 1 and comparing sampling from these numbers to sampling from a set of truly random numbers. We simulated random sampling from the data consisting of the numbers 1–100 (Fig. 2.1a and b) and from the non-random TPE data set (Table 2.1; Fig. 2.1c and d). Both data sets were sampled using a small sample size (Fig. 2.1a and c) as well as a large sample size (Fig. 2.1b and d). The figure shows the resulting four distributions of sample means.

As expected, the frequencies for trials (a) and (b) are centred on 50.5, illustrating zero bias, and the frequencies for (c) and (d) are centred to the left of 50.5, illustrating negative bias. The spread of frequencies for trials (a) and (c) is wider than in (b) and (d), illustrating less precision for (a) and (c) than for (b) and (d) due, as we shall see below, to the smaller sample size. Obviously, the sampling strategy for Fig. 2.1b gives the most accurate results, combining zero bias and comparatively high precision.

EXAMPLE 3: SELECTION BIAS WHEN SAMPLING STONES. Bias can be studied in more practical instances, as was done by Yates (1960). In one study, he laid out on a table a collection of about 1200 stones (flints) of various sizes with an average weight of 54.3 g. He asked 12 observers to choose three samples of 20 stones each, which would represent as closely as possible the size distribution of the whole collection of stones. The mean weight over all 36 samples proved to be biased upwards: the average weight of all 36 samples was 66.5 g. Ten of the observers chose stones whose average was greater than 54.3 g, and of their 30 samples only two had sample

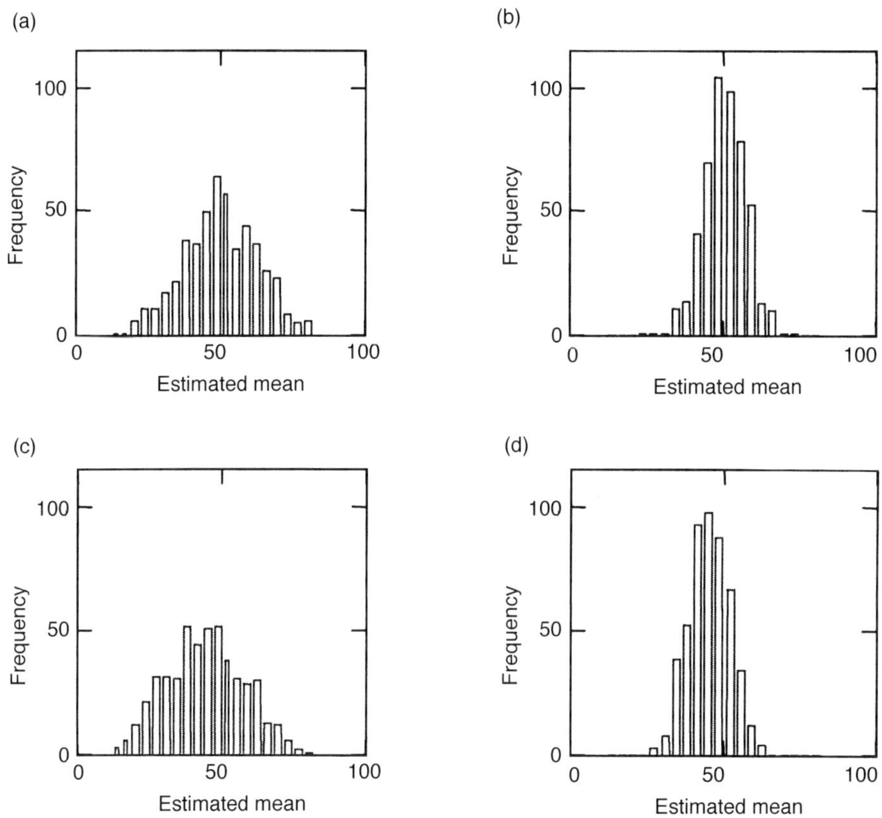

Fig. 2.1. Frequency distributions of 500 sample means based on random and non-random sampling of the numbers 1–100. Samples consist of: (a) five randomly chosen numbers, low precision, no bias; (b) 20 randomly chosen numbers, high precision, no bias; (c) five numbers chosen randomly from Table 2.1, low precision, bias; (d) 20 numbers chosen randomly from Table 2.1, high precision, bias.

means less than 54.3 g. All three of the samples of one of the two other observers had means less than 54.3 g. These results suggest that the non-random method of selecting stones by observers resulted in unrepresentative samples, leading to biased estimates of the average weight.

EXAMPLE 4: SELECTION BIAS WHEN COLLECTING 'ADDITIONAL' SAMPLE UNITS. In a second study, Yates describes how a carefully designed and tested sampling protocol for estimating wheat growth was, on two occasions, 'adjusted' at sampling time by an observer because the situation that he faced had not been envisaged in the protocol. Following the protocol as far as he could, he realized that he would collect only 192 measurements of plant height, rather than the required 256, so he added 64 of his own 'randomly chosen' plants, fortunately keeping the values separate. On the first date, the additional observations tended to be larger than the standard ones, while on the latter date they tended to be smaller. Yates suggests why these differences might have arisen: on the first occasion, the plants were only half grown, and the observer would have been more inclined to sample larger plants; on the second occasion, the plants had come into ear, and the observer may have over-compensated by looking for plants closer to the average.

These examples exemplify what is called *selection bias*. Yates (1960) lists four possible causes of selection bias:

1. Deliberate (but unsuccessful) selection of a representative sample.
2. Selection of sample units on the basis of a characteristic which is correlated with the properties to be investigated by sampling.
3. Omission of sample units that are (more) difficult or tedious to collect or inspect than others.
4. Substitution of omitted or rejected sample units by more readily observable units.

Selection bias is an ever-present danger in pest sampling work when sample units are not selected in a truly random manner.

A second source of bias is called *enumeration (counting) bias*. Enumeration bias occurs when counts on sample units are systematically less than or greater than the actual number on the sample unit. This is obviously of concern when organisms are very small, but may also be an issue when organisms are mobile or are difficult to recognize. Not only can a procedure which has bias be bad in itself, but different people may act in such a way that they have their own personal amounts of bias. Different users of a sampling plan might then arrive at different conclusions, even though they may be sampling from the same location and at the same time.

2.4 Dealing with Bias

To minimize selection bias, the choice of sample units should be codified so that the opportunity for bias to creep into the sampling process is minimized. If cost and time were of no concern, each possible sample unit could be numbered and sample units drawn by generating random numbers. Of course, this is impossible in practice, so a compromise must be made between selecting sample units entirely at random and making the sample process reasonable. Sampling procedures should be carefully codified to prevent any preferential selection of sample units which in some way or other are related to their pest status. We illustrate two common situations in which care is needed:

1. *Pests or pest injury is readily apparent on potential sample units.* Leaf-rolling caterpillars are often regarded as pests in apple orchards. These insects create feeding sites by wrapping a leaf or leaves with silken threads. These feeding sites are readily visible, even when the caterpillars are very small, so there is great danger in using a supposedly random collection of branch terminals or of clusters of leaves as a sample set. Observers might preferentially select or avoid leaves with feeding damage. To

circumvent this, a whole branch can be taken as the sample unit and all growing tips on the branch examined for caterpillars. In this case, the sample unit itself is redefined to minimize the chance of selection bias influencing the outcome.

2. *The local distribution of pests among potential sample units is known*. Even when the pest is minuscule and not readily visible, bias can arise because of the way it is distributed among potential sample units. A phytophagous mite on apple, the European red mite (*Panonychus ulmi*) is often more numerous on older leaves than on younger leaves. The density on leaves of intermediate age is most representative of the population in the whole tree. Therefore, to circumvent possible bias, leaves in the middle of fruit clusters or mid-way along an expanding terminal are used as sample units.

Codifying the way in which sample units are collected will not affect the other source of bias – inaccurate enumeration. In some cases, enumeration bias cannot be avoided. If this source of bias is consistent among observers then it can be largely ignored, other than being sure that estimates take into account either the under- or over-counting. If, however, enumeration bias differs greatly among observers, efforts must be taken to discover the source of the bias and, if possible, remove it. Otherwise, this source of bias can result in incorrect pest management decisions.

A procedure that is suspected to have bias is not necessarily to be avoided. One of the most important criteria in pest management is cost, and it frequently happens that procedures with less bias are more costly. In such a situation, good arguments need to be made if a less biased procedure is to be adopted. For example, a standard procedure for selecting plants in a field from which to take sample units is to follow some more or less straight paths across the field, taking samples at more or less equal intervals. This can easily introduce selection bias, either because of the choice of paths or the actual selection of sample units. However: (i) even if a complete list of all plants in the field could be envisaged, and a truly random sample taken, walking through the field to reach the plants would take a lot of time and would probably damage the crop too; and (ii) the protocol to select the sample unit cannot afford to be too complicated. It has been found in practice that the simple path method, along with close-to-random unit selection (e.g. tossing a stick, or taking, say, the tenth unit after arriving at a particular sampling location) is usually acceptable.

A somewhat different way in which bias can occur is when a sampling plan is proposed which is good on paper, but is hard to comprehend or difficult to implement. A pest manager may try very conscientiously to make the plan work, but it inevitably runs into problems. It is then almost inevitable that data errors will creep in, through exhaustion or frustration. Such a sampling plan is almost useless.

2.5 Criteria for a Trustworthy Sample

A trustworthy sampling plan should meet the following criteria:

1. The sample information should be *representative* of the actual pest density. This is simply another way of saying that unaccountable bias should be avoided. If the

design of the sampling programme is such that you *know* that the expected value of the sample mean is, say, 20% too low, you can easily adjust your estimate, but not otherwise.

2. The sample information should be *reliable*. It should not depend on the person collecting the data. Reliability goes beyond this though, in that the sample information should also not be affected by exogenous, uncontrolled variables such as the weather, or by the possible diurnal behaviour of the pest.

3. The sample information should be *relevant*, meaning that the estimate of pest abundance should have a reasonable relationship to crop yield and loss.

4. Sampling in the proposed way must be *practically feasible*; that is, the procedure must be simple enough to appeal to users and not be misunderstood; the time it takes to collect a sample, and the moment at which it must be collected must fit in the agenda of the user.

Meeting these criteria must take into account the specifics of the pest–crop system, its scale of space and time, and how and by whom it is managed. In this book, we will assume that a trustworthy sampling plan can be set up. We take it from there by outlining how the performance of proposed plans, as characterized for instance by probability of decision (or OC) functions and average sample number (ASN) functions, can be determined, modified and optimized. In the remainder of this book, we will often assume that the criteria of representativeness, reliability, relevance and practicality have been met in principle. When they are not met, looking at performance indicators such as the OC and ASN functions is of little use.

You might now recognize that we have not included precision as a criterion for a good sample. A sample protocol that delivers estimates with less variability than another protocol provides more precise estimates. Usually, obtaining more precise estimates requires more work either in terms of the type of information collected from each sample unit or in terms of the number of sample units examined. Depending on the context, added precision may or may not be important. We will discuss this repeatedly throughout the book, but what is important to understand now is that attaining high precision, or attaining some specified level of precision, cannot be used as a criterion of a good sample without considering to what end the sample information will be used. There will be instances when a moderately imprecise estimate of pest abundance is sufficient and other situations in which a more precise estimate is needed.

2.6 Spatially Explicit Data versus Data Lists

When we collect sample data, we do so in three-dimensional space. During most applications in pest management, this is ignored. Some sort of selection procedure is used to choose samples, but once pest counts are made on the sample units, the locations of the units are forgotten, and a summary such as the average count is all that is used in the decision-making process. On some occasions, however, the location of the sample units is important both for selection and for analysis. For example, when a known predictable edge effect exists, it may be wise to consider whether these areas should be managed differently. Possibly they alone should be sampled to act as an indicator of potential pest problems, or the edge and interior of a site should be managed differently.

Another instance in which all three space dimensions need to be considered is when the structure of the host, the behaviour of the pest and/or time considerations make it more convenient to select sample units hierarchically. For example, some methods for sampling the lucerne weevil on lucerne include protocols for initially sampling the stems and then sub-sampling leaves on these stems, counting pests on only the selected leaves. This type of sampling in more than one 'stage' presents further problems, and is discussed in detail in Chapter 8.

A final case in which all three space dimensions must be considered is when maps of pest abundance are desired. In such situations, the purpose of sampling is not just to estimate the average abundance of a pest over some sample space, but to quantify the spatial pattern of abundance. This is done so that control is only executed on those portions of the management unit where pests exceed some critical density. This type of sampling for pest management decision-making is not widely practised and is beyond the scope of this book. Interested readers are referred to Schotzko and O'Keeffe (1989, 1990) for an introduction to the subject.

2.7 Simulation of the Sampling Process

Throughout this book, we will use simulation of sampling processes as a tool to address questions about the behaviour and performance of sampling plans. Simulation is often the only practical way to do a particular analysis or to obtain the desired information. In the example discussed below, mathematical statistics could be used, but using simulation provides a different insight.

Simulation mirrors field sampling, and consists of four steps:

1. The population from which samples are to be drawn is described.
2. Sample observations are selected from the population at random, or using some other specified method.
3. The sample data are summarized; for example, by the sample mean.
4. Steps 2 and 3 are repeated many times and the results summarized so that the nature of the sample information can be studied.

Steps 1–3 are just the same as in field sampling. The fourth step is used to generate a 'long-term average' description of the properties of the sampling protocol. We have already used simulation to produce Fig. 2.1. We use sampling from the 'almost' random numbers 1–100 in Table 2.1 as an example.

One way of selecting a sample from the population of numbers summarized in Table 2.1 is to write them in an array underneath an array of serial numbers (s.n.) which go from 1 to the total number of sample units, as follows:

s.n.	1	2	3	4	5	6	7	8	9	10	11	etc.	184	185
Data	1	1	2	2	3	3	3	4	4	4	5	...	99	100

A simple computer program can be written to select random numbers in the range 1–185. Thus, 20 random numbers in this range can be chosen, and the sample then consists of the data values corresponding to each of the 20 randomly selected serial numbers. For example, if the 20 serial numbers were to include 5, 10, 5 and 8, representing the 5th (twice), the 8th and the 10th units in the population, then the sample data would include the data values 3, 4, 3 and 4.

In this example, we have allowed the 20 random numbers to include repetitions, but we can specify that no repetitions are allowed. We should then need to keep generating random numbers (often many more than 20 would be needed) until 20 distinct sample units are found. The former type of sampling is called sampling *with replacement*, and the latter is called sampling *without replacement*. Sampling without replacement should not be done in this way, unless the sample size, n, is small (see below).

A rationale for using with replacement sampling in this example is that the data are assumed to be typical of data that would be provided by any group of human beings. In other words, a much larger data set derived from many more people would have essentially the same characteristics as the data in Table 2.1. Removing sample units as they are chosen (i.e. sampling without replacement) would alter these characteristics, especially if the sample size were large relative to the total number of units, 185. On the other hand, if the data values represent a specific physical entity whose properties we need to estimate, sampling without replacement may be more sensible. For example, simulating sampling of grain cargoes which might contain a quarantine pest would be better done using 'without replacement' methods.

Another simulation method, which is easier in some circumstances, is to use the frequency counts. The data values are arrayed alongside the frequencies, the cumulated frequencies (c.f.) and cumulative probabilities (c.p.: c.p. = c.f./total number of sample units):

Data	1	2	3	4	5	6	7	8	etc.	99	100
Frequency	2	2	3	3	6	1	5	3	...	3	1
c.f.	2	4	7	10	16	17	22	25	...	184	185
c.p.	0.011	0.022	0.038	0.054	0.086	0.092	0.119	0.135	...	0.995	1.000

A random choice is made by getting the computer to provide a random number between 0 and 1, and finding where this number lies among the c.p. values. For example, if the random number were 0.090, you would find where it lies (between 0.086 and 0.092), look above to the top line, and choose the data value on the right to get the selection '6'. If the random number were 0.999, the selection would be '100'; and if it were 0.005, the selection would be '1'. If the number is exactly equal to one of the c.p. values, the selection is the data value directly above that c.p. value. If sampling with replacement is being simulated, this procedure is repeated 20 times.

Both of these methods can be used to simulate sampling with replacement. In each instance, the more convenient one should be used. For sampling without replacement, however, neither method is recommended. A sample of n units should be simulated by generating a random permutation of the sample units, and selecting the first n (Page, 1967).

With this introduction to simulation, let us now examine how to summarize the information collected from a sample and how to view the value of this summary.

2.8 The Variance of Sample Means

Beall (1939) described counts of Colorado potato beetles taken from a field in Ontario, Canada. Forty-eight contiguous rows of potatoes were subdivided into 48 lengths of 2 feet (71 cm) each, and the sample unit was defined as one 2 foot length. The numbers of Colorado potato beetles were noted for all sample units. A pictorial impression of the data is shown in Fig. 2.2. The numbers of sample units found with no beetles, the numbers with one beetle, with two beetles and so on are shown as a *frequency distribution* in Fig. 2.3a. Sampling from this field (as represented by the grid of $48 \times 48 = 2034$ units) can be simulated by selecting sample units at random 'without replacement' using the computer and summarizing the data to get the sample mean.

If the collection process is repeated (simulated) many times, the set of estimated means can be displayed as another frequency distribution, which can be compared with the frequency distribution representing the original data. Five hundred simulations were done of a sample protocol that takes 25 sample units and determines the average number of beetles in each set of 25 sample units. The frequency distribution is shown in Fig. 2.3b. Two things are evident from a comparison of the frequencies of the original data and of the sample means: the means of the two distributions are close to each other, but the spread of frequencies in the sample means is much less than in the original counts. Because each of the simulated samples represents a possible sample mean, you can see that you are more likely to be near the true mean with a sample of size 25 than if you were just to take one plant at random. What is more, statistical theory provides a guide that predicts how much closer you can expect to be, depending on your sample size.

The measure of spread that has been found most useful in theory and practice is the variance, σ^2. It is defined as an average of the squared difference of each data value from the true mean. The variance of sample values is defined as

$$\text{sample variance, } V = \frac{\sum_{1}^{n}\left(X_j - m\right)^2}{n-1} \tag{2.1}$$

where n is the number of sample units selected (the sample size) and m is the estimated mean. The division by one fewer than n, rather than n itself, is for theoretical reasons dealing with mathematical bias.

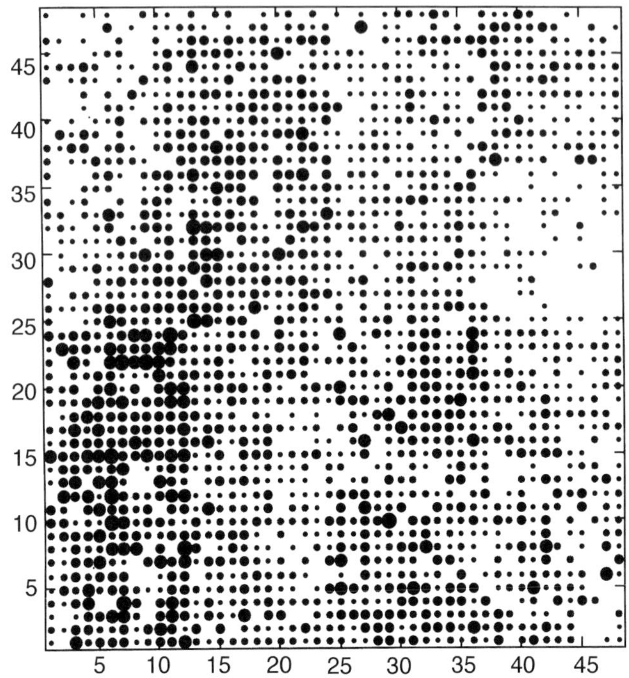

Fig. 2.2. A pictorial impression of counts of Colorado potato beetles in a field. Counts on sample units are classified into the size intervals 0, 1, 2–3, 4–6, 7–12, 13–20 and 21+. Increasing numbers of beetles per sample unit are indicated by circles of increasing size according to this classification, with a blank representing the zero class.

The guide provided by statistical theory is that the variance of the distribution of sample means in Fig. 2.3b is a simple function of σ^2 and n, and can be estimated from V and n:

$$\text{true variance of the sample mean} = \sigma^2 / n \tag{2.2}$$

It is estimated by

$$V_m = V / n \tag{2.3}$$

By taking n sample units, we have reduced the variance of the sample mean by a factor of $1/n$. As is evident from Equation 2.3, the variances of sample means decrease as sample size increases. Thus, the precision of a sample mean increases with increasing sample size. This can be better visualized by using the square root of the variance, and using standard deviation, standard error and so on (Table 2.2). A concept which is very useful in a variety of circumstances is the coefficient of variation (CV): it combines the variance with the mean, so that variability can be viewed as a percentage of the mean. This and other notations are presented in Table 2.2.

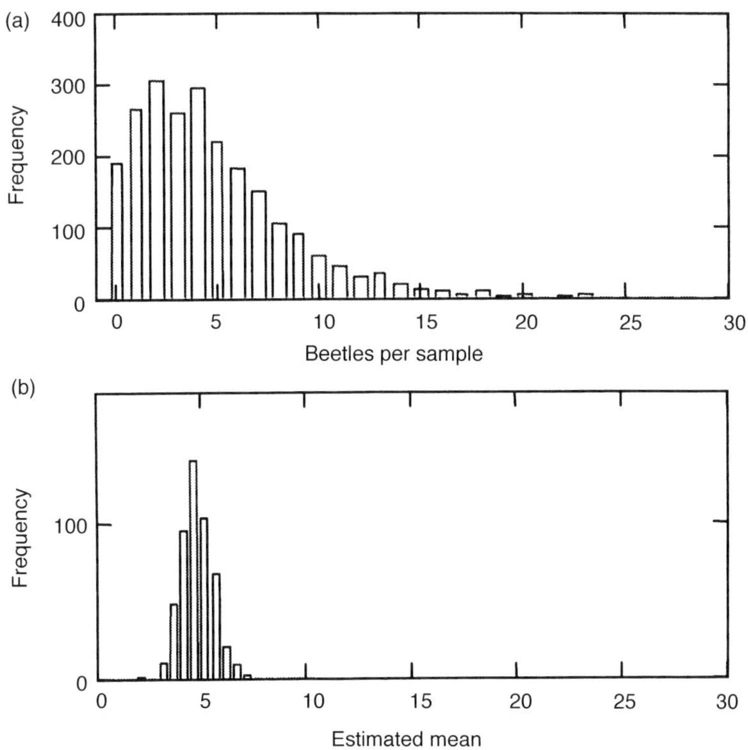

Fig. 2.3. (a) A frequency diagram for counts of Colorado potato beetles; mean = 4.74. (b) A frequency diagram for 500 means estimates from samples of 25 units each; mean = 4.75.

2.9 The Distribution of Sample Means – the Central Limit Theorem

A second useful result that we can take from statistical theory is the *Central Limit Theorem*. As the sample size increases, not only does the variance of estimated means decrease, but the shape of the distribution of means standardizes to the shape of what is called the *normal distribution*. The normal distribution has a bell-like shape, and two of its properties are that about two-thirds of it lies within one standard deviation of its mean and about 95% lies within two standard deviations of its mean. Therefore, we have a heuristic idea of what happens when sample sizes increase:

1. With about 66% probability, the sample mean, m, is within $\sigma/\sqrt{n}$ of the true mean, μ.
2. With 95% probability, the sample mean, m, is within $2\,\sigma/\sqrt{n}$ of the true mean, μ.

σ can be replaced by the estimate, $\sqrt{V}$ (Fig. 2.4).

Table 2.2. Notation for various statistical quantities based on the mean and the variance. There is some inconsistency in the literature (for example, s and $s/\sqrt{n}$ are often referred to as standard deviations). In this book we use the definitions noted here.

True values	
μ	Mean
σ^2	Variance
σ	Standard deviation (*sd*)
$\frac{\sigma}{\sqrt{n}}$	Standard deviation of a sample mean (sd_m)
$\frac{\sigma}{\mu}$	Coefficient of variation
$\frac{\sigma/\sqrt{n}}{\mu}$	Coefficient of variation of a sample mean
Sample estimates	
m	Sample mean
V	Sample variance, often written as s^2
$\sqrt{V}$ or s	Sample *sd*, also called standard error (*se*)
$\sqrt{\frac{V}{n}}$ or $\frac{s}{\sqrt{n}}$	Standard error of a sample mean (*sem*)
$\frac{\sqrt{V}}{m}$ or $\frac{s}{m}$	Estimated coefficient of variation
$\frac{\sqrt{V/n}}{m}$ or $\frac{s/\sqrt{n}}{m}$	Estimated coefficient of variation of a sample mean

The properties of the normal distribution have been studied and many of them described in detail; they can be found in statistical texts. In particular, the *cumulative distribution function* of the normal distribution has been tabulated. The cumulative distribution function defines the probability that a random variable takes on values less than or equal to a specified constant. We can find in published tables (available in many textbooks) the theoretical probability of a sample mean being, say, 1.338 standard deviations away from the true mean – or any other number of standard deviations. The only thing that we need to be careful about is that the sample size is large enough for us to assume the normal distribution. There is much folklore on this subject, and no globally true criterion exists. However, most workers in pest management are prepared to assume a normal distribution if their sample size is 25–30 or more.

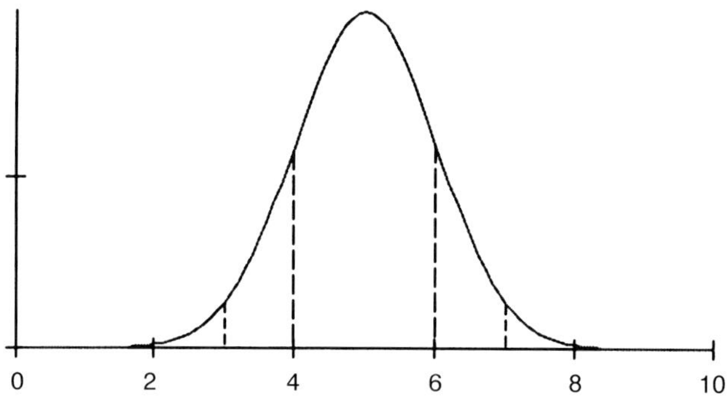

Fig. 2.4. A typical normal distribution, with mean equal to 5 and standard deviation equal to 1. The range from 4 to 6 contains about two-thirds of the distribution, and the range from 3 to 7 contains about 95% of the distribution.

Therefore we have two special reasons to be grateful to statistical theory:

1. For the formula which tells us how the variance is reduced when the sample size increases.
2. For defining the shape of the distribution of sample means when the sample size is large.

These are illustrated in Exhibit 2.1.

Exhibit 2.1. *Variance of sample means and the Central Limit Theorem*

Sampling was simulated on the computer by randomly selecting sample observations from Beall's record of Colorado potato beetle counts. Three sample sizes were used: 5, 25 and 50 sample units. For each sample size, the sampling process was simulated 500 times, thereby resulting in 500 estimates of the mean. The variance, σ^2, of the original counts was 14.995 and the mean, μ, was 4.74. The mean and variance of the 500 simulated sample means were calculated and graphed as functions of the sample size. The theoretical variance of the sample (Equation 2.2) was also calculated and graphed as a function of the sample size. The results are shown in Figs 2.5 and 2.6.

The squares in Fig. 2.5 represent the mean and variance of the original counts, while the circles are the average of the sample means (the mean of the means) and variances of the sample means. The means are essentially the same for the original counts and for all three sample sizes. The variances of the sample means decrease with increasing sample size and closely follow the theoretical variance (solid line).

In Fig. 2.6, the 500 estimates of the sample means are arranged as frequency distributions and are compared to frequencies based on the normal distribution (lines). Frequencies are shown for an interval rather than a single value of the mean (e.g. 0.2–0.4 versus 0.3), because the estimated means are continuous; and in order to calculate a frequency, the number of means in an interval must be tallied. As the sample size increases, the distribution of means is more like a normal distribution.

Continued

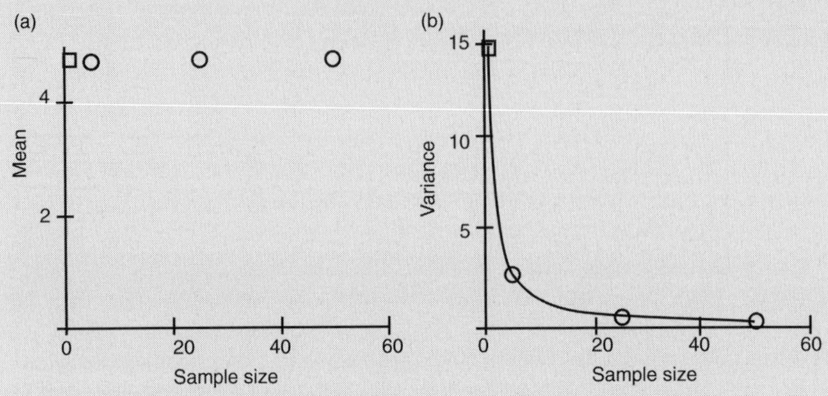

Fig. 2.5. The (a) mean and (b) variance of the original count of Colorado potato beetles (□) and of means and variances of means (○) for sample sizes 5, 25 and 50. The solid line in (b) is the theoretical variance of the sample mean, according to Equation 2.3.

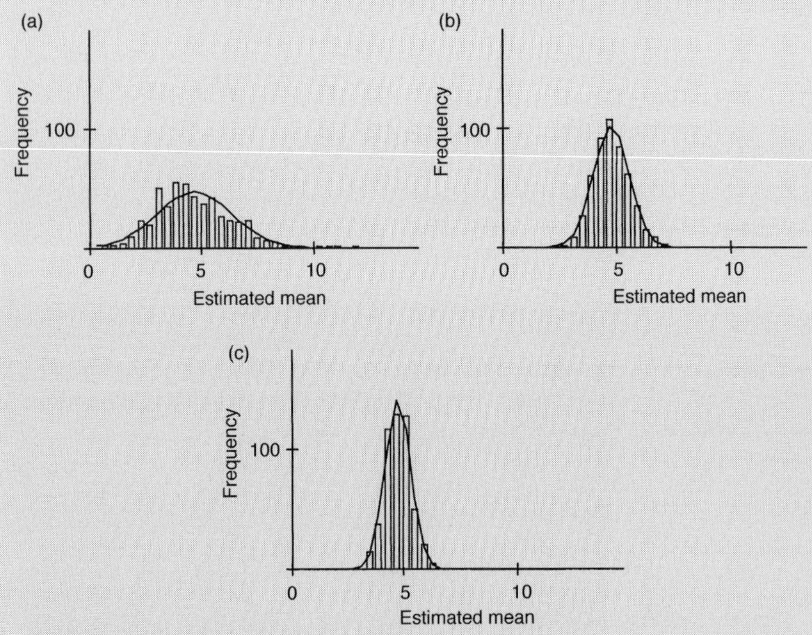

Fig. 2.6. Frequency distributions of 500 sample means for sample sizes of (a) $n = 5$, (b) $n = 25$ and (c) $n = 50$. Solid lines are theoretical frequencies based on the normal distribution.

2.10 What Does this Mean for Decision-making?

The purpose of sampling in pest management is to gather information on pest abundance so that a decision can be made on the need for some control action. Using the results discussed above, we can make inferences based on sample data which allow us to make informed decisions. For example, suppose that, following the principles laid out in this and the preceding chapter, we have decided that the critical density (*cd*) for Colorado potato beetles in fields such as Beall's, is 3.5 beetles per sample unit, and that 25 sample units are sufficient. In practice, this would mean that the management protocol is to inspect 25 sample units, calculate the average number of beetles per unit, and recommend a control action if the sample mean is greater than 3.5. Typical values for the sample mean (obtained by 1000 simulations) are displayed in Fig. 2.7, along with an indication of *cd*.

By looking carefully at Fig. 2.7a, we can count that about 30 of the 1000 sample means were less than 3.5, equal to a proportion 0.030, or 3%. It is tedious and hard on the eyes to do this for each sample protocol, but a good approximation is available based on the central limit theorem. For this, it is most convenient to transform Fig. 2.7a into Fig. 2.7b, by changing the *x*-axis from simple average counts of beetles per sample unit to their standardized form:

$$z = \frac{m - \mu}{sd_m}, \quad \text{where } sd_m = \frac{\sigma}{\sqrt{n}} \tag{2.4}$$

The transformed value, z, of the mean, m, is normally distributed (approximately), and has a mean value equal to zero and a variance equal to one. It is much easier to use published tables of the normal distribution using z than using m, but to do that, we need to transform *cd* in the same way:

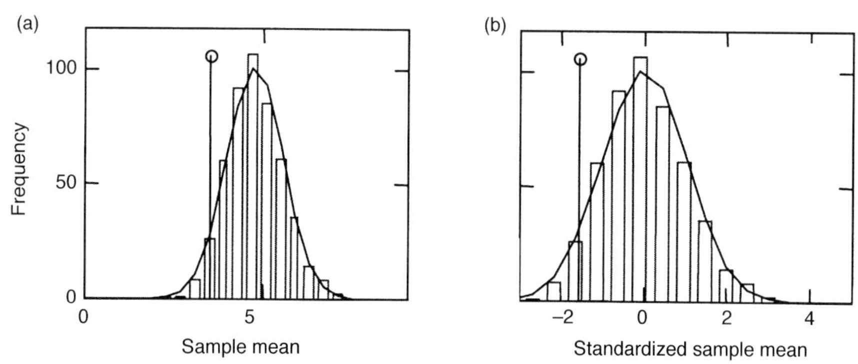

Fig. 2.7. Frequency distributions for means based on 25 sample units each and 1000 simulations; the critical density, *cd*, is shown by a tethered balloon. (a) Unadjusted sample means; (b) standardized sample means, based on Equation 2.4.

$$\text{transformed } cd = \frac{cd - \mu}{sd_m} \tag{2.5}$$

$$\text{calculated as} = \frac{3.5 - 4.736}{\left(\sqrt{14.995}\right)/5} = \frac{-1.236}{0.774} = -1.596 \tag{2.6}$$

The goal of all these calculations is to estimate the probability of a mean of 25 sample units being less than or equal to 3.5, which is (approximately) equal to the probability of a standardized normal deviate (z) being less than or equal to -1.596 (see Fig. 2.7b), and this is something that we can look up in tables. Note that z is only notionally calculated: it is the fact that it is a standardized normal deviate that is important, leading to the validity of looking up a table to obtain the probability. This probability is 0.055, which can be compared with the counted result, 0.030. The difference between these two numbers is the result of $n = 25$ not being large enough for the Central Limit Theorem to hold exactly, and because 1000 simulations may not be enough to give accurate enough results. For most workers, such an approximation is usually good enough, given all the other imponderables relating to pest management decision-making.

Thus, we can be reasonably sure that the probability of a sample mean based on 25 sample units is less than or equal to 3.5 is about 0.05. What this means for decision-making is that the probability of not intervening, if such a sample protocol were to be used for Beall's field, is about 0.05.

2.11 Calculating a Probability of Decision (OC) Function

These calculations can be generalized to apply to any field of potatoes which is sampled in a similar way to Beall's. Before using any sample protocol, we need to know what sort of decision recommendation it will give for whatever field may be sampled. We need to calculate the probability of the sample data recommending one or other management action. In Chapter 1 we introduced probability of decision and OC functions as summaries of these probabilities. The OC function is the probability, based on a field sample, of classifying pest density as less than or equal to some specified *cd* and thus the probability of deciding not to intervene.

The OC is the probability that m is less than or equal to *cd*, given a value for μ. This probability can be determined, as above, provided that we know the true variance, σ^2, of the sample counts, and provided that the sample size, n, exceeds some minimal level for the Central Limit Theorem to be relevant. Under these circumstances, we can calculate z (Equation 2.3) for any value of μ, and compare it with the similarly transformed *cd* (Equation 2.4) to obtain a probability based on the normal distribution. As above, z is not actually *calculated* – it is the fact that it is approximately a standardized normal deviate that is used to estimate probabilities. Thus we can calculate the OC function for a sampling protocol based on n equal to 25. Some values of the function are given in Table 2.3.

By now, we can predict what will happen to the OC function as the sample

Table 2.3. OC function values when an estimated mean is used to classify density with respect to a critical density, *cd,* equal to 3.5. A sample size of 25 is used, the sample mean is assumed to be normally distributed and $sd_m = 0.774$.

	μ								
	2	2.5	3	3.25	3.5	3.75	4	4.5	5
$\frac{cd - \mu}{sd_m}$	1.94	1.29	0.65	0.32	0	−0.3	−0.7	−1.3	−1.9
$OC = P\left(z \le \frac{cd - \mu}{sd_m}\right)$	0.97	0.9	0.74	0.63	0.5	0.37	0.23	0.1	0

size, n, is increased. As the sample size is increased, the standard deviation of the mean, sd_m, decreases. As a result, the standardized form of *cd*, $(cd - \mu)/sd_m$, increases in absolute value (unless $\mu = cd$). This causes the spread of values in the second line of Table 2.2 to become larger, which in turn causes the spread of probabilities in the third line to shrink. The end result is that OC function becomes steeper about *cd* as the sample size increases. This is illustrated in Exhibit 2.2.

The OC function for the sampling plan used to classify beetle densities with respect to *cd* can also be determined using simulation. To simulate the OC function, the following procedure is used:

1. A range of true means (μ's) is specified for which OC values are to be generated.
2. For each value of μ, a random variate is generated from a normal distribution with mean μ and standard deviation $\sigma / \sqrt{n}$, where σ is the population standard deviation. Each of these random variates represents a sample mean, m.
3. If $m \le cd$, μ is classified as less than or equal to *cd* and the recommendation is to do nothing; otherwise, μ is classified as greater than *cd*.
4. Steps 2 and 3 are repeated several times and the proportion of times μ is classified as less than or equal to *cd* is determined. This is the simulated OC value for that value of μ.
5. Steps 2–4 are repeated for each μ in the range of true means.

How close the simulated OC is to the true OC depends on the variability in the sample estimate and the number of times step 4 is repeated. This is illustrated in Exhibit 2.2.

Exhibit 2.2. *Simple OC functions based on the Central Limit Theorem*

For sampling plans that classify the population density on the basis of an estimate of the mean, the OC is the probability of a sample mean, normally distributed with mean μ and standard deviation $\sigma / \sqrt{n}$, being less than *cd*. This probability can be found by determining the normal cumulative probability distribution value for these parameters. Because $\sigma / \sqrt{n}$ decreases with increasing sample size, the OC function

Continued

becomes steeper as the sample size is increased. This is illustrated in Fig. 2.8, in which OC functions are shown for three sample sizes, 25, 50 and 100, when $cd = 3.5$ and $\sigma^2 = 15$.

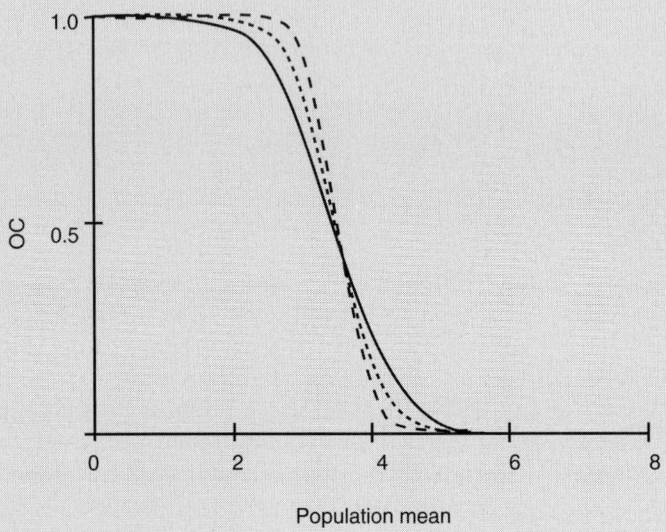

Fig. 2.8. OC functions computed using a normal cumulative distribution function for three sample sizes when $cd = 3.5$ and $\sigma^2 = 15$: ——, $n = 25$; ----, $n = 50$; - - -, $n = 100$.

OC functions can also be calculated using simulation. The normal distribution with the parameters μ and $\sigma/\sqrt{n}$ provides a model for sample means each calculated from a set of n sample units. Many such random numbers can be generated and the proportion of times these numbers are less than cd is the OC value for a particular μ. Simulation for a range of μ's results in a set of proportions that constitute an OC function. When simulation is used to determine OC values, the number of times sampling is simulated for each value of μ determines how accurate (smooth) the OC function will be. Three OC functions, one calculated analytically and two using simulation, for $n = 25$ and $\sigma^2 = 15$ are shown in Fig. 2.9. Those determined by simulation used either 25 or 250 simulation replicates for each OC value. Note that when 250 simulation replicates were used, the simulated OC is close to the analytical result. The number of simulation replicates required to obtain a smooth OC function varies depending on the variability in the sample data.

The accuracy of a simulated OC function can be predicted to a certain extent. Equation 2.2 can be used to estimate the variance of any one simulated OC value, if the variance of the true value (p) is known. We shall show in Chapter 4 that a simulated OC value follows a binomial distribution with expectation p and variance $p(1 - p)/sr$, where sr is the number of simulation replicates. It follows from the Central Limit Theorem applied to the simulation process that the standard error of a simulated OC value is

$$\text{simulation } se = \sqrt{\frac{p(1-p)}{sr}} \qquad (2.7)$$

and, for two-thirds of the time, the simulated OC should be within one standard error of the true OC, as noted above (Section 2.9). This is illustrated for the OC function in Fig. 2.9, which is based on 25 simulation replicates: the simulated values were mostly within one standard error of the true values (Fig. 2.10). When OC functions, estimated by simulation, are compared with each other, their accuracy should be noted before bold statements are made.

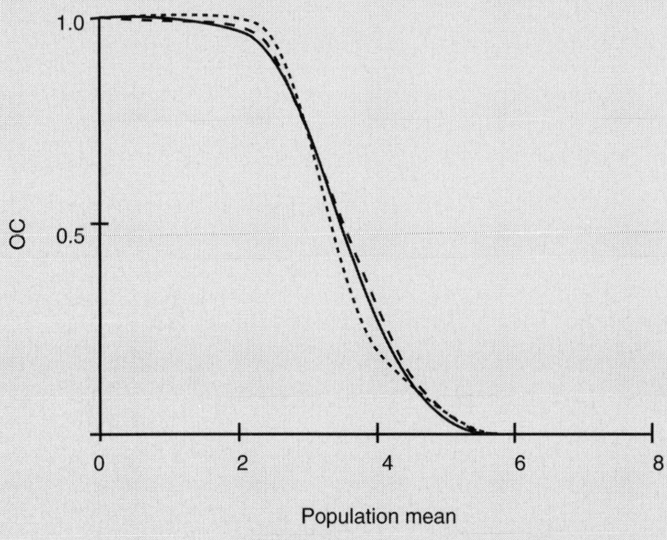

Fig. 2.9. OC functions computed using a normal cumulative distribution function and via simulation for $n = 25$, $cd = 3.5$ and $\sigma^2 = 15$: determined using the normal distribution function (——), using simulation with 25 replicates (------) and with simulation using 250 replicates (- - - -).

Continued

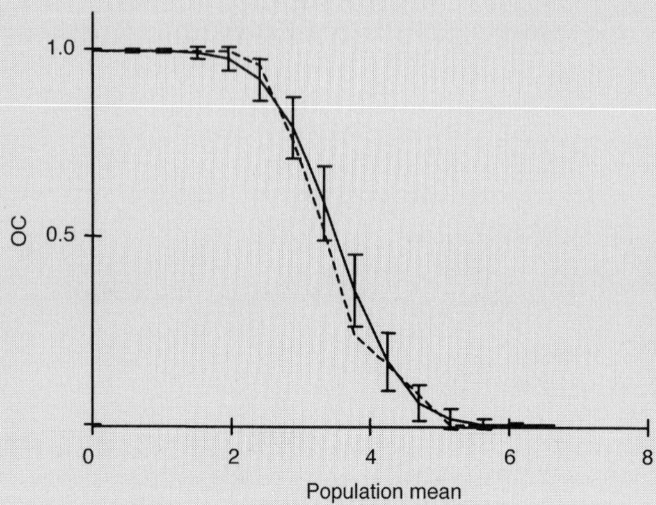

Fig. 2.10. Predicting the accuracy of the simulated OC functions in Fig. 2.9. The OC function is computed using the normal distribution (——), with error bars indicating one standard error around it based on 25 replicates, and the simulated OC function is based on 25 replicates (- - - -).

For readers who have worked with statistical hypothesis testing and statistical significance, the OC function may not be new. An experiment could be set up to examine whether the mean of a test population or process (e.g. the survival rate of an insect under stress) is, or is not, less than a certain specified value, C. We plan to assume a normal distribution for the data, with standard deviation equal to σ, and use a 5% significance test (note that for a standardized normal distribution, the probability of getting a value less than 1.645 is equal to 0.05). What this means is that after the experiment, we intend to state that the true mean of the test population is significantly less than C if the experimental mean is less than $C - 1.645\sigma/\sqrt{n}$. Depending on the true mean, the experiment will have varying probabilities of finding significance. A plot of these probabilities against true mean values is called the *power function* of the experiment.

Making a pest management decision on the basis of a sample mean, as in Exhibit 2.2, is similar to doing such an experiment. If

$$cd = C - 1.645\sigma/\sqrt{n}$$

the OC function of Exhibit 2.2 and the power function for such an experiment are the same. The plans discussed in Exhibit 2.2 can be regarded as designs for experiments to test whether the true mean of pests per sample unit is, or is not, less than C; it depends on how you view what is going on. Looked at the other way, there are many statisticians who advocate looking at the entire power function for a proposed experiment, rather than at the probabilities of accepting or rejecting one specific (often artificial) null hypothesis. Tukey (1950) drew attention to the similarity of OC functions and power functions.

2.12 Summary

Five key ideas have been presented in this chapter:

1. Bias is an omnipresent factor when sampling for pest management. It is important to consider data collection carefully, so that bias does not unwittingly influence the sample outcome.
2. A trustworthy sample satisfies four criteria: it should be representative, reliable, relevant and practically feasible.
3. The precision of an estimated population parameter increases as the sample size increases. This is because the variance of an estimated parameter decreases with increasing sample size.
4. The distribution of sample means is described by a normal distribution function provided that a minimum sample size is used. Often, $n > 25$ is sufficient for the normal approximation to be acceptable.
5. OC functions for fixed samples that classify density based on an estimated mean can be calculated using the normal cumulative distribution function. OC functions for these sampling plans can also be generated by simulation.

References and Suggested Reading

Beall, G. (1939) Methods of estimating the population of insects in a field. *Biometrika* 30, 422–439.

Brenner, R.J., Focks, D.A., Arbogast, R.T., Weaver, D.K. and Shuman, D. (1998) Practical use of spatial analysis in precision targeting for integrated pest management. *American Entomologist* 4, 79–101.

Cressie, N.A.C. (1993) *Statistics for Spatial Data.* John Wiley, New York, 900 pp.

Dent, D. (ed.) (1995) *Integrated Pest Management.* Chapman & Hall, London, 356 pp.

Guttierez, A.P. (1995) Integrated pest management in cotton. In: Dent, D. (ed.) *Integrated Pest Management.* Chapman & Hall, London.

Legg, D.E., Shufran, K.A. and Yeargan, K.V. (1985) Evaluation of two sampling methods for their influence on the population statistics of alfalfa weevil (Coleoptera: Curculionidae) larva infestation in alfalfa. *Journal of Economic Entomology* 78, 1468–1474.

Milikowski, M. (1995) *Knowledge of Numbers: a Study of the Psychological Representation of the Numbers 1–100.* University of Amsterdam, The Netherlands, 135 pp.

Page, E.S. (1967) A note on generating random permutations. *Applied Statistics* 16, 273–274.

Schotzko, D.J. and O'Keeffe, L.E. (1989) Geostatistical distribution of the spatial distribution of *Lygus hesperus* (Heteroptera: Miridae) in lentils. *Journal of Economic Entomology* 82, 1277–1288.

Schotzko, D.J. and O'Keeffe, L.E. (1990) Effect of sample placement on the geostatistical analysis of the spatial distribution of *Lygus hesperus* (Heteroptera: Miridae) in lentils. *Journal of Economic Entomology* 83, 1888–1900.

Tukey, J.W. (1950) Chairman's closure. In: *Acceptance Sampling, a Symposium*. American Statistical Association, Washington, DC, pp. 150–155.

Yates, F. (1960) *Sampling Methods for Censuses and Surveys*, 3rd edn. Griffin, London, 440 pp.

3 Classifying Pest Density

3.1 Introduction

In this chapter we return to basics: the choice between sampling for estimation and sampling to make decisions. We begin by contrasting estimation and classification, and state why classification of pest density is more appropriate than estimation for pest management decision-making. With this in mind, we examine how sample costs can be reduced by collecting samples in batches, deciding after each batch whether more samples are needed to come to a decision. Setting up such plans requires knowledge of the variance of sample data, especially for mean pest densities near the critical density. We present variance–mean models which have been used as formulae to estimate variances for any pest density, and we discuss the usefulness and limitations of these models.

3.2 Classification versus Estimation

The goal of integrated pest management is to protect the crop from pests in an economical way, relying first on preventative and environmentally benign practices, such as crop rotation and sanitation, plant resistance, and naturally occurring natural enemies and antagonists, and then, if the need arises, on the judicious use of corrective control tactics such as pesticides or inundative natural enemy releases. Many factors come into play when determining the need for pest control, such as predicted weather conditions, closeness to harvest, other plant stresses and crop value. A critical piece of evidence that is always required is whether pressure on the crop from pests is too high.

At first glance, one might think that a really good estimate of pest density would be the ideal piece of evidence. A grower could then weigh all the information together, and come to a decision on the need for control. No one would argue with this in principle, but in practice no reliable objective procedures for combining all relevant information have yet been developed (although some serious attempts have been made). One must therefore rely on a few key pieces of information which, when coupled with practical experience and understanding, can indicate a management action. One key piece of information is whether pest density is above or below a level which is likely to cause unacceptable damage (see Chapter 1). This is in principle a matter of classification. Precise knowledge about the true density is unnecessary, except as a guide as to whether the true density is above or below the critical density. Thus, because the information contained in a precise estimate of density can rarely be fully used, density estimation seems to be less appropriate for decision-making than density classification.

But are estimation and classification really so different? In both procedures, samples must be collected, pests organisms are counted, and on many occasions a sample mean is calculated (i.e. an estimate is found). It is only here that a vital difference is detected: in classification, the goal is to find out if pest density is too high, and an estimate (if it is calculated at all) is only useful as an intermediate step to reach that goal. At this point, therefore, we must tie down the differences between sampling for estimation and sampling for classification, and note typical criteria for designing good sampling plans for either purpose.

3.2.1 Estimation

Suppose that we want to relate pest density to crop yield, or we want to determine how the presence of natural enemies influences pest density. In circumstances such as these, we need estimates of pest numbers per sample unit or per area. These estimates – whether high, low or moderate – must be precise enough to give useful results after they had been used in mathematical calculations or models (possibly very complex ones). We want to be sure that the results of our calculations will be acceptable to others and not easily brushed aside as, for example, 'too vague'. We want to be able to use variances and the Central Limit Theorem, perhaps in sophisticated ways, to delimit how close our sample estimates probably are to the true values.

This problem can usually be solved by relying on the *representativeness* (no unaccountable bias) and *reliability* (no effect of uncontrolled variables) of the sample protocol (see Section 2.5), and also on the two important statistical results presented in Chapter 2:

1. Variance of the estimate of the mean, m, based on n sample units is equal to σ^2 / n and is estimated by V / n.
2. For large n, the distribution of m is normal.

Representativeness and reliability ensure that the sample mean, m, is an unbiased estimate of the true pest density, μ. In other words, the *mathematical expectation* of m is equal to μ. Because of the two statistical results, the standard deviation of the sample mean is $\sigma / \sqrt{n}$, and, if n is large enough, the shape of the distribution of m is known (it is close to normal). These can all be combined to show that the distribution of all hypothetical sample means, m, is centred on the true mean density, μ, and the distribution is shaped like a normal distribution with standard deviation equal to $\sigma / \sqrt{n}$, which can be estimated by $\sqrt{V / n}$. Because all that we have is one value of m, we want to turn this statistical result around to allow ourselves to infer a value for μ.

In Chapter 2 we noted some properties of the normal distribution, one of which is that 95% of it lies within two standard deviations of the mean. Retaining our assumption that the distribution of all hypothetical sample means in the above discussion is normal, then 95% of these means are no more than two standard deviations away from the true mean. This is the basis of what are called confidence intervals. We can assume that our single value m is one of the 95% which are within two standard deviations of the true mean μ , and say that μ must therefore be no further than $2\sigma / \sqrt{n}$ from the sample mean, m. We indicate that this is an assumption by saying that the range

$$m - 2\sigma / \sqrt{n} < \mu < m + 2\sigma / \sqrt{n} \tag{3.1}$$

is a 95% *confidence interval*. We can replace $\sigma / \sqrt{n}$ by $\sqrt{V / n}$, to obtain

$$m - 2\sqrt{\frac{V}{n}} < \mu < m + 2\sqrt{\frac{V}{n}} \tag{3.2}$$

Confidence intervals are useful ways of describing where true means are. For example, the press often quotes survey results in words such as 'the percentage support for the governing party is between 37% and 43%, with a 1 in 20 chance of error'. This corresponds to a 95% (95/100 = 1 − 1/20) confidence interval:

37% < true support < 43%

Naturally, we are not stuck with a 95% confidence interval. We can choose any other percentage, but then the factor 2 in Equation 3.2 must be changed accordingly. For example, if a 90% confidence interval were used, 2 would be replaced by 1.645. This result can be checked by looking up tables of the normal distribution. The number 2 is an approximation for the 95% interval; a more exact value is 1.96. The general formula for a 100(1 − α)% confidence interval based on the normal distribution is

$$m - z_{\alpha/2}\frac{\sigma}{\sqrt{n}} < \mu < m + z_{\alpha/2}\frac{\sigma}{\sqrt{n}} \tag{3.3}$$

where $z_{\alpha/2}$ is defined by the following statement: 100(1 − α)% of a standard normal distribution (i.e. mean = 0, variance = 1) is within $z_{\alpha/2}$ of the mean. A 95% interval has α = 0.05, because if you are unlucky and your estimate (m) happens to be among the 5% which are outside the two standard deviation interval around μ, it could equally well be on either side (Fig. 3.1a): α is the overall probability of 'getting it wrong'; that is to say that the interval does *not* contain the true mean. Convention divides this error probability equally on both sides of the interval (hence $\alpha/2$ and $z_{\alpha/2}$), but this is not absolutely necessary. In the extreme case you may decide that you want a one-sided interval, with all the uncertainty put on one side of the estimated value m (Fig. 3.1b).

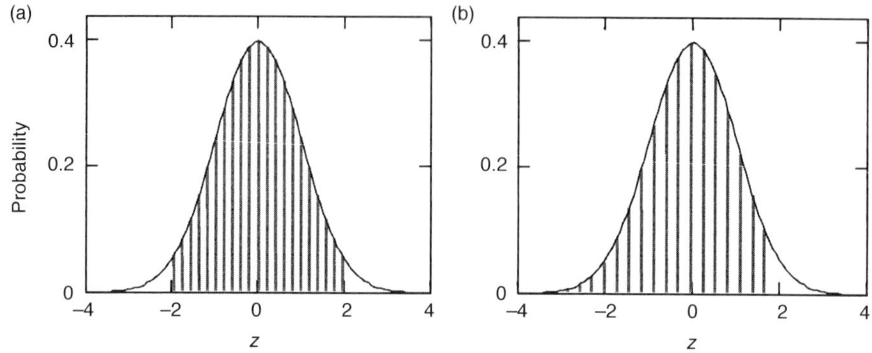

Fig. 3.1. Areas under the standardized normal distribution (mean = 0, variance = 1) indicated by shading. (a) Two-sided interval, −1.960 to 1.960, contains 95% of the total probability; (b) one-sided interval, less than 1.645, contains 95% of the total probability.

3.2.1.1 *Choosing the sample size,* n, *to achieve specified precision*

Typically, a confidence interval is used after data have been collected, but if a reasonable estimate, V, of σ^2 is available beforehand, Equation 3.3 can be turned around to indicate how many sample units are required to achieve a desired precision. The half-width of the interval, substituting V for σ^2, is

$$w = z_{\alpha/2}\sqrt{\frac{V}{n}} \tag{3.4}$$

which can be turned into an expression that defines n:

$$n = V\left(\frac{z_{\alpha/2}}{w}\right)^2 \tag{3.5}$$

If we wished the 95% confidence interval ($z_{\alpha/2} = 1.96$) to have a half-width equal to 2.5 when $V = 60$, then $n = 60(1.96/2.5)^2 = 36.9$, so 37 sample units would be required.[1]

The coefficient of variation (see Table 2.2) is often used to indicate precision after data have been collected. It also can be turned around to indicate how many sample units are required to achieve a desired precision, provided again that a reasonable estimate, V, of σ^2 is available beforehand. The formula for CV, the coefficient of variation of the sample mean, is (from Table 2.2) as follows:

$$\mathrm{CV} = \frac{\sqrt{V/n}}{\mu} \tag{3.6}$$

[1]Strictly speaking, when σ is estimated, z based on the normal distribution should be replaced by a number based on another distribution (the *t*-distribution). However, if the value of n is large enough to satisfy the Central Limit Theorem, the difference should be ignorable.

so

$$n = V\left(\frac{1}{\mu \times CV}\right)^2 \tag{3.7}$$

For example, to obtain a CV equal to 25% when $\mu = 5$ and $V = 60$, then $n = 60(1/(5 \times 0.25))^2 = 38.4$, so 39 sample units would be required.

Equations 3.5 and 3.7 are similar, but a critical difference is in how the required precision is specified. In Equations 3.4 and 3.5, w is a constant, but in Equations 3.6 and 3.7 CV is proportional to μ. In practical terms, this means that n defined by Equation 3.7 changes as μ changes, but n defined by Equation 3.5 does not. The two methods can be made equivalent if w and CV are both specified either as constants or as proportional to μ.

3.2.2 Classification

Clearly, any procedure which gives an estimate of μ can be used for classification. Either of the Equations 3.5 or 3.7 can be used to determine n for a sampling plan, data could be collected and the sample mean could be obtained. This mean would then be compared with a critical density to make the classification:

$$\mu \le cd \quad \text{if } m \le cd \qquad \text{or} \qquad \mu > cd \quad \text{if } m > cd \tag{3.8}$$

We can measure how good the classification is by turning around the confidence interval (Equation 3.2) and determining whether m is inside or outside the classification interval

$$cd \pm z_{\alpha/2}\sqrt{\frac{V}{n}} \tag{3.9}$$

If m is outside the classification interval, the (confidence) probability of the classification (Equation 3.8) being correct is $1 - \alpha$, but if m is inside the classification interval, the (confidence) probability of the classification being correct is less than $1 - \alpha$. The confidence probability, $1 - \alpha$, can be regarded as the probability of getting it right. For pest management purposes, we are interested in the 'correctness' of the classification. What we want in general terms is that the sample size, n, is adjusted so that m is outside such a classification interval because a decision may then be made with confidence.

For readers with some knowledge of statistics, it is worth noting that the above is linked to hypothesis testing. If m is outside the classification interval (Equation 3.9), then what is called a statement of statistical significance can be made: μ is significantly different from cd, at the $100\alpha\%$ probability level. What we want for classification is to be able to make a statement such as this whenever we take a sample for decision-making. We should like a sampling plan that automatically tries to adjust n accordingly.

The key distinction between estimation and classification lies in the purpose to which the sample information is directed. With estimation, we begin by specifying how close we wish an estimate to be to a true value. With classification, we begin by specifying how frequently we wish to make correct classifications, or how frequently we will tolerate incorrect classifications. In both cases, sample sizes can be adjusted to achieve the specified objective. Clearly, concepts pertinent to estimation and classification are related, but the purposes to which these concepts are applied are quite different. In this chapter, we begin to show how concentrating on sampling for classification produces its own optimal values for sample size. These optimal values will turn out to be radically different from those for estimation.

3.3 Reducing Sampling Costs: Classification by Sampling in Batches

In Chapter 2 we described how classification could be done after collecting a fixed number of sample units and, using the normal distribution, we showed how OC functions could be calculated. Fixed sample size plans have certain advantages: simplicity and the relative assurance that a representative sample can be obtained. However, when classification with respect to a critical density is the objective, intuition suggests that fewer samples are needed to make a correct classification when pest abundance is very different from the critical density than when it is close to it. By reducing the number of samples required to make a correct classification, if the data allow so, sampling costs are reduced, which is obviously an important consideration.

Statisticians have proposed ways of cutting short sample collection if the evidence overwhelmingly supports one of the possible classifications. The simplest approach is to make one preliminary sample to get a quick idea of the situation, and then take a second one if more information is needed. This process is called double sampling. It may appear to be an obvious procedure, and it is, but working out the probabilities of decision (operating characteristic (OC) function) and the corresponding average sample number (ASN) function is not easy.

3.3.1 *Two batches*

A simple double sampling plan is to take an initial sample ('batch') of n_B sample units, and calculate the sample mean, m, and the ends of a classification interval (Equation 3.9) around critical density (cd):

$$\begin{aligned} L &= cd - z_{\alpha/2}\sqrt{\frac{V}{n_B}} \\ U &= cd + z_{\alpha/2}\sqrt{\frac{V}{n_B}} \end{aligned} \tag{3.10}$$

If the sample mean is greater than U, a decision is made to intervene, and if it is below L, the decision is not to intervene. If it is between these two values, a second sample of n_B units is taken and the sample mean for all $2n_B$ units is calculated, and a final decision is made, however small the difference between *cd* and the sample mean. If the sample mean is greater than *cd*, intervene; otherwise, do not intervene. No further sample units are taken. This approach is an attempt to ensure $1 - \alpha$ confidence in the final classification (see Section 3.2.2).

In practice, it is not necessary to calculate the sample mean each time – the cumulative total count of organisms in the samples can be used. This reduces calculations during the sampling. The equations for L and U are then

$$L = n_B\left(cd - z_{\alpha/2}\sqrt{\frac{V}{n_B}}\right)$$
$$U = n_B\left(cd + z_{\alpha/2}\sqrt{\frac{V}{n_B}}\right) \tag{3.11}$$

The pair (L, U) at n_B sample units, and *cd* (or $2n_Bcd$ if using Equation 3.11) at $2n_B$ sample units together define a primitive *stop boundary*, often referred to simply as a *boundary*. A stop boundary specifies termination points in a graph or chart at which points representing the sample data are plotted. Each point is plotted as data are collected, with the number of sample units along the horizontal axis (x-axis) and the current sample mean, or total, on the vertical axis (y-axis). The position of each new point is compared with the stop boundaries. If a boundary is reached or crossed, a final classification is made based on the location of the final point (classifying μ as greater or less than *cd*); otherwise, more sample units are collected. This is illustrated in Fig. 3.2.

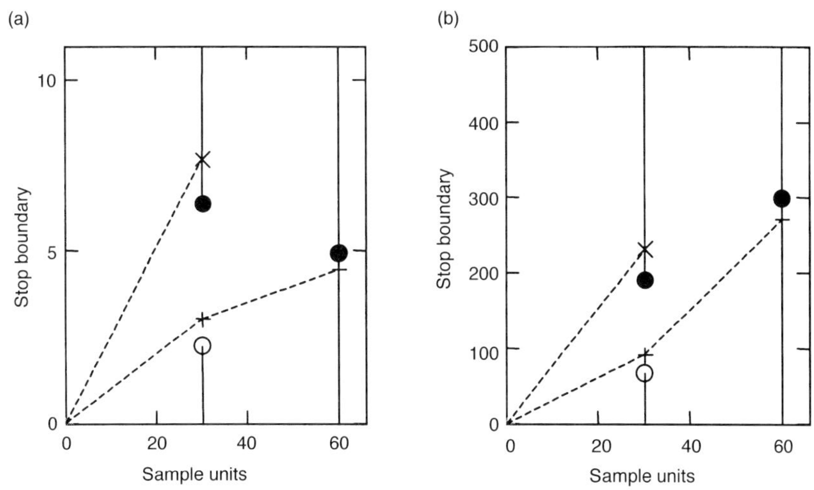

Fig. 3.2. Stop boundaries for a batch sampling plan with two batches; boundary points are represented by open (L) and closed (U) circles. (a) Stop boundaries with the average count per sample unit as dependent variable; (b) stop boundaries with the total count as dependent variable. Two sample runs are illustrated: one stopping after the first batch (×), and the other stopping after the second batch (+).

It is not easy to calculate the probability of decision (OC) or ASN functions for double sampling plans analytically: the probabilities of intervening or not intervening based on the initial n_B sample units are easy enough, but the overall function involves conditional probabilities and double integrals, which are beyond the scope of this book. However, simulation can be used to get approximations. The basic idea behind the simulation approach is described in Chapter 2, and is the method used in this book to estimate OC and ASN functions.

When we simulate the sampling process to estimate properties of a sampling plan, we need to use the true variance in generating the simulated sample data. Hence, to evaluate a sampling plan, we must take on two roles, one for setting up the parameters of the sampling plan and the other for simulating and estimating OC and ASN functions. For the first, we assume the role of the user who does not know σ^2 and, for the second, we assume the role of an all-knowing being who does. The steps in the simulation are as follows:

1. Set up the sampling plan: calculate the stop boundaries L and U for sample totals.
2. Specify a range of true means (μ's) for which OC and ASN values will be generated.
3. For each value of μ, do *sr* simulations of the sampling plan:
3a. Mimic sampling by generating a random variable from a normal distribution with mean $n_B\mu$ and standard deviation $\sigma\sqrt{n_B}$. Note that the true variance, σ^2, and not its estimate, V, is used.
3b. Compare the random variable with U and L (note that U and L are based on the estimated variance, V). If it is less than L, classify the density as less than or equal to *cd* and proceed to step 3d. If it is greater than U, classify the density as greater than *cd* and proceed to step 3d. Otherwise …
3c. Generate another random variable from the same distribution and compare the total of the two random variables to $2n_Bcd$ and make a classification accordingly. Go to step 3d.
3d. Repeat steps 3a, 3b and 3c *sr* times.
4. Calculate the number of times density is classified as less than or equal to *cd*, and divide this by *sr*. This is the estimated probability of not recommending intervention; it can be plotted against the corresponding value of μ as part of the OC function.
5. Calculate the average sample size as S/sr, where S is the sum of the number of instances when the first sample was outside the interval (L, U) multiplied by n_B, and the number of instances when the first sample was inside this interval multiplied by $2n_B$. This can be plotted against the corresponding value of μ as part of the ASN function.
6. Return to step 2 and repeat for another value of μ.

Changes in the parameters of a double sampling plan have some predictable general effects on the OC and ASN functions, but the magnitude of these effects, and the interactions among parameters, are not easily predicted. The OC and ASN functions relate the true mean, μ, to two properties of the sampling plan, as exemplified in the simulation scheme above. The OC function represents the probability of deciding not to intervene, and the ASN function represents the expected number of sample units. The effects of different parameter changes in a double sampling plan, which are explained and illustrated in Exhibit 3.1, are described in Table 3.1:

Table 3.1. The effects of different parameter changes in a double sampling plan.

Change	Consequence
Increase error probability, α (decrease z_α)	The classification interval (L,U) is narrowed, which in turn decreases the ASN function. The OC function is made flatter near *cd*, but the magnitude of the change may be slight
Increase n_B	The OC function is made steeper and the ASN function is increased
Increase *cd*	The OC and ASN functions are shifted to the right on the pest density axis
Increase the true variance, σ^2, above the estimated variance, V	The OC function is made flatter. The ASN function is (usually) decreased near *cd*, but increased far from *cd*

Exhibit 3.1. *Batch sequential sampling with two batches*

Here we illustrate how changes in parameters of double sampling plans influence OC and ASN functions. Double sampling plans were used to classify density about a *cd* of 4.0. The estimated variance of sample counts (V) was 50. The base parameters for the sampling plan were $\alpha = 0.2$ and batch sample size (n_B) = 20. Sample means were assumed to be distributed normally, and simulation based on the normal distribution was used to generate OC and ASN functions. Unless otherwise stated, the true variance, σ^2, was taken to be equal to V. Each OC and ASN estimate was calculated using 1000 simulation replicates. Parameters were changed one at a time to investigate the influence of α, n_B, *cd* and σ^2 on the OC and ASN.

The effect of α The value of α influences the points L and U that define the intermediate boundary. Increasing α narrows the interval (L,U), with the effect that sampling stops more often at n_B. This reduces the number of times when $2n_B$ sample units are needed, so the ASN is decreased and more misclassifications occur (OC is flattened), especially near pest densities corresponding to L and U. Simulations were done for $\alpha = 0.05$, 0.2 and 0.4. The resulting OC and ASN functions are shown in Fig. 3.3. As α was increased, the ASN decreased and the slope of the OC function decreased. Changes in the OC functions were slight, while for population means close to *cd* the ASN was nearly halved.

The effect of n_B When n_B is increased, the precision of the sample mean at n_B and at $2n_B$ is increased (see Equations 2.2 and 2.3), and the classification error is decreased (see Section 3.2.2). Therefore, the OC function is made steeper, at the expense of a high ASN, as can be exemplified by simulation. Larger values of n_B resulted here in increased ASN functions, and had a notable effect on the OC curves (Fig. 3.4). Increasing n_B had a greater proportional effect on OC than had

Continued

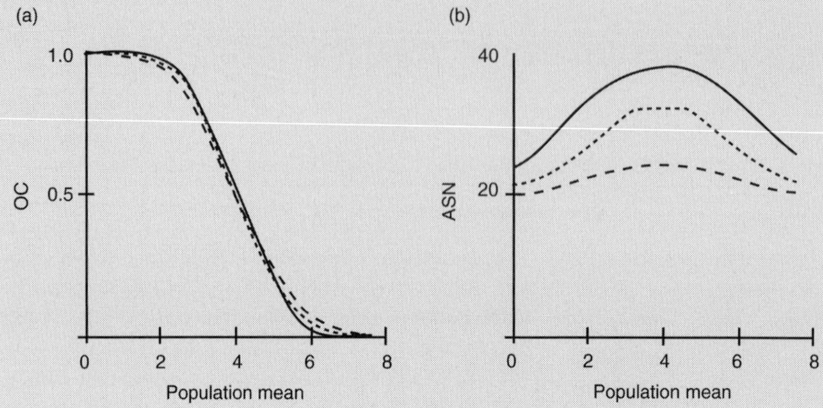

Fig. 3.3. The effect of α. OC (a) and ASN (b) functions for double sampling plans with $n_B = 20$, $cd = 4$ and $V = 50$: $\alpha = 0.05$ (——), 0.2 (-----) and 0.4 (- - - -).

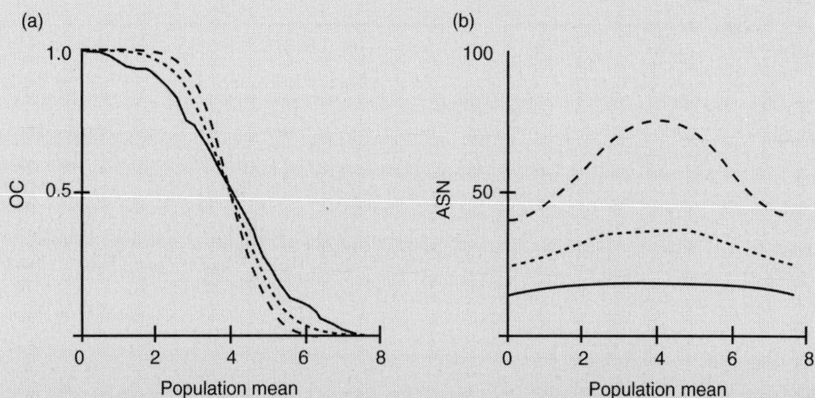

Fig. 3.4. The effect of n_B. OC (a) and ASN (b) functions for a double sampling plans with $cd = 4$, $V = 50$ and $\alpha = 0.05$: $n_B = 10$ (——), 20 (-----) and 40 (- - - -).

increasing α in the previous paragraph. Both α and n_B influence the intermediate boundary (L,U), but n_B also affects the precision of the sample means.

The effect of cd Increasing *cd* always shifts the OC and ASN functions to the right and decreasing *cd* shifts the functions to the left. The effects of a 10% change in *cd* either way are shown in Fig. 3.5. Changes in the OC function are easily seen; the effect is less easily seen in the ASN function, but it is there nonetheless. Because *cd* is often only crudely estimated in practice, the effect of this parameter on the OC function is always worth checking. Different pest managers and growers may feel inclined to adjust a *cd* to their own needs, especially if a value for *cd* has

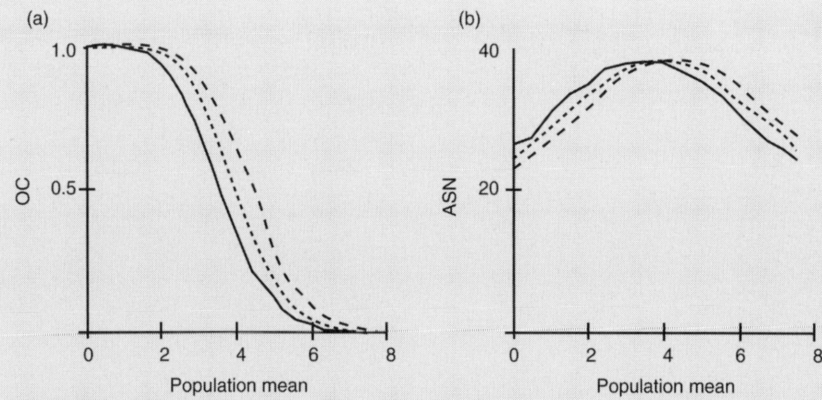

Fig. 3.5. The effect of *cd*. OC (a) and ASN (b) functions for a double sampling plans with $n_B = 20$, $V = 50$ and $\alpha = 0.05$: $cd = 3.6$ (——), 4.0 (-----) and 4.4 (- - - -).

been obtained elsewhere and imported into their area, but this raises questions that should be addressed following principles described in Chapter 1. It is possible that, despite the effort involved, more effort should be put into estimating critical densities in practice.

The effect of σ^2 An estimate, V, of the true variance of sample observations, σ^2, is required to construct the boundaries for a double sampling plan. However, V is rarely identical to σ^2. The relationship between σ^2 and V will be frequently addressed in this book, although from a more complex perspective. A first insight can be gained by studying the simple situation in which the estimated variance is constant but the true variance varies. Boundaries were constructed using $V = 50$, but different values of σ^2 (25, 50 and 100) were used to simulate the OC and ASN functions. The results are shown in Fig. 3.6.

When $\sigma^2 > V$, the OC becomes flatter, indicating greater errors (reduced accuracy) in classification. This is understandable because the (simulated) sample observations are more variable than expected, resulting in a greater number of incorrect decisions. The ASN function is reduced for means close to *cd*, because the increased variability at n_B sample units increases the probability of stopping after the first batch: the sample mean is more likely to be outside the interval (L,U) of Equation 3.10. The ASN function is increased for means much greater or smaller than *cd*, because the increased variability at n_B sample units now decreases the probability of stopping after the first batch. The effects on OC and ASN are opposite when $\sigma^2 < V$.

The changes in variances used in this example are extreme – a halving or doubling of the nominal value. However, they illustrate an effective approach to deciding how well σ^2 needs to be estimated, and can reduce considerably the fieldwork spent estimating σ^2. Based on a preliminary estimate (V), results as shown in Fig. 3.6 can be obtained. σ^2 must be estimated by V well enough so that:

Continued

1. The OC and ASN corresponding to extreme values of σ^2 are acceptable.
2. More extreme values of σ^2 are unlikely to occur in practice.

Considerations such as these can reduce the amount of effort put into estimating σ^2, at the expense of merely doing some simulations.

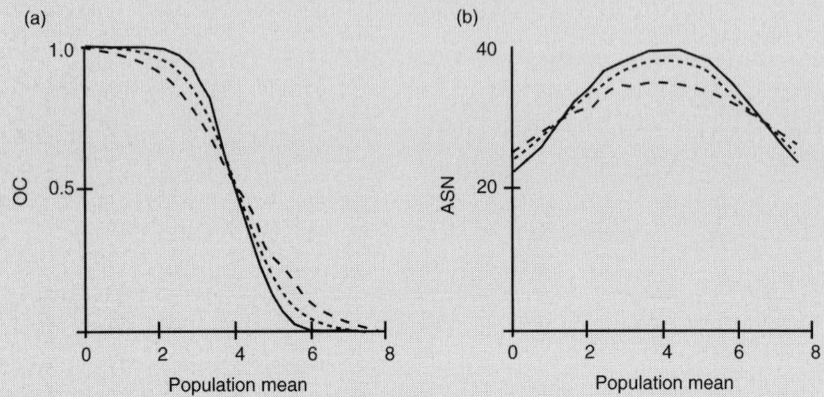

Fig. 3.6. The effect of σ^2. OC (a) and ASN (b) functions for a double sampling plans with $n_B = 20$, $cd = 4$, $V = 50$ and $\alpha = 0.05$: $\sigma^2 = 25$ (——), 50 (-----) and 100 (- - - -).

3.3.2 More than two batches: batch sequential sampling plans

If double sampling seemed like a good idea, why not extend the idea to more batches of samples? For example, if the sample sizes at each batch in the above plan were divided into two batches of 10 each (or even 40 batches of one each!), the maximum number of samples would still be 40, but there might be a chance of stopping sampling with fewer sample units examined. When three or more batches are used, the resulting plans are called *batch sequential sampling plans*. In general, when the maximum number of possible sample batches is I, the general formulae for the lower and upper stop boundaries for sample totals at each intermediate batch, i, are as follows:

$$\left.\begin{aligned} L_i &= i \times n_B\left(cd - z_{\alpha/2}\sqrt{\frac{V}{i \times n_B}}\right) \\ U_i &= i \times n_B\left(cd + z_{\alpha/2}\sqrt{\frac{V}{i \times n_B}}\right) \end{aligned}\right\} \tag{3.12}$$

and

$$L_I = U_I = I \times n_B cd \tag{3.13}$$

If batch number I is reached, a final decision is made, however small the difference is between *cd* and the sample total.

As the maximum number (I) of batches is increased and n_B is reduced (keeping the same maximum number of sample units, In_B), there is more opportunity for sample counts to cross the boundaries earlier, especially when the true mean is far from *cd*, resulting in a lower ASN. Other things being equal, OC functions for lower ASN plans are flatter than those for higher ASNs, meaning that there are fewer correct classifications. The general rule is that the higher the ASN, the steeper is the OC function, although improvements in the OC function are often disappointing relative to increases in the ASN. We have noticed on numerous occasions that large changes in the ASN can result in meagre changes to the OC: the gain in OC appears to be not worth the pain of a greatly increased ASN. However, as we noted in Chapter 1, whether an OC and ASN pair is acceptable, or better than another pair, must be decided in a wider context.

Because we are assuming that the Central Limit Theorem applies to sample means estimated from each set of samples, we cannot reduce the batch sample size, n_B, much below 20 and still be reasonably confident the assumption is adequate. We address this subject more fully in later chapters. For now, we impose the limitation that the batch sample size is sufficient to ensure that the Central Limit Theorem applies.

We can now begin to realize the wide choice that is available for designing a batch sequential sampling plan. The patterns outlined for double sampling also apply to batch sequential sampling with the addition that increasing the number of batches (while holding the overall maximum sample size constant) generally reduces ASN and flattens the OC. When designing a classification sampling plan, the most important thing to bear in mind is that the objective is to achieve an acceptable OC and ASN subject to the sample information being representative and reliable. We shall illustrate this design process shortly (Exhibit 3.2), but first we must address one important technical difficulty.

3.4 Variance–Mean Relationships

A perfectly usable decision-making plan can be set up even if σ^2 is completely unknown. All that is needed is a value for *cd*. An estimate of σ^2 at *cd* is necessary for setting up batch sampling plans as above (Equations 3.10–3.12), but a simple plan can do without it (see e.g. Section 2.11). However, an estimate of σ^2 is critical for assessing the properties of the decision guide; in particular, how often correct decisions are obtained, as expressed in the OC function, and how many sample units are needed, as expressed in the ASN function. Without estimates of the OC and ASN functions, there is no way to judge the decision guide. Until now, we have assumed that σ^2 is a constant. Unfortunately, this is much too simplistic. It is well known empirically that the variance of counts of biological organisms tends to increase with the mean count. There has been much dispute over a period of years about indices and models to describe the relationship between the variance and the mean (Taylor, 1984), but the dust has settled somewhat, and two entirely different empirical formulae have emerged. One is referred to as Iwao's variance–mean relationship (IVM) and the other as Taylor's Power Law (TPL: 'Power Law' is used to emphasize the mathematical formulation):

$$\text{IVM:} \quad \sigma^2 = (a + 1)\mu + (b - 1)\mu^2 \tag{3.14}$$

$$\text{TPL:} \quad \sigma^2 = a\mu^b \tag{3.15}$$

The interpretations of the two parameters, a and b, are different from one model to the other. Both TPL and IVM have been used on many occasions to estimate the variance, given the mean, but TPL is more commonly used than IVM. In this book, we shall use TPL. The interpretation of the TPL parameters is discussed below.

To use TPL in practice, it must have been estimated beforehand. This is done by fitting the model to data that consist of sets of estimated means and variances. TPL can be estimated well only if data are available from a wide range of pest densities, and if the estimated means and variances are reasonably precise. A general rule of thumb is that at least 40 sample observations are required to estimate each sample mean and variance, although smaller numbers have been successfully used (see, e.g. Taylor *et al.*, 1988). If the sample sets used to estimate TPL are small, the scatter about the relationship will be large, and the fit will be poor. The range of means must cover the critical density and the densities which might occur in management practice. No clear guidance can be given on the number of data points that are required. It depends on the spread but, evidently, the more points the better.

Given good data, it is usually sufficient to estimate the regression of the logarithmic form of Equation 3.15, that is, the regression of $\ln(\sigma^2)$ on $\ln(\mu)$:

$$\ln(\sigma^2) = \ln(a) + b\ln(\mu) \tag{3.16}$$

Strictly, the regression technique is inappropriate because the values used for σ^2 and μ in the regression (namely, V and m) are estimates with random error, so b will tend to be underestimated (Perry, 1981). However, where the range for m is wide and the precision of the points representing the sample sets is good, the underestimation should be ignorable. All it requires is a lot of fieldwork!

Once the parameters of TPL have been estimated, the variance is a function of the mean and can be calculated for any value of the mean. Thus equations such as Equations 3.10, 3.11 or 3.12 can be written in the form:

$$\begin{aligned} L &= cd - z_{\alpha/2}\sqrt{\frac{a \times cd^b}{n_B}} \\ U &= cd + z_{\alpha/2}\sqrt{\frac{a \times cd^b}{n_B}} \end{aligned} \tag{3.17}$$

TPL can also be incorporated into the simulations for the OC and ASN functions.

The real value of a variance–mean relationship lies in its *portability*. Many empirical relationships work well in the context of the data that generated them, but break down when tried under different conditions. A portable relationship is one that works under a wide range of conditions. TPL has been shown to have this property in many instances: the parameter b is often regarded as being species-specific, while a depends primarily on the sampling method. There has been considerable controversy on this over the years (Taylor, 1984; Trumble *et al.*, 1989). It is often possible to show differences in a or b due the effects of plant cultivar or variety, the presence or absence of natural enemies, pesticide application or geographical location. However, there is often much similarity in the parameter estimates, even among different species of pests which have similar biologies and for which sample units are similar (e.g. phytophagous mites on temperate fruit trees). Modest differences in the parameters for TPL can have small or even ignorable effects on OC and ASN functions. Luckily, variability in TPL parameters is easy to investigate using simulation, and practitioners are advised to study the effects of potential variability before spending considerable effort gathering data with which to estimate TPL parameters. This is illustrated in Exhibit 3.2.

Another type of variability in TPL can also be checked by simulation. Even if portability is accepted, there remains the fact that there is variation about the regression model Equation 3.16. In practice, this variability looks approximately normal, so a more realistic model than Equation 3.16 would be:

$$\ln(\sigma^2) = \ln(a) + b\ln(\mu) + z(0,\sigma_\varepsilon), \quad \text{or} \quad \sigma^2 = a\mu^b e^z \tag{3.18}$$

where $z(0,\sigma_\varepsilon)$ is a normally distributed random variable with mean 0 and standard deviation σ_ε. In other words, each time a sampling plan is used to classify a particular density during a simulation, Equation 3.17 could be used to generate the variance of the population being sampled. If an OC determined using simulation is to be based on 500 simulation runs, then 500 values of $\ln(\sigma^2)$ could also be determined, one for each run. As will be demonstrated in later exhibits, the general effect of including this idea is to reduce the slope of the OC function. The square root of the mean square error for the regression used to estimate TPL can be used as an estimate of σ_ε.

Exhibit 3.2. *Adjusting batch sampling parameters to achieve a specified goal for the OC and ASN functions*

Here we illustrate how the parameters of batch sequential sampling plans can be adjusted to achieve design criteria specified for the OC and ASN functions. The targets specified for the OC here are very stringent and possibly unrealistic for practical purposes. However, they provide a useful sequence of problems which can be solved by simulation techniques.

The objective chosen here is to classify density about a critical density of 5, so that the OC when $\mu = 4$ is greater than 0.9 (written as OC(4) > 0.9), and also that OC(6) < 0.1. The average sample size is to be as small as possible subject to the constraint that at least 25 samples are taken (to achieve a representative sample). The expected variance of sample observations can be described as a function of the mean using TPL with parameters $a = 3.0$ and $b = 1.5$. Upper confidence limits

Continued

(95%) for the TPL parameters are 3.5 and 1.7; lower 95% confidence limits are 2.5 and 1.3. A further requirement is that the OC specifications be met when the 95% confidence limits for the TPL parameters are used. As we shall see, the design process passes through a series of steps in which the sampling plan is modified to improve its properties (by changing the boundary specifications). Simulations are done after each modification to evaluate the effect on the performance of the plan in terms of OC and ASN.

Sampling plan: batch sample size, n_B, *and maximum number of batches,* maxb A starting point in the design process is to examine the effect of batch sample size and maximum number of batches on the OC and ASN. Three sampling plans were examined with n_B and *maxb* equal to 100 and 1, 50 and 2, and 25 and 4. Note that the maximum sample size for all three plans is 100. Boundaries were calculated using $\alpha = 0.2$, and TPL parameters $a = 3.0$ and $b = 1.5$. The OC and ASN functions were determined using simulation with 1000 iterations for each OC and ASN value. The same TPL values used to calculate the boundaries were used in the simulations. The simulated OC and ASN functions are shown in Fig. 3.7. The OC functions are all alike, and all three plans have OC(4) > 0.9 and OC(6) < 0.1. Plan 3 ($n_B = 25$, $maxb = 4$) has the lowest ASN.

Testing: TPL parameters As shown previously in Exhibit 3.1, using variances in the simulations different from those used to construct the boundaries can affect the OC and ASN functions in several ways. If larger variances are used, the OC becomes flatter; whereas if smaller variances are used, the OC becomes steeper. In terms of TPL, if larger parameter values are used, the OC becomes flatter, and conversely when smaller parameters are used. When the upper confidence limits for *a* and *b* (3.5 and 1.7) were used, none of the three plans described above had

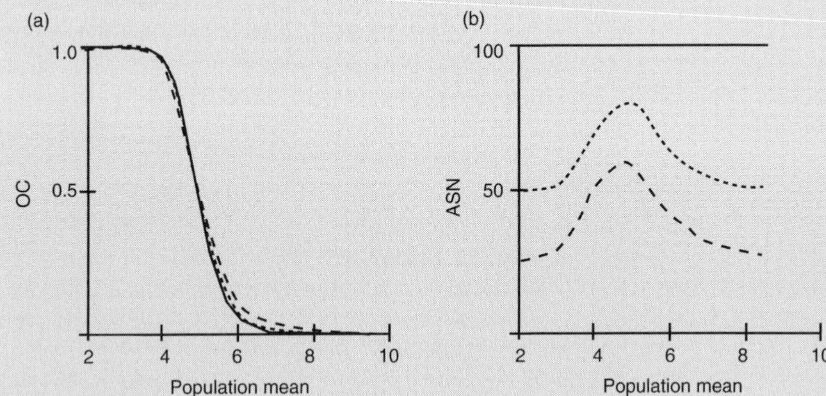

Fig. 3.7. OC (a) and ASN (b) functions for three batch sequential sampling plans with stop boundaries based on $cd = 5$, $\alpha = 0.2$ and TPL *a* and *b* = 3.0 and 1.5. The batch sample size (n_B) and maximum number of batches (*maxb*) were varied. For plan 1 (——) $n_B = 100$ and $maxb = 1$; for plan 2 (-----) $n_B = 50$ and $maxb = 2$; and for plan 3 (- - - -) $n_B = 25$ and $maxb = 4$.

OC(6) < 0.1 (for plans 1, 2 and 3, the values were 0.125, 0.135 and 0.196, respectively). OC and ASN functions for the plan with $n_B = 25$ and $maxb = 4$ were investigated in more detail by using extreme values (upper and lower confidence limits) for the TPL parameters in the simulations. The results are shown in Fig. 3.8. Although there are clear differences among the OC functions, the specified target OC(6) < 0.1 is not met, so further design changes are needed. Specifically, the OC functions must be made steeper and/or moved along the pest density axis.

Sampling plan: TPL parameters and α One strategy for making the OC function steeper is to increase the width of the stop boundaries. This can be accomplished by decreasing α and/or increasing the variance (increasing TPL a and b) used to calculate the boundary. Both strategies were explored by determining OC and ASN functions for three sampling plans, all with $n_B = 25$ and $maxb = 4$:

plan 1: $\alpha = 0.2$ TPL $a = 3.5$ TPL $b = 1.7$
plan 2: $\alpha = 0.05$ TPL $a = 3.0$ TPL $b = 1.5$
plan 3: $\alpha = 0.05$ TPL $a = 3.5$ TPL $b = 1.7$

The simulations used the upper confidence limits for the TPL values ($a = 3.5$, $b = 1.7$). The OC functions did become steeper, but one of the design criteria was still not met: values for OC(6) were 0.153, 0.121 and 0.116 for plans 1, 2 and 3, respectively. The OC and ASN functions are shown in Fig. 3.9, and it can be seen that relatively large changes in the ASN function can lead to only modest changes in the OC function.

Sampling plan: cd Another approach to meeting the design criteria is to shift the entire OC function to the left by reducing the *cd*. Shown in Fig. 3.10 are OC

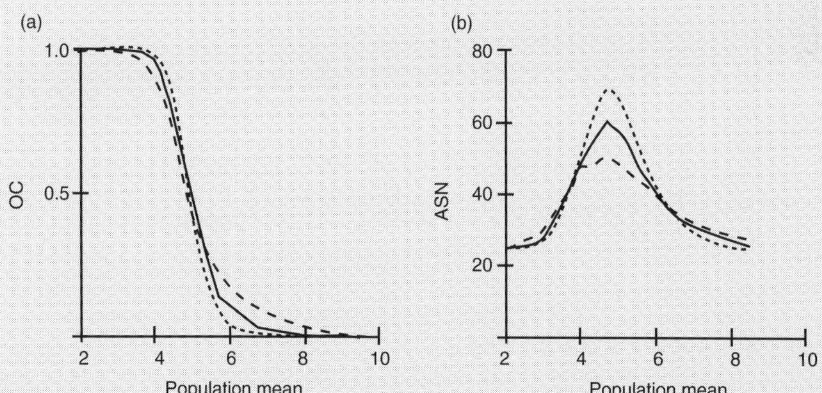

Fig. 3.8. OC (a) and ASN (b) functions for three batch sequential sampling plans with stop boundaries based on $cd = 5$, $\alpha = 0.2$, TPL a and $b = 3.0$ and 1.5, $n_B = 25$ and $maxb = 4$. For plan 1 (——) simulations were done using $a = 3.0$ and $b = 1.5$; for plan 2 (-----) simulations used $a = 2.5$ and $b = 1.3$; and for plan 3 (- - - -) simulations used $a = 3.5$ and $b = 1.7$.

Continued

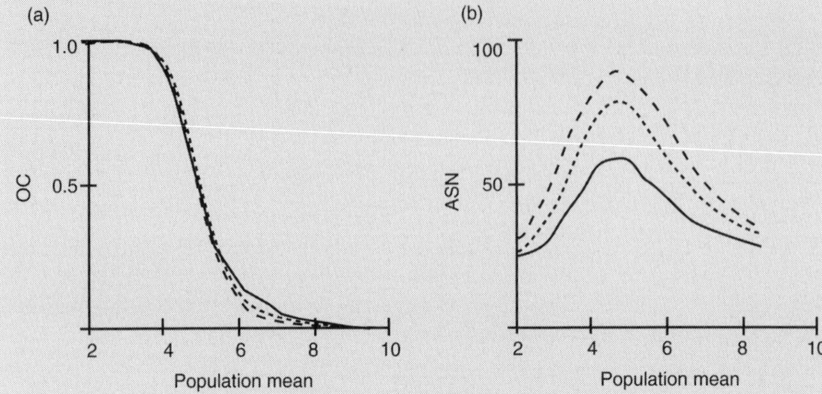

Fig. 3.9. OC (a) and ASN (b) functions for three batch sequential sampling plans with *cd* = 5, $n_B = 25$, and *maxb* = 4, and with the following α and TPL *a* and *b* values: $\alpha = 0.2$, $a = 3.5$, $b = 1.7$ (——); $a = 0.05$, $\alpha = 3.0$, $b = 1.5$ (-----); $\alpha = 0.05$, $a = 3.5$, $b = 1.7$ (- - - -). Sampling was simulated using TPL values $a = 3.5$ and $b = 1.7$.

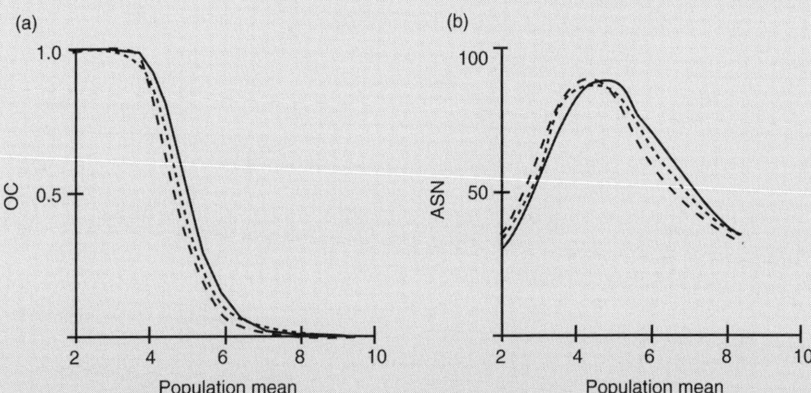

Fig. 3.10. OC (a) and ASN (b) functions for three sampling plans with $\alpha = 0.05$, $a = 3.5$ and $b = 1.7$, *cd* = 5 (——), 4.8 (-----) and 4.6 (- - - -). Simulations were performed using $a = 3.5$ and $b = 1.7$.

and ASN functions for three sampling plans set up with $\alpha = 0.05$, $a = 3.5$, $b = 1.7$ and three values for *cd* (5.0, 4.8 and 4.6). The simulations used TPL values $a = 3.5$ and $b = 1.7$. As the *cd* value was decreased, the OC and ASN functions were shifted to the left. OC(4) values for the three plans were 0.947, 0.903 and 0.835. OC(6) values were 0.128, 0.096 and 0.055. The second plan therefore meets the design criteria.

Testing: TPL parameters Although the design criteria have finally been met, we must remember that the sampling plan was constructed using upper confidence interval values for TPL *a* and *b*, and that these parameter values will usually over-predict the variance of counts from the population being sampled. The consequences of this can be evaluated by determining OC and ASN for lower values of TPL *a* and *b*. Shown in Fig. 3.11 are OC and ASN functions when stop boundaries were based on $\alpha = 0.05$, $a = 3.5$, $b = 1.7$ and $cd = 4.8$. The following variants were used for the simulations:

plan 1: TPL $a = 3.5$ TPL $b = 1.7$ (solid lines)
plan 2: TPL $a = 3.0$ TPL $b = 1.5$ (small dashed lines)
plan 3: TPL $a = 2.5$ TPL $b = 1.3$ (long dashed lines)

By using larger values of *a* and *b*, greater classification precision is achieved. The differences among the ASN functions are modest, with at most a 20% increase in the average number of samples required to make a classification.

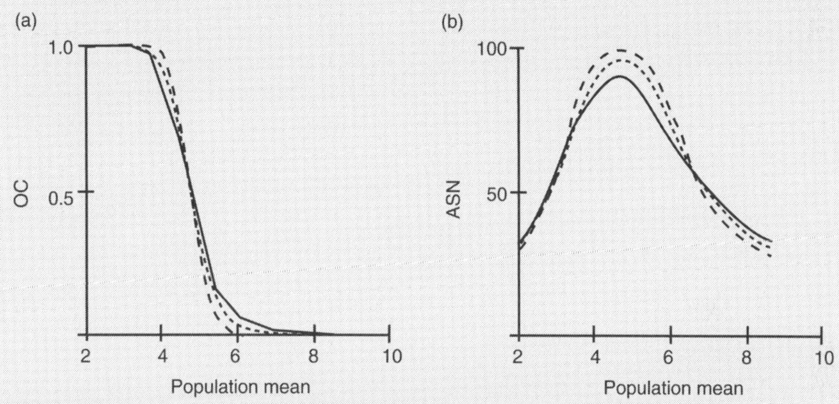

Fig. 3.11. OC (a) and ASN (b) functions for three sampling plans with $\alpha = 0.05$, $a = 3.5$, $b = 1.7$ and $cd = 4.8$. Simulations were done using $a = 3.5$ and $b = 1.7$ (——), $a = 3.0$ and $b = 1.5$ (-----) and $a = 2.5$ and $b = 1.3$ (- - - -).

Before ending the discussion on variance–mean relationships, we need to address the following question: Why is the variance not estimated at the time of sampling? Would this allow us to forego the reliance on a formula (e.g. TPL) which may not be perfectly applicable to our situation? Estimating the variance at time of sampling is not recommended for essentially three reasons. Unless sample sizes are very large, variance estimates are notoriously imprecise (e.g. Snedecor and Cochran, 1967). The variance used in the formulation of stop boundaries (e.g. L_i and U_i) is the variance at the *cd*; replacing it with an estimated variance implies a different basis for batch sampling (not necessarily a bad thing!). Using an estimated variance in the stop boundary formulae for each decision sample would be complicated and impractical. In practice, the vast majority of practical sampling plans for pest management decision-making use variance–mean models, or other tools based on prior knowledge about variance, to determine expected variability.

3.5 Summary

1. Sampling to estimate pest density and sampling to classify pest density are different. They have different goals. Criteria for a good estimation sampling plan are not necessarily the same as those for a good classification sampling plan.
2. For pest management decision-making purposes, basing a sampling plan on criteria related to classifying pest density is usually more appropriate than basing it on estimation criteria. This is because pest management decision-makers are mainly interested in whether pest density (or incidence) exceeds some critical level. Sampling for classification provides, through the OC function, a direct assessment of the effectiveness of the sampling plan.
3. By collecting data in batches and noting whether a good decision can be made after each batch, sampling costs can be reduced. This is implemented by setting up stopping boundaries for cumulative sample data.
4. The variance of sample counts of biological organisms is rarely constant. This variance can be modelled as a function of the sample mean. Variance–mean models are often portable.
5. The parameters of batch sequential sampling plans can be manipulated to achieve specific classification criteria, defined in terms of OC and ASN functions. Of course, it is possible that certain specifications are unattainable.
6. Stochastic simulation of the sampling process is a powerful method for estimating OC and ASN functions of sampling plans. Comparing OC and ASN functions for different sampling plans can be used to design good sampling plans.

References and Suggested Reading

Dodge, H.F. and Romig, H.G. (1944) *Sampling Inspection Tables – Single and Double Sampling*. John Wiley, New York, 106 pp.

Iwao, S. (1968) A new regression method for analysing the aggregation pattern of animals. *Researches on Population Ecology* 10, 1–20.

Perry, J.N. (1981) Taylor's power law for dependence of variance on mean in animal populations. *Applied Statistics* 30, 254–263.

Snedecor, G.W. and Cochran, W.G. (1967) *Statistical Methods*, 6th edn. Iowa State University Press, Ames, Iowa, 593 pp.

Taylor, L.R. (1984) Assessing and interpreting the spatial distributions of insect populations. *Annual Review of Entomology* 29, 321–359.

Taylor, L.R., Perry, J.N., Woiwod, I.P. and Taylor, R.A.J. (1988) Specificity of the spatial power-law exponent in ecology and agriculture. *Nature* 332, 721–722.

Trumble, J.T., Brewer, M.J., Shelton, A.M. and Nyrop, J.P. (1989) Transportability of fixed-precision level sampling plans. *Researches on Population Ecology* 31, 325–342.

Distributions

4

4.1 Introduction

In this chapter we describe how spatial patterns in the field are related to observed frequency distributions of sample counts. We describe four probability distributions which are tools for characterizing the observed frequency distributions of sample observations. These distributions are the Poisson distribution, the negative binomial distribution, the binomial distribution and the beta-binomial distribution. We show how these probability distributions can be helpful for assessing the properties of sampling plans for decision-making.

When a random pattern of points is overlaid with a regular grid of sample units, the resulting frequency distribution of classes with 0, 1, 2, ... points conforms to the Poisson probability distribution. If the spatial pattern is not random but is aggregated, the resulting frequency distribution is usually more 'spread out' and has larger variance than the Poisson variance. The negative binomial distribution can often describe such a long-tailed frequency distribution.

Sometimes it is possible to observe only whether or not a sample unit contains or is affected by a pest. The binomial probability distribution provides a model for characterizing sampling results expressed as the proportion of affected sample units. When sample units are taken in clusters, and the spatial pattern of the pest is aggregated, the beta-binomial probability distribution can be used to characterize the frequency distribution of the number of affected sample units in a cluster.

We explain how the parameters of these distributions can be fitted to data, using the maximum likelihood method, and how the goodness of fit can be assessed with the χ^2 statistic.

4.2 Spatial Pattern, Frequency Distribution and Probability Distribution

The concept of spatial pattern[1] is fundamental for understanding pest populations and for planning methods aimed at assessing their potential for crop damage. Each

[1] Following Pielou (1977), we use the expression 'spatial pattern' rather than 'spatial distribution'.

biological organism has its own preferred habitat, and that is where it is most likely to be found. The Colorado potato beetle, for example, feeds on potato plants, so to describe its spatial pattern, plants must be examined, the beetles counted, and the data displayed as in Fig. 2.2. The spatial pattern of a biological organism is the description of how the organism is dispersed (or disperses itself) in space. The key aspect of a spatial pattern is that the locations of the organisms are retained in the description. Although Fig. 2.2 is a valid picture of a spatial pattern, other equally valid ones are possible, such as a description of the numbers of beetles on potato stems, or on individual leaves. Ultimately, one can even ignore the host and envisage a spatial pattern of individuals as points in space, although this could be misleading if the hosts are not evenly distributed. Such a concept (spatial pattern as points in space) is useful for understanding the theory behind various sampling strategies for pest management. It is especially useful for describing the link between spatial pattern, frequency distributions and probability distributions.

A frequency distribution can be extracted from a spatial pattern, provided that a sample unit is defined. If a sample unit forms part of the spatial structure (as in Fig. 2.2), it is simple enough to count the number of sample units where there are no individuals, one individual, two individuals and so on, as in Fig. 2.3. If there is no obvious sample unit, for example, with soil-borne organisms, a sample unit must be chosen before a frequency distribution can be calculated.

The difference between a spatial pattern and a frequency distribution is huge. They are altogether different things. The spatial pattern is what we might see in a field. The frequency distribution of number of specimens per sample unit is a less information-rich piece of information, abstracted from the spatial pattern. You cannot restore the spatial pattern from knowledge of a frequency distribution, because all the information on spatial coordinates is lost. Therefore, you neither know how specimens were arranged within the units; nor do you know how the more densely and less densely populated units were arranged with respect to one another. Spatial information is very important for the design of the sampling pattern; for example, the transect to follow through a field and the distances at which to take samples. We shall discuss this further in Chapter 8.

Other infestations by the Colorado potato beetle would have different spatial patterns from that shown in Fig. 2.2. However, it has been found in practice that, for a wide variety of species, when the same sample unit is used to examine the same species in similar environments and the overall mean density is similar, the frequency distributions are also similar to one other. So although spatial patterns may be somewhat different from one time or place to another time or place, the frequency distributions are often similar. This indicates that there is a role for theoretical probability distributions. Theoretical probability distributions are models for expected frequency distributions, based upon assumptions relating to the biological process involved.

A frequency distribution and a probability distribution have much in common. Both can be presented as bar charts with classes on the horizontal axis (often indicating the number of pests per sample unit) and frequencies or probabilities on the vertical axis. A frequency distribution is generally graphed with absolute numbers of occurrences on the vertical axis, while a probability distribution (the name indicates it) is always graphed with probabilities on the vertical axis. A probability distribution is a model for a frequency distribution in the sense that the expectation (or long-term average; see Section 2.2) of each frequency is proportional to the corresponding probability:

$$E(f_i) \propto p_i \tag{4.1}$$

and the constant of proportionality is the sample size, n:

$$E(f_i) = np_i \tag{4.2}$$

Of course, this is true only if the probability distribution, p_i, is the correct model for the frequency distribution. Even if Equation 4.2 is true, the frequencies are unlikely to be equal to the probabilities, because of the randomness inherent in the biological processes involved and in the sampling. The frequency distribution tends to be more jagged than the corresponding probability distribution, especially for small sample size, n. The values f_i/n which correspond to the p_i are often called *relative frequencies*.

There are three reasons why crop protectionists are interested in knowing probability distributions that describe the observed frequency distributions of sampled pests:

1. As summary descriptions of reality.
2. As components of simulation models which allow us to explore the behaviour of sampling methods, using a computer. Such distributions are often indispensable tools when calculating the expected performance of sampling methods.
3. As models to calculate variability among sample units, so that we may make a crop protection decision, solely based on comparing the average pest density to a threshold. However, descriptive variance–mean relationships, such as Iwao's (IVM) or Taylor's Power Law (TPL), (see Chapter 3), are better suited to this purpose.

In this chapter, we explore the relationship between a spatial pattern and the sample data taken from it, as well as how probability distributions can characterize the frequency distributions of sample observations.

4.3 Random Patterns and the Poisson Probability Distribution

In a field which is infested by a pest (an arthropod, disease, weed and so on), individuals are spread about in some manner over the area of the field. The simplest situation arises when the location of each specimen is completely independent of the location of others. Such a pattern is random. This might occur immediately following a wind-borne invasion of pest insects or of disease propagules which are deposited indiscriminately over a large crop area. With a computer, such a random pattern can easily be generated. Points are generated one by one. The x and y coordinates of each new point are drawn from a uniform probability distribution over the interval 0 to 1, irrespective of the location of points already generated. Thus each coordinate value in the interval (0,1) is equally likely. This procedure was used to generate Fig. 4.1(a), which can be overlaid with a regular grid of sample

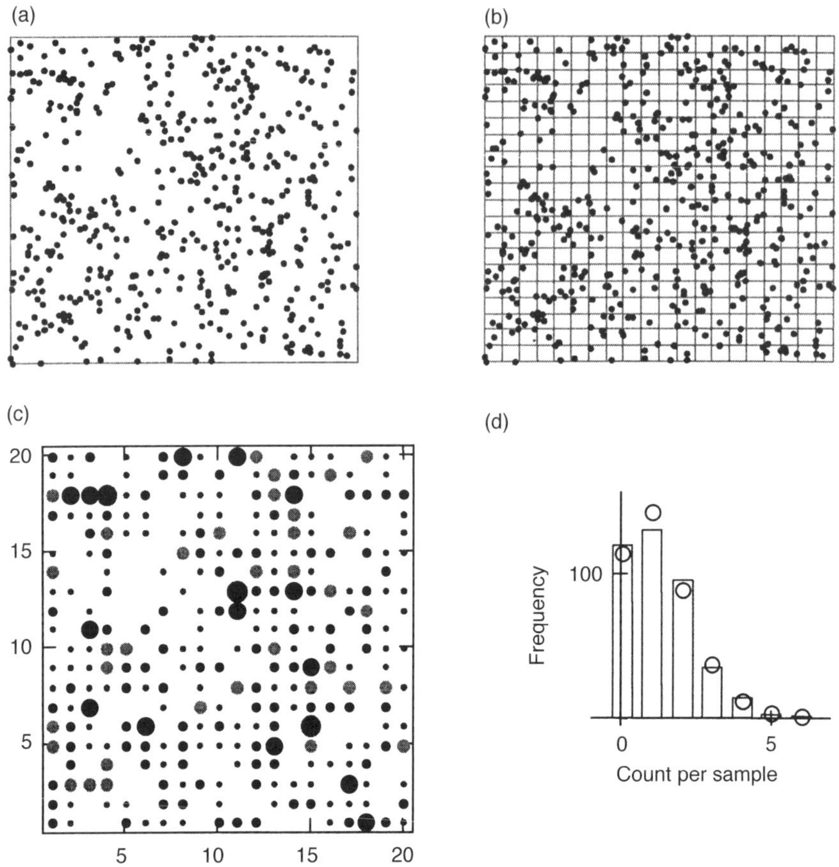

Fig. 4.1. (a) A random spatial pattern of points. (b) The random pattern in (a) with a 20 × 20 grid superimposed. (c) A spatial pattern derived from (b) as a summary of (a). (d) Frequencies derived from (b) (bars) and fitted Poisson frequencies (○).

units (Fig. 4.1b). The spatial pattern of numbers in sample units can be displayed graphically as dots whose sizes represents the numbers of individual organisms in the sample units (Fig. 4.1c). The frequency distribution of these numbers can be calculated as in Chapter 2 (Fig 4.1d). What kind of frequency distribution should one expect when points are spread out randomly like this?

Statistical theory shows that, when events occur at random locations in a region, the number of events which can be found in any specified area within the region follows a Poisson probability distribution. This implies that we can predict the shape of the frequency distribution if we know the rate at which the events occur or, equivalently, if we know the average number of events in a unit of area. This average number of events constitutes the only parameter of the Poisson distribution. It is called the mean and is denoted by the Greek symbol μ. The Poisson probabilities of $x = 0, 1, 2, \ldots$ events per sample unit are defined as follows:

$$p(x \mid \mu) = e^{-\mu} \frac{\mu^x}{x!} \tag{4.3}$$

where $x!$ denotes the product of the first x integers:

$$x! = 1 \times 2 \times 3 \times \ldots \times x$$

The notation $p(x \mid \mu)$ is a common way of writing formulae for probabilities. It should be understood as the probability of getting x individual organisms in a sample unit when the parameter is equal to μ. The vertical bar is a convenient way of separating the data (x) from the parameter (μ). Typical shapes of the Poisson distribution are shown in Fig. 4.2.

On the basis of Equations 4.2 and 4.3, we can compare the frequency distribution obtained by our simulated random process with the expected frequencies for a Poisson distribution. The parameter μ is 1.25 because, in Fig. 4.1, 500 points were placed in a field with $n = 400$ sample units. With $\mu = 1.25$, the expected frequencies can be calculated from Equation 4.3:

$$E(f_x) = np(x \mid 1.25) = 400e^{-1.25} \frac{1.25^x}{x!} \tag{4.4}$$

A comparison between the simulated results and theory is shown in Fig. 4.1d. The process of finding the theoretical model which is closest to the data is called model fitting. Here we have fitted the Poisson probability distribution to the observed frequencies. The fitting procedure is quite simple, because we need to calculate only the mean number of individuals per sample unit. This mean is an unbiased estimator of μ. We usually denote it by the arabic equivalent of μ, namely m.

How does this help us make decisions? From the decision-making perspective, the most important help it gives is that we now know the variance, σ^2, of the

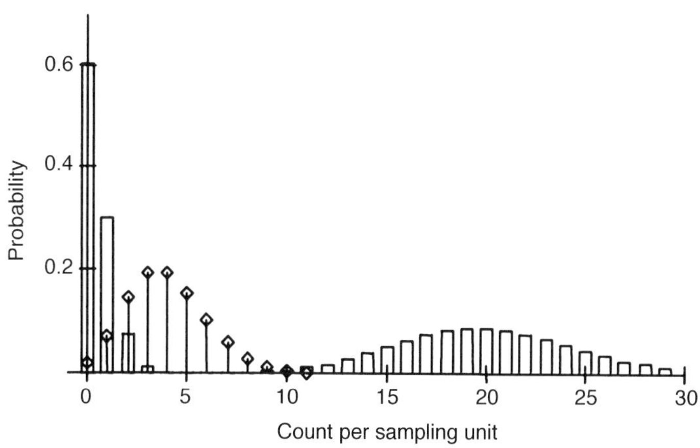

Fig. 4.2. Poisson distributions with $\mu = 0.5$ (first set of bars), 4 (◇), 20 (second set of bars).

number of individuals in a sample unit. For a Poisson distribution, the variance is equal to the mean: $\sigma^2 = \mu$. With this knowledge, we can predict the variance of the mean for samples of any size n as in Chapter 2. Therefore, assuming that we can use the normal distribution approximation as in Chapter 2, we can go as far as calculating probability of decision or operating characteristic (OC) functions.

The probability distribution that we have obtained depends on the size and shape of the sample unit. As a consequence, the OC function relates to the sample unit used in the preliminary work, and cannot be used directly for a different sample unit. For instance, when counting pests on a pair of adjacent plants, rather than on one plant, the mean, μ, changes to 2μ. Likewise, the variance, σ^2, changes to 2μ. This new mean and variance would have to be used for calculating the OC function. Therefore, if a Poisson distribution can be assumed for the new sample unit, these few mathematical calculations are all that is required to adjust the OC function.

4.4 Fitting a Distribution to Data Using the Maximum Likelihood Principle[2]

In the previous section, we used the sample mean to estimate the parameter μ of the Poisson distribution without indicating why this was a good idea. A motivation can be given by the maximum likelihood principle. According to this principle, the overall probability (or 'likelihood') of getting the data which were in fact obtained is calculated, and the model parameters for which this 'likelihood' is maximized are defined as the maximum likelihood estimates. These estimates have many desirable properties. For example, in most situations which might be encountered by pest managers, these estimates have the highest precision. Other properties are beyond the scope of this book, and are discussed in statistical textbooks. In general, the maximum likelihood principle forms the basis for estimating parameters of probability distributions. As an aside, it is also the principle that leads to least squares estimates in linear regression.

Suppose that we have sample data $X_1, X_2, \ldots, X_n$, and we want to estimate some kind of probability distribution $p(x \mid \theta)$ with parameter θ. Then we should look for the parameter that gives the highest value for the product

$$L(X_1,X_2,\ldots,X_n \mid \theta) = p(X_1 \mid \theta)\, p(X_2 \mid \theta) \ldots p(X_n \mid \theta) \tag{4.5}$$

This product, $L(X_1,X_2,\ldots,X_n \mid \theta)$, is called the likelihood function. Its value depends on the parameter (or parameters) θ and on the sample data. The parameter θ can be the mean, the variance or some other defining characteristic. For the Poisson distribution, the likelihood is

$$L\left(X_1,X_2,\ldots,X_n \mid \mu\right) = e^{-n\mu}\,\frac{\mu^{x_1+x_2+\ldots+x_n}}{X_1!X_2!\cdots X_n!} = e^{-n\mu}\,\frac{\mu^{nm}}{X_1!X_2!\cdots X_n!} \tag{4.6}$$

[2]This section is full of mathematics and may be skipped.

It is often easier to maximize the logarithm of the likelihood, so we do that. At the same time we drop the division by $(X_1!\,X_2!\,\cdots\,X_n!)$, which has no influence on our attempt to maximize L as a function of μ:

$$\ln(L) = -\,n\mu + nm\ln(\mu) \tag{4.7}$$

It is worth noting that the data enter this expression only through the sample mean, m, so that all we need to do is find the value of μ, in terms of m, which maximizes ln(L). This can be done by using calculus, and the value of μ so found is called the maximum likelihood estimate of μ, denoted by $\hat{\mu}$; that is, we put a 'hat' on μ. For the Poisson distribution, the maximum likelihood solution is always $\hat{\mu} = m = \Sigma X_i/n$. The log likelihood for the Poisson distribution with $\mu = 5$ is shown in Fig. 4.3. In this chapter, we shall use the 'hat' notation to denote a maximum likelihood estimate, as a reminder to the reader; but this will not be necessary in subsequent chapters, so we will drop the practice after this chapter.

4.5 Testing the Goodness of Fit of a Probability Distribution

There are objective ways of testing the goodness of the fit between a set of observed frequencies and a theoretical distribution fitted to them. We include here a brief

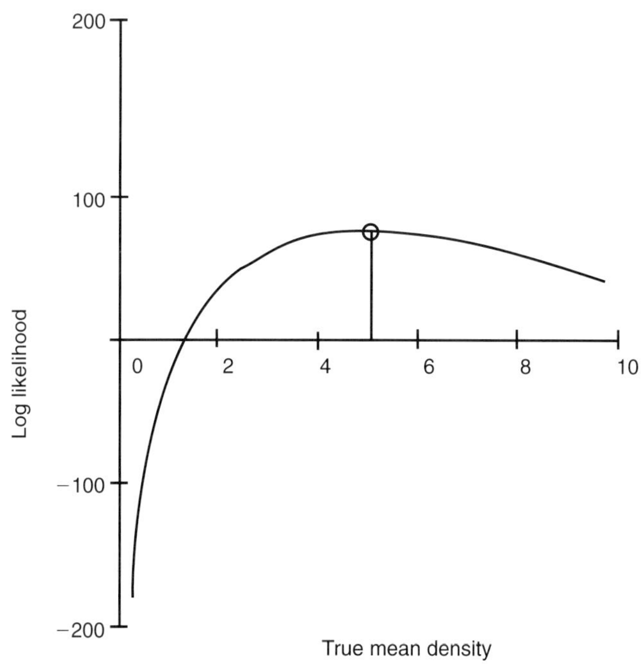

Fig. 4.3. A maximum likelihood example based on the Poisson distribution. The curved line represents the log-likelihood for sample size $n = 25$, as a function of μ. The maximum is at $\mu = m$, which here is equal to 5.

guide to one of them, the χ^2 test. In the simplest case, based on n sample units, the parameters of the probability distribution model are determined (usually by maximum likelihood), and the expected frequencies are calculated as $n\ \mathrm{p}(x_i\,|\,\hat{\theta})$, where $i = 1, 2, \ldots, c$ and $x_i = i - 1$. Note that here we are dealing with frequency classes x_i, whereas in Equations 4.5 and 4.6, and all of Section 4.3, X_j refers to the jth sample unit, and $j = 1, 2, \ldots, n$. The $n\ \mathrm{p}(x_i\,|\,\hat{\theta})$ are used in a comparison with the observed frequencies, f_i, to calculate a statistic X^2 :

$$X^2 = \sum_1^c \frac{\left(f_i - n\mathrm{p}\left(x_i\,|\,\hat{\theta}\right)\right)^2}{n\mathrm{p}\left(x_i\,|\,\hat{\theta}\right)} \tag{4.8}$$

where c is the number of frequency classes. Note that the capital letter, X, is used to distinguish the expression in Equation 4.8 from the theoretical χ^2 distribution. If the data truly come from the distribution $\mathrm{p}(x\,|\,\theta)$, with $\theta = \hat{\theta}$, X^2 follows what is called the χ^2 distribution (with a few provisos, as discussed below), whose properties are well documented. The χ^2 distribution has one parameter, its degrees of freedom. The number of degrees of freedom, df, is easily calculated as $c - 1 - n_p$, where c is the number of classes and n_p is the number of fitted parameters ($n_p = 1$ for the Poisson distribution). If the data do not come from the distribution $\mathrm{p}(x\,|\,\theta)$, or, for some reason or other, θ is not equal to $\hat{\theta}$, X^2 tends to be large, because it is basically the weighted sum of squared differences between observed (f_i) and expected ($n\ \mathrm{p}(x_i\,|\,\hat{\theta})$) frequencies. Therefore, large values of X^2 indicate that the fit to the distribution $\mathrm{p}(x\,|\,\theta)$ is not good.

How large is large? Because we know the properties of the χ^2 distribution, we can find the probability of getting a χ^2 value equal to or greater than the calculated value of X^2. Suppose that this probability is 0.2. What this means is as follows:

- either the data come from the distribution $\mathrm{p}(x\,|\,\theta)$ with $\theta = \hat{\theta}$, but an event with probability of occurrence equal to 0.2 occurred, or
- the data did not come from the distribution $\mathrm{p}(x\,|\,\theta)$ with $\theta = \hat{\theta}$

Our choice depends on how we view the probability 0.2 in connection with our preconceived notion that the data might conform to the distribution $\mathrm{p}(x\,|\,\theta)$. If this preconception is strong, then why not accept that we have just been unlucky and got a result which should happen only one time in five – that's not too unlikely. But what if the probability had been 0.02? Is our preconception so strong that a 1 in 50 chance is just unlucky? Or should we reject the possibility that the data conform to the distribution $\mathrm{p}(x\,|\,\theta)$? It has become standard usage that a probability equal to 0.05 is a borderline between: (i) accepting that we have just been unlucky and (ii) that we should have grave doubts about our preconceived notion.

The above procedure is an example of testing a hypothesis. Here the hypothesis is that the data conform to the distribution $\mathrm{p}(x\,|\,\theta)$. The data are used to estimate parameters, X^2 is calculated and a χ^2 probability (P) is found. The hypothesis is accepted if P is greater than some standard, often 0.05, and rejected otherwise. If there are many data sets, the proportion of data sets where the hypothesis is rejected can be compared with the standard probability that was used. If all the data sets really do conform to the distribution $\mathrm{p}(x\,|\,\theta)$, this proportion should be close to the standard probability itself. Testing statistical hypotheses in this way has become common in many branches of biological science, and some technical terms should be noted. X^2 is an example of a 'test statistic', and the χ^2 probability obtained from it is called a 'significance probability'. If the significance probability, P, is less than 0.05, the hypothesis is said to be rejected at the 5% level. This test is often called the χ^2 goodness-of-fit test.

Use of the χ^2 test implies that the observed frequencies can be regarded as following a probability model, with the class probabilities given by $p(x_i \mid \hat{\theta})$. However, a few conditions must be fulfilled before the results can be analysed by the χ^2 test:

1. None of the expected frequencies, $n\ p(x_i \mid \hat{\theta})$, should be small. Traditionally, 5 has been regarded as the minimum value, but some statisticians think that 1 is often large enough. If the tail class frequencies are too small, the classes should be grouped until all classes have acceptable expected frequencies. Some arbitrariness is unavoidable here. Note that if there is grouping, the number of degrees of freedom, df, must be recalculated, because the number of classes, c, has changed.[3]

2. The number of sample units, n, should not be small. This requirement is connected with the first because if n is small many of the expected frequencies will be small, and too much grouping is not good. Another reason for n to be large (generally, at least 100) is that the estimated parameters and the significance test will be more reliable.

These concepts are illustrated in Exhibit 4.1.

Exhibit 4.1. *Sampling a random pattern produces a Poisson distribution*

A random spatial pattern can be generated as in Fig. 4.1a. Sample units can be defined by superimposing regular grid lines, as in Fig. 4.1b. Sample units of different shapes and sizes can be defined by using different grids. Four types of sample unit were created for the random spatial pattern of 500 points shown in Fig. 4.1a (see Fig. 4.4):

- 25 sample units in a 5 × 5 array (five sample units along each of the horizontal and vertical axes)
- 100 sample units in a 10 × 10 array
- 400 sample units in a 20 × 20 array
- 100 sample units in a 20 × 5 array.

The means and variances were as follows:

Number, n_x, of sample units along the x-axis	5	10	20	20
Number, n_y, of sample units along the y-axis	5	10	20	5
Grid ($n_x \times n_y$)	5 × 5	10 × 10	20 × 20	20 × 5
Number of sample units ($n = n_x \times n_y$)	25	100	400	100
Mean per sample unit	20	5	1.25	5
Calculated variance per sample unit	29.1	5.88	1.41	6.08

Continued

[3]There are some theoretical difficulties when classes are grouped relating to degrees of freedom (Chernoff and Lehmann, 1954), but in the present context they should be ignorable.

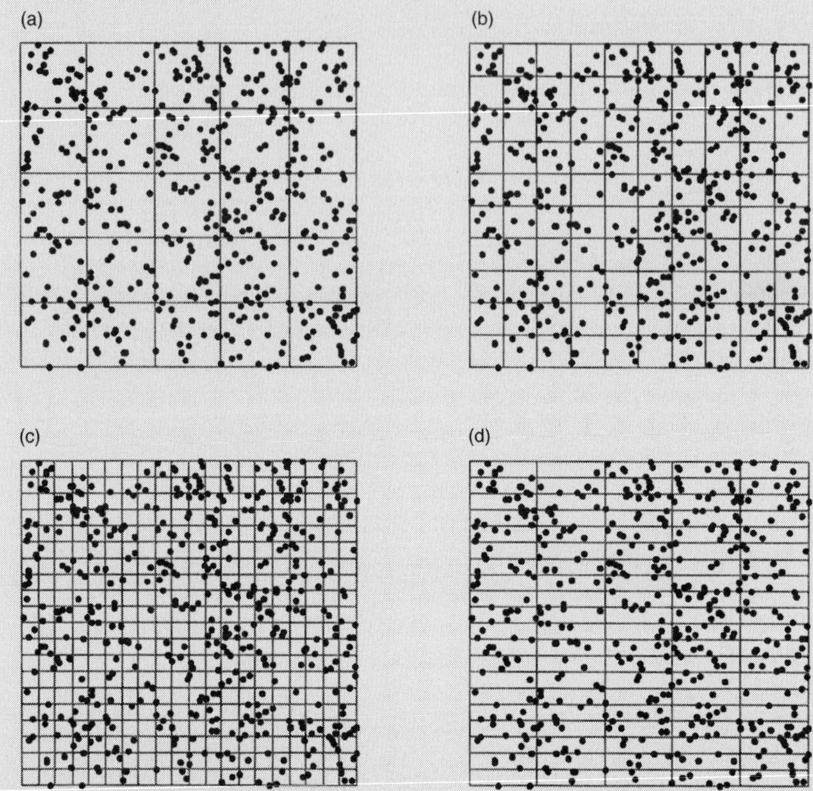

Fig. 4.4. A random spatial pattern of 500 points, with superimposed grids of sample units. (a) 5 × 5 grid, mean = 20, variance = 29.1; (b) 10 × 10 grid, mean = 5, variance = 5.88; (c) 20 × 20 grid, mean = 1.25, variance = 1.41; (d) 5 × 20 grid, mean = 5, variance = 6.08.

Although not exactly equal to the means, the variances were close to the means for all shapes and sizes of sample unit. The frequency distributions were calculated for each sample unit, and the Poisson distribution was fitted to each by maximum likelihood (Fig. 4.5). As noted above, the maximum likelihood estimate of μ for the Poisson distribution is always equal to the mean, m. The χ^2 significance probabilities, P, were calculated using the grouping criterion that the expected frequencies for the tail classes should be at least equal to 1:

Grid ($n_x \times n_y$)	5 × 5	10 × 10	20 × 20	20 × 5
X^2	19.9	7.8	6.4	9.8
df = $c - 1 - n_p$ ($n_p = 1$)	14	9	4	9
Significance probability, P	0.13	0.55	0.17	0.36

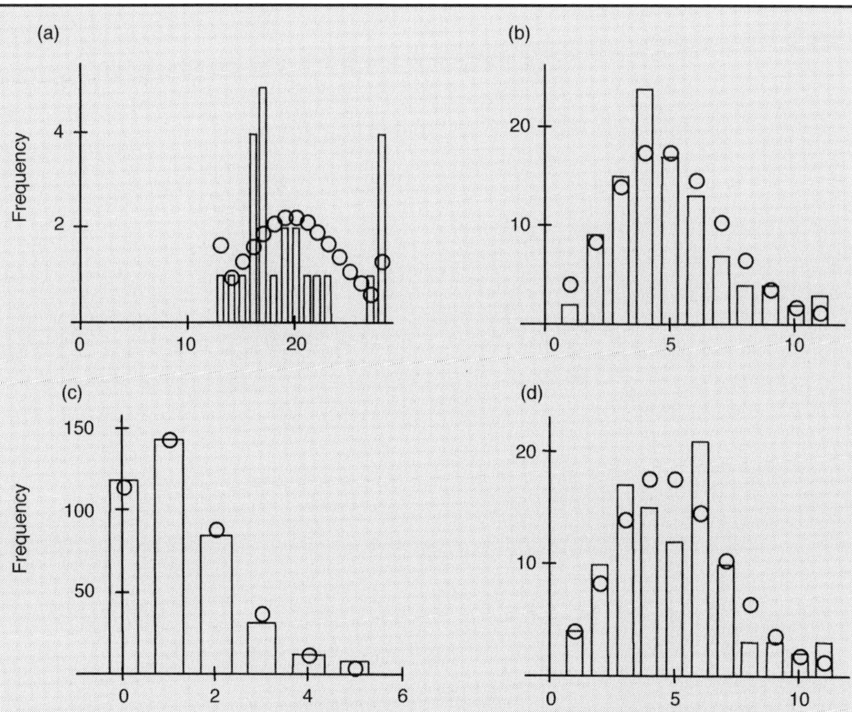

Fig. 4.5. Observed frequency distributions for the four grid types shown in Fig. 4.4, with a Poisson distribution fitted to each (○) with grouping criterion equal to one. (a) 5 × 5; (b) 10 × 10; (c) 20 × 20; (d) 5 × 20.

Because these are all greater than the standard value, $P = 0.05$, we would be justified in assuming a Poisson distribution in each case.

Despite the non-significant P-value for the 5 × 5 grid, Fig. 4.5a is not convincing: the frequencies do not really look like the theoretical values. This is mainly because $n = 25$ is too few sample units for a good test. There are people who are reluctant to accept the results of a statistical significance test if the data do not look convincing. It is unwise to lean too heavily on a statistical crutch. Statistical tests are most useful when they summarize something that is not difficult to swallow.

4.6 Aggregated Spatial Patterns and the Negative Binomial Distribution

We know from experience that pests are often not distributed randomly. They tend to be found in aggregated spatial patterns. There are many biological mechanisms that lead to aggregation: the existence of 'good' and 'bad' resource patches, settling of offspring close to the parent, mate finding and so on. Likewise, there are many ways in which models for such processes can be combined to obtain probability distributions to account for the effect of spatial aggregation.

What kind of frequency distribution should we expect when we sample an aggregated pattern? It has been found in practice that, where sample units are collected at random and the numbers of pests are spatially aggregated at the scale of the sample units, the variance tends to be greater than the mean. There are many theoretical distributions (in one way or another related to the Poisson distribution), which allow more variability among numbers in sample units: their variances are greater than their means. The probability distribution that is most often used in describing sampling distributions of pests is the negative binomial probability distribution. Because of its mathematical versatility, this distribution has been found to be a powerful workhorse for matching the frequencies of a wide variety of pest distributions in the field. The negative binomial distribution has two parameters, the mean μ, and a parameter k, which is generally called the *exponent* or *clustering parameter* of the distribution. The variance is as follows:

$$\sigma^2 = \mu + \frac{\mu^2}{k} \tag{4.9}$$

and the formulae for calculating the negative binomial probabilities are as follows:

$$\left.\begin{aligned} &\mathrm{p}(0 \mid \mu,k) = \left(\frac{k}{\mu+k}\right)^k \\ &\mathrm{p}(x+1 \mid \mu,k) = \mathrm{p}(x \mid \mu,k)\frac{k+x}{x+1}\frac{\mu}{\mu+k} \\ &\text{for } x = 0, 1, \ldots \end{aligned}\right\} \tag{4.10}$$

If k increases while μ remains constant, the variance decreases. For very large k, the negative binomial distribution is practically indistinguishable from the Poisson distribution. The effects of changing the parameters is illustrated in Fig. 4.6. When $\mu = 5$ (Figs 4.6a and c), the distribution looks like a Poisson distribution if $k = 10$, but if $k = 0.1$, the tail of the distribution is very long: although μ is still equal to 5, the probability of getting a count equal to 20 or greater is 0.07 (Fig. 4.6c, box). These figures (Figs 4.6a and c) illustrate how decreasing k increases the variance per sample unit (Equation 4.9): for $\mu = 5$,

k	0.1	1	10
Variance	255	30	7.5

The effect of varying μ while k remains constant is easier to understand (Figs 4.6b and d). As μ increases, the probability of getting a zero count decreases and the distribution is stretched out to the right. Again, these figures illustrate how increasing the mean also increases the variance (Equation 4.9): for $k = 1$,

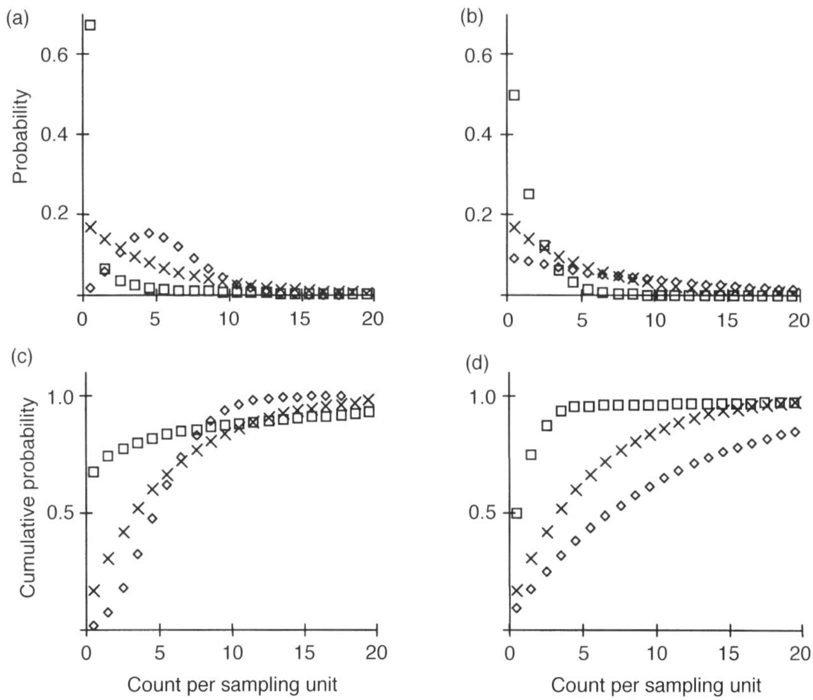

Fig. 4.6. Negative binomial distributions, showing the effect of changing one parameter while keeping the other one fixed. (a) and (b) show probabilities; (c) and (d) show corresponding cumulative probabilities. (a,c) $\mu = 5$ and $k = 0.1$ (□), 1 (×) and 10 (◇); (b,d) $k = 1$ and $\mu = 1$ (□), 5 (×) and 10 (◇).

μ	1	5	10
Variance	2	30	110

The versatility of the negative binomial distribution allows it to model Beall's data very well, where the Poisson distribution fails badly (Fig. 4.7). The mean and variance of the data are 4.74 and 15.00, respectively (see Chapter 2, Exhibit 2.1). The fitted values and tests (grouping so that expected class frequencies ≥ 1) are as follows:

	$\hat{\mu}$	$\hat{k}$	X^2	df	P
Poisson	4.74	–	6220	12	0.00
Negative binomial	4.74	2.21	30.6	25	0.20

The value of $\hat{k}$ is an indication of aggregation at the spatial scale of the sample unit, and for μ near 4.74. If the sample unit is changed or the mean density is different, there is no guarantee that k will remain the same. In fact, there is much

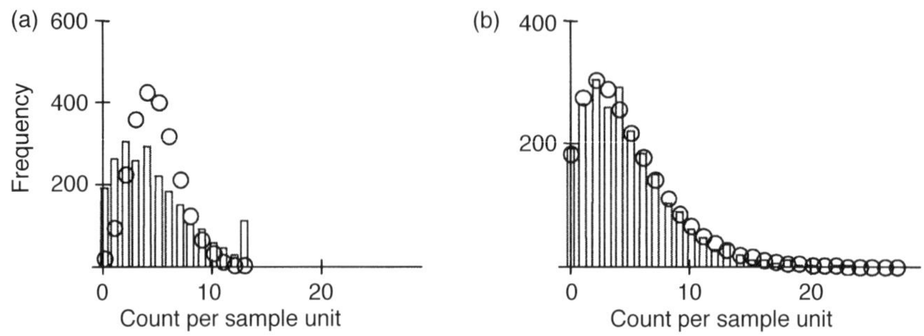

Fig. 4.7. Fitting probability distributions to Beall's data (Fig. 2.2). (a) Poisson distribution (○), $\hat{\mu} = 4.74$; (b) negative binomial (○), $\hat{\mu} = 4.74$, $\hat{k} = 2.21$.

evidence in the literature that k changes with sample unit and mean density (see, e.g. Taylor, 1984).

Computers can be made to generate aggregated spatial patterns in a number of ways (for details, see the appendix to this chapter). A computer-generated aggregated spatial pattern of points is shown in Fig. 4.8a. When a 36 × 36 grid is superimposed on this 'field' to create 1296 sample units, the aggregation among sample units can be easily seen (Fig. 4.8b). We can fit the Poisson and negative binomial distributions to these data by maximum likelihood (Fig. 4.9). We do not need a statistical test here to infer that the points are not randomly distributed – that is to say, not at the spatial scale of the chosen sample unit – or that the negative binomial distribution is a plausible model for the frequency distribution. Using the spatial pattern in Fig. 4.8a as a basis, we illustrate the effect on k of changing the size and shape of the sample unit.

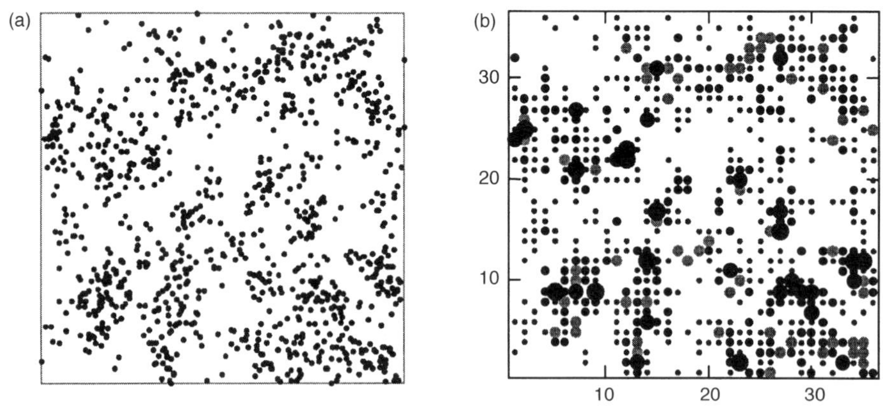

Fig. 4.8. (a) A computer-generated spatial pattern. (b) A representation of the pattern in terms of a grid (36 × 36) of sample units, where each small dot represents the presence of one point in the sample unit, and each increase in size represents the presence of one more point.

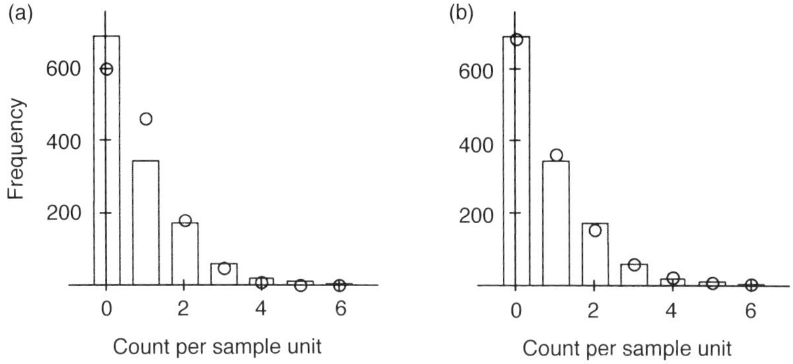

Fig. 4.9. Fitting distributions to the data shown in Fig. 4.8(b). (a) Poisson; (b) negative binomial. Fitted frequencies are shown as circles (○).

Exhibit 4.2. *The effect of sample unit size and shape on the negative binomial* k

The spatial pattern of points displayed in Fig. 4.8a was subdivided into sample units by grids of various shapes and sizes. The negative binomial distribution was fitted to each frequency distribution in turn.

Initially, all sample units were square, with equal numbers of sample units (n_x and n_y) along each axis:

Grid ($n_x \times n_y$)	$n = n_x \times n_y$	$\hat{\mu}$	$\hat{k}$	X^2	df	P
9 × 9	81	12.4	3.09	35.1	31	0.28
18 × 18	324	3.10	2.01	17.2	10	0.07
27 × 27	729	1.37	1.74	8.15	6	0.23
36 × 36	1296	0.77	1.70	5.3	4	0.25
48 × 48	2304	0.43	1.77	3.58	3	0.31

It is always a good idea to check where a test has produced a significance probability near the critical value (we use $P = 0.05$ here as the critical value). Therefore, before proceeding further, we need to check on '18 × 18' ($P = 0.07$). The frequency distribution and the fitted frequencies are shown in Fig. 4.10a. No wild departure is evident (the worst fits are for counts 5 and 8), so we can proceed to use the above results. The value of k increases as the size of the sample unit increases. Therefore, the aggregation decreases as sample unit size increases. A plausible reason for this is that a larger sample unit contains more centres of aggregation in the spatial pattern, partially 'averaging out' the aggregation effect. Although there is no general guarantee that increasing the size of the sample unit will increase the value of k, it often occurs in practice.

Continued

What happens when the shape of the sample unit is changed? We can investigate this by keeping the size of the sample unit constant, but adjusting the grids to make rectangular sample units. Based on the sample unit size for '18 × 18', we find the following:

Grid ($n_x \times n_y$)	$\hat{\mu}$	$\hat{k}$	X^2	df	P
3 × 108	3.09	4.12	12.2	9	0.20
4 × 81	3.10	4.86	5.76	8	0.67
6 × 54	3.09	3.39	10.6	10	0.39
12 × 27	3.09	2.04	10.1	11	0.52
18 × 18	3.10	2.01	17.2	10	0.07
27 × 12	3.09	2.15	13.8	11	0.24
54 × 6	3.09	3.19	15.2	9	0.09
81 × 4	3.09	4.11	7.31	10	0.70
108 × 3	3.09	7.06	9.3	8	0.32

Again, we must check where near-significance occurred. A plot of the frequencies and the fitted values of '54 × 6' are shown in Fig. 4.10b. There is no obvious feature which should make us want to reject the fit, so we can continue. There appears to be a trend for k to increase as the shape of the sample unit becomes more elongated, in either direction. When the pattern was originally generated, no purposeful 'directional' pattern was imposed. A plausible reason for increasing k is that as the length of the sample unit grows, the sample unit itself contains parts of more and more centres of aggregation, thus attaining, in part, some of the attributes of a larger sample unit. From the previous discussion on the effect of sample unit size on k, we might therefore expect a larger k as the rectangular shape of the sample unit becomes more elongated.

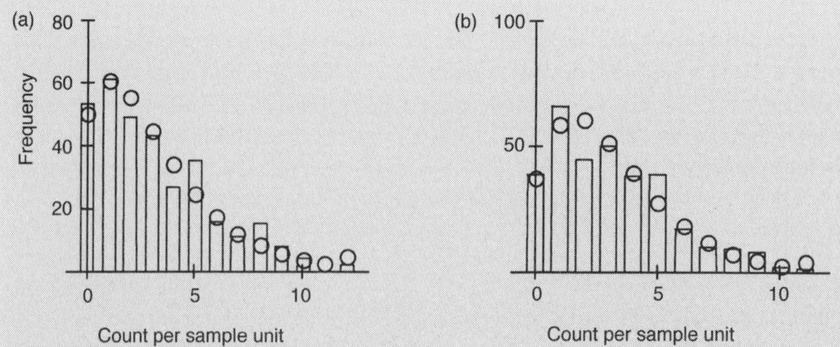

Fig. 4.10. Fitting the negative binomial distribution to the pattern in Fig. 4.8a subdivided into different sized sample units. (a) 18 × 18 superimposed grid; (b) 54 × 6 superimposed grid. Fitted frequencies are shown as circles (○).

Exactly how the value of *k* changes as the sample unit size changes depends on the type and degree of aggregation in the spatial pattern. Certainly, *k* depends on the spatial aggregation and on the size and shape of the sample unit. Estimating aggregation, *per se*, is beyond the scope of this book. For further discussion on measures of aggregation see, for example, Pielou (1977).

4.6.1 Estimating the negative binomial k from a variance–mean relationship

At first sight, the results presented in Exhibit 4.2 should appear disturbing. On the one hand, the negative binomial distribution is versatile enough to fit the frequency distributions of many species of interest to pest managers. On the other hand, the parameter k can vary depending merely on the size and shape of whatever sample unit is used. Have we taken one step forward and one step backward? The answer is that we have not, because once we have decided on a sample unit (based on concepts discussed in Chapter 2) we are generally not interested in changing its size and shape. If there are several sample units which satisfy the criteria noted in Chapter 2, we can always compare their precisions and choose the best. But once the sample unit has been agreed upon, we are then only concerned about how changes in the mean per sample unit, μ, might affect k. Here we can use one of the variance–mean relationships discussed in Chapter 3, namely

$$\text{IVM:} \quad \sigma^2 = (a+1)\mu + (b-1)\mu^2 \tag{4.11}$$

$$\text{TPL:} \quad \sigma^2 = a\mu^b \tag{4.12}$$

Using what is called the 'method of moments', we can relate either of these equations to Equation 4.9, to obtain an estimate of k:

$$\begin{aligned} &\text{IVM: } (a+1)\mu + (b-1)\mu^2 = \mu + \frac{\mu^2}{k}, \\ &\qquad \text{leading to } k = \frac{\mu}{a+(b-1)\mu} \end{aligned} \tag{4.13}$$

or

$$\begin{aligned} &\text{TPL: } a\mu^b = \mu + \frac{\mu^2}{k}, \\ &\qquad \text{leading to } k = \frac{\mu^2}{a\mu^b - \mu} \end{aligned} \tag{4.14}$$

Once the parameters of either IVM or TPL are estimated, we can estimate k for any value of μ, the mean per sample unit, and we do not need to worry about finding models to justify it (Binns, 1986). Therefore, if we have a good estimate of IVM or of TPL, we can estimate probability of decision (OC) functions for a range of mean densities using the negative binomial distribution. We shall use Equation 4.14 frequently throughout the book.

4.7 The Effect of Sample Size on Fitting Distributions

In our examples so far, we have fitted the Poisson and negative binomial distributions to all of the data. That is, we have used the data from all the sample units superimposed on a spatial pattern. In a sense, therefore, we have been working with the 'true' values in the field. In practice, however, only some of the sample units are collected and analysed. Can we compare the fits that we obtain from all the data with those we might obtain when we collect only some of the sample units?

Exhibit 4.3. *Using all sample units, or only a subset*

If we collect only a few sample units, we cannot expect to obtain a good representation of the true frequencies. How much do we gain in precision by taking more than a minimum number of sample units? Here we start with the spatial pattern and grid of sample units displayed in Fig. 4.8b, and compare the fits to the negative binomial distribution when fewer than the total number (1296) of sample units are chosen and analysed.

Fifty sample units were collected at random from the total number of units (1296) in Fig. 4.8b. Poisson and negative binomial distributions were fitted to these data (Fig. 4.11):

	$\hat{\mu}$	$\hat{k}$	X^2	df	P	P (based on all 1296 sample units)
Poisson	0.780	–	3.08	2	0.22	< 0.001
Negative binomial	0.780	1.57	1.16	2	0.56	0.25

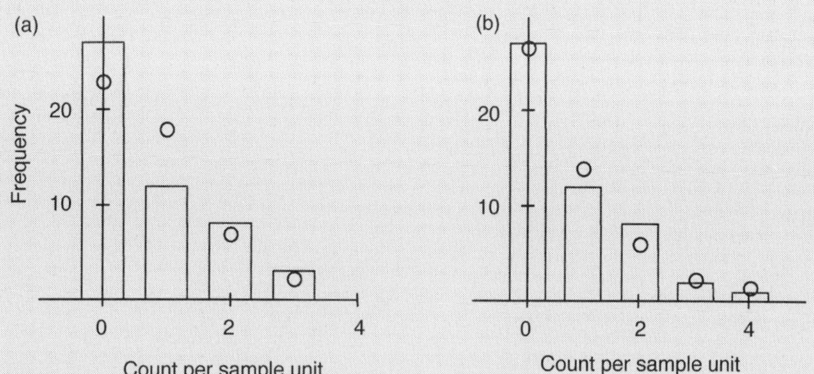

Fig. 4.11. Fitting Poisson and negative binomial distributions to a subsample of the total number of sample units shown in Fig. 4.8b. (a) A frequency distribution of 50 randomly selected sample units and the fitted Poisson distribution. (b) A frequency distribution of 50 randomly selected sample units and the fitted negative binomial distribution.

Based on these 50 sample units alone, it could be inferred (wrongly) that the Poisson distribution is an adequate probability model. Clearly, we need to take a reasonable number of sample units if we are going to fit distributions and make inferences from the fits. Another two samples were taken with 100 and 200 sample units each. This time, the inferences were like those using all the data (Table 4.1). In this instance, 50 sample units were too few and 100 were enough.

We can compare the (approximate) standard errors (se) of the maximum likelihood estimates (Table 4.1) with curves analogous to Equation 2.3, namely

$$V_m = \frac{V}{n} \quad \text{or} \quad sem = \sqrt{\frac{V}{n}} \tag{4.15}$$

These curves are

$$se(\mu)_n = se(\mu)_{1296}\sqrt{\frac{1296}{n}} \quad \text{and} \quad se(k)_n = se(k)_{1296}\sqrt{\frac{1296}{n}} \tag{4.16}$$

so they are equal to the data values at sample size $n = 1296$. We can plot the standard errors of the estimates of μ and k (Table 4.1) along with these curves (Fig. 4.12). It is not surprising that the $se(\hat{\mu})$ values follow the curves, because Equation 4.15 refers specifically to means, but the same relationship holds (approximately) for $se(\hat{k})$ also. We can generally assume the same effect of sample size on the variance of anything that we estimate, provided that all other aspects of the sampling remain the same.

It is important to collect enough sample units to be able to decide whether a specified probability distribution really does fit a given spatial pattern, at the level of that sample unit. We have found that 100 sample units is usually enough, given the fact that many different fields need to be sampled before a general conclusion can be drawn about what is an adequate probability model.

Table 4.1. Fitting subsamples of the complete data set shown in Fig. 4.8b to the negative binomial distribution. Fitted parameters ($\hat{\mu}$ and $\hat{k}$) and their standard errors for all 1296 sample units, and for random subsamples.

	$\hat{\mu}$	$se(\mu)$	$\hat{k}$	$se(\hat{k})$	P_P[1]	P_{nb}[2]
All data ($n = 1296$)	0.772	0.029	1.70	0.259	0.00	0.25
$n = 200$	0.780	0.075	1.78	0.712	0.00	0.58
$n = 100$	0.690	0.103	1.313	0.679	0.04	0.80
$n = 50$	0.710	0.145	1.57	1.168	0.62	0.27

[1] Significance probability for fitting the Poisson distribution.
[2] Significance probability for fitting the negative binomial distribution.

Continued

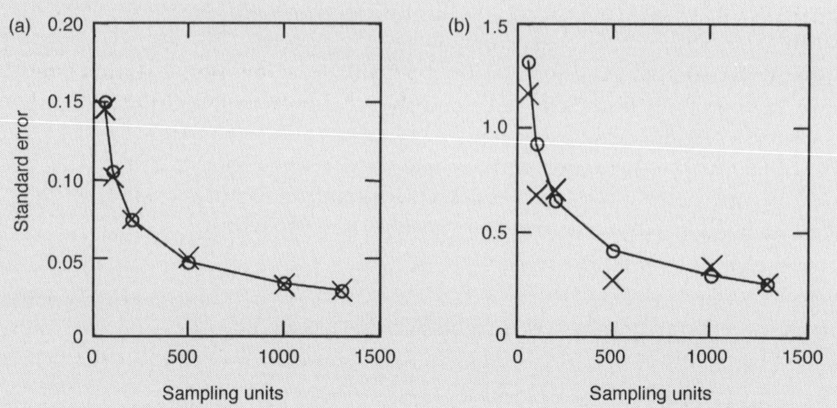

Fig. 4.12. The relation between the standard errors of estimates of negative binomial parameters and sample size: estimated values and a theoretical curve (Equation 4.16). The base data are the sample units illustrated in Fig. 4.6b. (a) Estimates of μ; (b) estimates of k.

4.8 The Mechanics of Fitting Distributions to Data[4]

In the previous section we stated that the parameters of the negative binomial distribution were fitted to the observed frequencies without being explicit on how this was done, except to note that the principle of maximum likelihood was used. The problem is to find the maximum value of an expression (such as Equations 4.5 or 4.7) that represents the combined probability of getting the data as a function of one or more unknown parameters. Occasionally, as with the Poisson distribution, there is an analytical solution (as we have demonstrated), but usually there is no analytical solution and we have to use a numerical optimization algorithm.

Such an algorithm works something like the following:

1. An initial parameter set is determined, either automatically by the algorithm or given by the user.
2. An area of search around the initial parameter set is determined based on user input or internal calculations.
3. The algorithm searches the likelihood function by calculating the likelihood at various values of the parameters in the search area, using built-in knowledge of what the shape of the likelihood ought to be, and calculates a parameter set which is likely to be nearer the solution.
4. The likelihood is calculated at this new parameter set.
4a. Based on a criterion, often given by the user (such as a desired small value for the distance between the new and old parameter sets), the algorithm decides

[4]This section is full of mathematics and may be skipped.

whether it has reached a solution. If it has, the algorithm ends with the current parameter set.

4b. If the stopping criterion is not satisfied, a new iteration is started (step 5), unless a predetermined maximum number of iterations has already been done, in which case the algorithm stops.

5. Initiate the next iteration.

5a. If the new parameter set gives a better value for the likelihood, the new parameter set replaces the old set, and the algorithm returns to step 3, possibly defining a smaller area of search.

5b. If the new parameter set does not give a better value, most algorithms retain the previous parameters set and return to step 3 with new search parameters (for example an enlarged or differently shaped area of search).

To create even a reasonable algorithm requires a very good understanding of numerical methods and the underlying mathematics. When no more than two parameters are used, a very simple algorithm can be used. Starting from an initial parameter set (e.g. μ and k of the negative binomial distribution), calculate a 5×5 array of parameter values using values of δ_1 and δ_2, which are determined as 'small' with respect to μ and k, respectively:

$\mu - 2\delta_1, k + 2\delta_2$	$\mu - \delta_1, k + 2\delta_2$	$\mu, k + 2\delta_2$	$\mu + \delta_1, k + 2\delta_2$	$\mu + 2\delta_1, k + 2\delta_2$
$\mu - 2\delta_1, k + \delta_2$	$\mu - \delta_1, k + \delta_2$	$\mu, k + \delta_2$	$\mu + \delta_1, k + \delta_2$	$\mu + 2\delta_1, k + \delta_2$
$\mu - 2\delta_1, k$	$\mu - \delta_1, k$	μ, k	$\mu + \delta_1, k$	$\mu + 2\delta_1, k$
$\mu - 2\delta_1, k - \delta_2$	$\mu - \delta_1, k - \delta_2$	$\mu, k - \delta_2$	$\mu + \delta_1, k - \delta_2$	$\mu + 2\delta_1, k - \delta_2$
$\mu - 2\delta_1, k - 2\delta_2$	$\mu - \delta_1, k - 2\delta_2$	$\mu, k - 2\delta_2$	$\mu + \delta_1, k - 2\delta_2$	$\mu + 2\delta_1, k - 2\delta_2$

The likelihood can be calculated at each of the 25 sets of parameters, and the maximum value can be determined, thus finding the next pair of values for the parameters. These are then placed in the middle of another array, defined by smaller values of δ_1 and δ_2 than those used in the previous iteration. Repeating this a number of times, a solution can be reached.

Two critical features of this, or of any search for a likelihood solution, remain: which parameters to use, and how to obtain initial values. The first of these may not appear to be a problem. Why not use the parameters used in the mathematical definitions (e.g. μ and k for the negative binomial)? The answer lies at least partly in the initial values. It has now become fairly well known that parameters that reflect easily identified properties of the data are best when searching for a maximum likelihood solution (see, e.g. Ross, 1970, 1990). One reason for this is that initial values are then easy to calculate directly from the data and, furthermore, these values should not be far from the solution. For fitting distributions, it is easy to calculate the mean and variance of a set of data, and if the parameters are chosen to be closely related to these, the search for the solution should be relatively easy. For example, the parameters of the negative binomial distribution are the mean itself (μ) and the parameter k, which is closely related to the variance and the mean: $k = \mu^2/(\sigma^2 - \mu)$.

4.9 The Binomial Distribution

In many instances, the only information that is obtainable, or even needed, is whether a sample unit contains, or does not contain, a pest. Note that 'contain a pest' includes, for example, 'infected by a pathogen' and, for convenience, we use the expression 'infected sample unit'. If the probability of a sample unit being infected is p, the probability of x infected sample units out of the total n sample units selected at random from a field is given by the binomial distribution:

$$\mathrm{p}(x \mid n, p) = \binom{n}{x} p^x (1-p)^{n-x} \tag{4.17}$$

The theoretical mean and variance of x are as follows:

$$\text{mean number of infested sampling units} = np, \quad \text{with} \quad \text{variance} = np(1-p) \tag{4.18}$$

and the theoretical mean and variance of the incidence, x/n, are

$$\text{mean incidence} = p, \quad \text{with} \quad \text{variance} = \frac{p(1-p)}{n} \tag{4.19}$$

Typical shapes of the binomial distribution are shown in Fig. 4.13. If economic injury is related to pest incidence, the binomial distribution can be used directly to

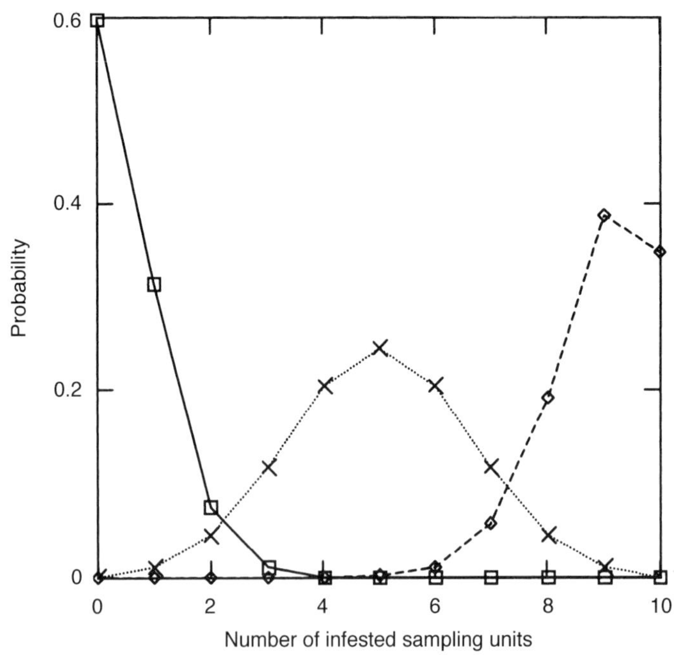

Fig. 4.13. Binomial distributions with $n = 10$ and $p = 0.05$ (□),0.5 (×) and 0.9 (◇).

develop sampling plans for decision-making in crop protection. If economic injury is related to density, however, but the sample data are proportions of plants containing a pest, the problem arises of relating incidence to density. This is discussed in detail in Chapter 7.

A striking feature of the binomial distribution is that, in a sense, it always holds when sampling is random. For any spatial pattern which is sampled by random selection of sample units, the proportion of sample units containing a pest (or that are infected) can be characterized by the binomial distribution.

The binomial distribution is relevant also for simulated OC functions. Each simulation of a sampling plan provides an estimate of the true probability of making a decision not to intervene, whatever the mean density. This is analogous to collecting binomial samples in the field. The mean of *sr* simulation replicates is an estimate of the true probability of a decision not to intervene, and is therefore distributed exactly as the mean of *sr* samples from a binomial distribution. This is important for comparing simulated OC functions.

4.10 The Beta-binomial Distribution

When sampling to estimate incidence, it is sometimes practical to take a *cluster* of sample units at each location instead of only one sample unit. The main argument for this would be to save time and effort. However, this immediately destroys the inevitability of the binomial distribution. If the pest is aggregated at the scale of the sample cluster, the number of infected sample units per cluster is no longer binomial. In particular, the variance is increased. Such sampling data can be analysed using the *beta-binomial* distribution.

The beta-binomial distribution describes the number of infected units (x) in a cluster of R sample units and has three parameters: R, α and β, although R is known (of course!). The formulae for the beta-binomial probabilities are formidable, involving the gamma function:

$$p\,(x\,|\,R,\alpha,\beta) = \binom{R}{x} \frac{\Gamma(\alpha+\beta)}{\Gamma(\alpha)\,\Gamma(\beta)} \frac{\Gamma(\alpha+x)\,\Gamma(\beta+R-x)}{\Gamma(\alpha+\beta+R)} \tag{4.20}$$

It may be comforting to know that the gamma function is just an extension of the factorial function (!): for integer arguments i, $\Gamma(i)$ equals $(i - 1)!$ For example, $\Gamma(3) = 2! = 2$, and $\Gamma(4) = 3! = 6$. For non-integer arguments $(i + u)$, where i is an integer and $0 < u < 1$, $\Gamma(i + u)$ lies between $(i - 1)!$ and $i!$ For example, $\Gamma(3.1) = 2.198$ and $\Gamma(3.9) = 5.299$.

When n clusters of R sample units are collected, the mean and variance of the proportion of infected sample units, under the beta-binomial distribution, are as follows:

$$\left.\begin{aligned} &\text{mean incidence} = p \\ &\text{variance} = \frac{p\,(1-p)}{R\,n}\,[1+\rho(R-1)] \\ &\text{where } p = \frac{\alpha}{\alpha+\beta},\quad \rho = \frac{1}{1+\alpha+\beta} \end{aligned}\right\} \tag{4.21}$$

and where p is the overall probability of a sample unit being infected, and ρ is the intra-cluster correlation. Typical shapes of the beta-binomial distribution for $p = 0.05$, 0.4, 0.9 and for $\rho = 0.02$, 0.1, 0.33, 0.7 are shown in Figs 4.14a–d. The corresponding mathematical parameters α and β are as follows:

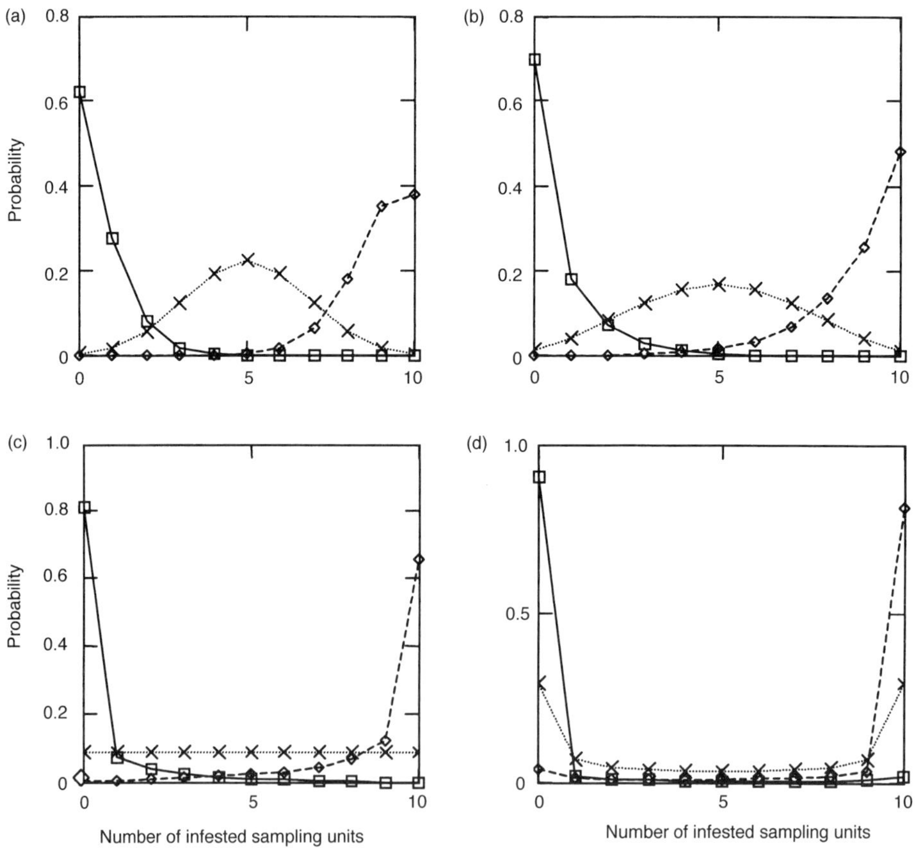

Fig. 4.14. Beta-binomial distributions with cluster size equal to 10. Each graph presents distributions for $p = 0.05$ (□), 0.5 (×) and 0.9 (◇). (a) $\rho = 0.02$; (b) $\rho = 0.1$; (c) $\rho = 0.33$; (d) $\rho = 0.7$.

	ρ	$p=$	0.05	0.5	0.9
(a)	0.02	$\alpha=$	2.45	24.5	44.1
		$\beta=$	46.55	24.5	4.9
(b)	0.1	$\alpha=$	0.45	4.5	8.1
		$\beta=$	8.55	4.5	0.9
(c)	0.33	$\alpha=$	0.102	1.015	1.827
		$\beta=$	1.929	1.015	0.203
(d)	0.7	$\alpha=$	0.021	0.214	0.386
		$\beta=$	0.407	0.214	0.043

Note that when ρ is small ($\rho = 0.02$), the distributions (Fig. 4.14a) are not very different from the corresponding binomial distributions (Fig. 4.13). When ρ is large, the distribution becomes U-shaped, even for small or large values of p (Fig. 4.14d).

Exhibit 4.4. *The description of the number of virus-infected plants per cluster in sugar beet*

As part of a study on the spread of beet yellows virus in experimentally inoculated fields of sugar beet in The Netherlands, symptoms were noted on all plants in 21 rows of one of the fields. The degree of aggregation was estimated in terms of the intra-cluster correlation, ρ.

The disease was clearly aggregated (Fig. 4.15). Plants were allocated to clusters of 10 consecutive individual plants within rows (Fig. 4.16). The binomial and beta-binomial distributions were fitted to the numbers of affected plants in each cluster. The binomial distribution is clearly unable to model the data, but the beta-binomial distribution provides a good fit (Fig. 4.17). The fitted parameters were as follows:

	p	ρ	α	β	X^2	df	P
Binomial	0.195	0	–	–	1417	5	< 0.001
Beta-binomial	0.193	0.291	0.472	1.968	9.6	8	0.29

The beta-binomial distribution fits the data with different cluster sizes also. The value of p does not change much, but the value of ρ does:

Cluster size, R	p	ρ
1×5 ($R = 5$)	0.195	0.363
1×10 ($R = 10$)	0.193	0.291
1×15 ($R = 15$)	0.195	0.270
3×5 ($R = 15$)	0.197	0.294

The effect of cluster size on the value of ρ is investigated in more depth in the next exhibit.

Continued

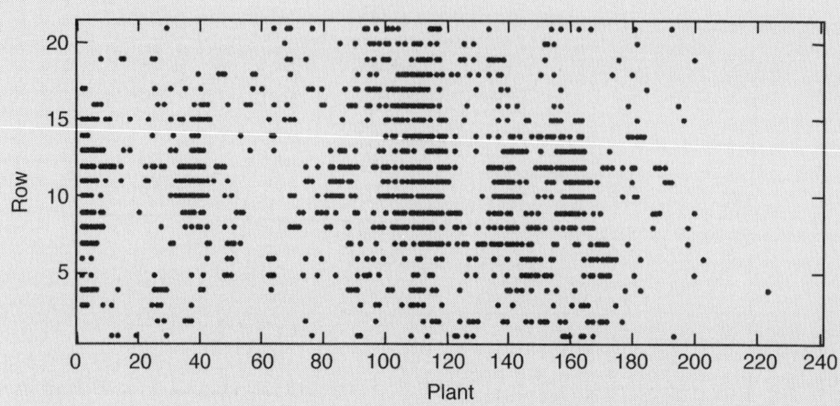

Fig. 4.15. The incidence of beet yellows virus on individual sugarbeet plants in 21 rows of 240 plants.

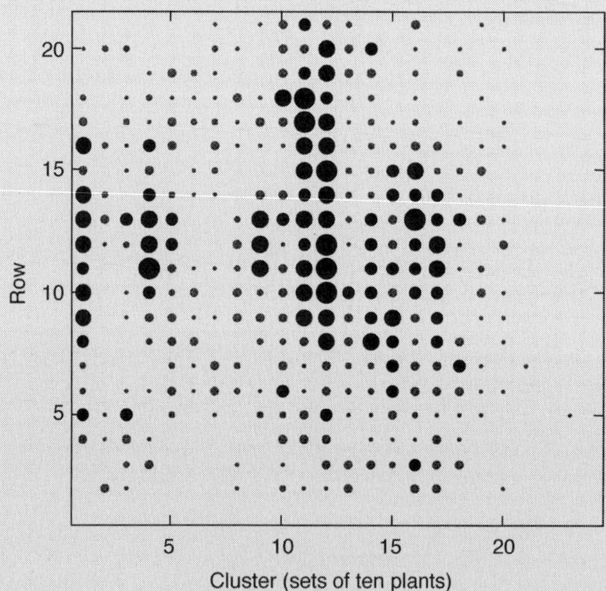

Fig. 4.16. The numbers of diseased sugarbeet plants in clusters of 10 plants in each row of the field depicted in Fig. 4.15. Each small dot represents a cluster with one diseased plant, and each increase in size of the dot represents one or two more diseased plants in a cluster: 1, 2, 3 or 4, 5 or 6, 7 or 8, 9 or 10.

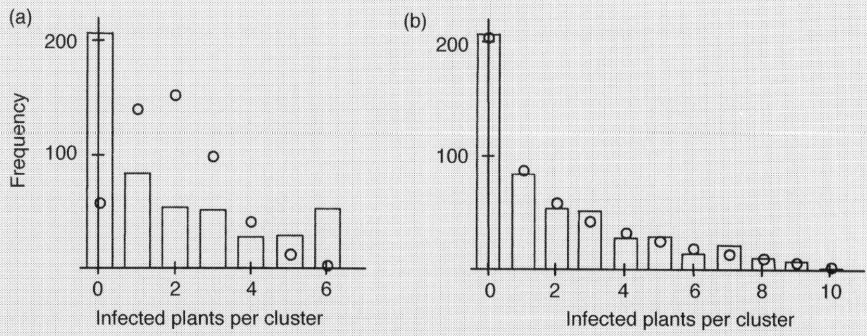

Fig. 4.17. Fitting the data shown in Fig. 4.16 to the binomial (a) and the beta-binomial (b) distributions, using a grouping criterion equal to 1. Fitted frequencies are shown as circles.

The intra-cluster correlation, ρ, is not estimated during routine pest management sampling, but is important in defining the variance of the incidence estimate (Equation 4.21). Intuitively, the average correlation among sample units in a cluster should decrease as the cluster size increases, because of distance. This attenuation of the correlation as cluster size increases is illustrated in the following exhibit.

Exhibit 4.5. *The effect of cluster size and shape on the intra-cluster correlation*

Under random sampling, any aggregation among sample units is immaterial, and the binomial distribution is appropriate. If for any reason sampling is not random, then the binomial distribution cannot be assumed to hold. In particular, if sample units are chosen in clusters of adjacent sample units, the beta-binomial distribution should be used. The grid of sample units shown in Fig. 4.8b was rendered into a pattern of incidence by allocating 0 to units with 0 or 1 points and 1 to units with at least two points (Fig. 4.18). The effect of cluster size and shape on the intra-class correlation, ρ, was studied for this spatial pattern of incidence. The results are shown in Table 4.2.

The estimate of incidence was always near 0.20, but the estimate of ρ depended on the size and shape of the cluster. The larger clusters (12 sample units) tended to have lower values of ρ than the smaller clusters (six sample units). This is not surprising: the aggregation in the spatial pattern is local (as we noted above when we used the same data to study the negative binomial distribution), and the correlation between sample units decreases as the distance between them increases. As clusters became more elongated, ρ decreased. This result is similar to that noted in Exhibit 4.2, and the reason for the decrease is the same: clusters with longer edges cover more sources of aggregation, and thus average them out.

Continued

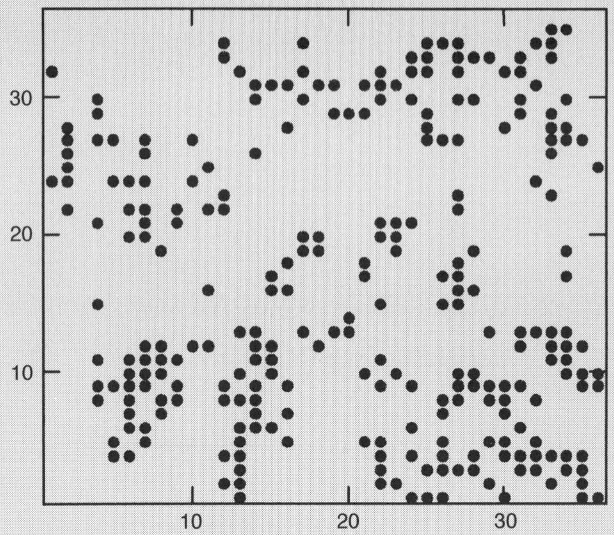

Fig. 4.18. The spatial pattern derived from Fig. 4.8b. Each sample unit in Fig. 4.8b containing two or more individual points is plotted here as a point. All other sample units are blank.

Table 4.2. Estimates of incidence and intra-cluster correlation for different sizes and shapes of clusters, used to sample the computer-generated spatial pattern in Fig. 4.18. All frequency distributions fitted the beta-binomial distribution, but none fitted the binomial distribution.

Number of sample units along the		Estimates and approximate standard errors			
x-axis	*y*-axis	*p*	*se*(*p*)	ρ	*se*(ρ)
4	1	0.200	0.014	0.166	0.032
2	2	0.201	0.014	0.217	0.035
1	4	0.201	0.013	0.156	0.032
6	1	0.201	0.014	0.098	0.026
3	2	0.200	0.015	0.174	0.031
2	3	0.200	0.015	0.165	0.030
1	6	0.202	0.014	0.113	0.026
12	1	0.201	0.016	0.103	0.024
6	2	0.201	0.017	0.113	0.026
4	3	0.199	0.018	0.149	0.029
3	4	0.200	0.018	0.136	0.028
2	6	0.200	0.016	0.107	0.025
1	12	0.202	0.015	0.075	0.021

> Note that we have not referred here to the defining mathematical parameters α and β of the beta-binomial distribution. The parameters p and ρ are easier to understand in practical terms than α and β. As can be seen from Equation 4.21, there are simple relationships among α, β, p and ρ.

As with the negative binomial distribution, we are left with the knowledge that the beta-binomial distribution is able to model incidence very well, but do we need to estimate ρ every time? A model relating the beta-binomial variance to the binomial variance has been proposed by Hughes *et al.* (1996) which can do for the beta-binomial distribution what TPL can do for the negative binomial distribution (Section 4.5.1). Hughes *et al.* (1996) showed that the relationship

$$\left.\begin{aligned} \text{beta-binomial variance} &= A(\text{binomial variance})^b \\ \text{where binomial variance} &= \frac{p(1-p)}{R} \end{aligned}\right\} \tag{4.22}$$

fitted a number of data sets, and proposed it as a general formula. We retain their original notation. On the basis of this formula and Equation 4.21, the intra-class correlation can be calculated as

$$\begin{aligned} \rho &= \frac{1}{R-1}\left(\frac{\text{beta-binomial variance}}{\text{binomial variance}} - 1\right) \\ &= \frac{1}{R-1}\left(\frac{AR^{1-b}}{[p(1-p)]^{1-b}} - 1\right) \end{aligned} \tag{4.23}$$

4.11 Relationships among the Distributions

For certain choices of parameters, the distributions in this chapter may be indistinguishable from each other.

The negative binomial distribution with mean μ approaches the Poisson distribution with mean μ as k increases. For example, compare Fig. 4.2 (diamond) with Fig. 4.6a (diamond).

When sampling to detect a rare event (binomial distribution with very small p), the total number of sample units where the rare event is observed (X_n out of a total of n sample units) is approximately Poisson with mean equal to np. This is exemplified in Fig. 4.19. When $p = 0.025$ and $n = 40$, the Poisson and binomial distributions are indistinguishable. When p is increased to 0.125 there is little difference between Poisson and binomial. When $p = 0.25$, there begins to be a noticeable difference, but it is still not great. The binomial distribution is more

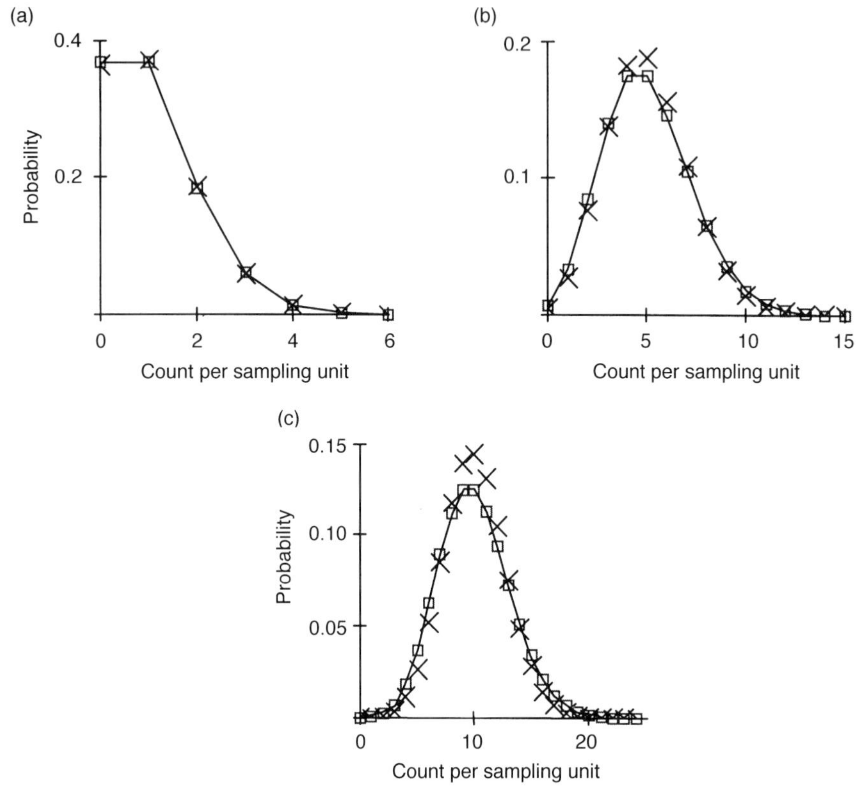

Fig. 4.19. A comparison of the Poisson (—□—) and binomial (×) distributions. (a) Poisson mean = 1, binomial n = 40, p = 0.025; (b) Poisson mean = 5, binomial n = 40, p = 0.125; (c) Poisson mean = 10, binomial n = 40, p = 0.25.

peaked than the Poisson, reflecting the fact that its variance is always less than its mean.

The beta-binomial distribution with mean probability p $(= \alpha/\{\alpha + \beta\})$ approaches the binomial distribution with mean probability p as both α and β increase (incidentally, forcing ρ to decrease).

The beta-binomial distribution is a binomial distribution when the cluster size, R, is equal to 1.

When R is large and p is small, the beta-binomial distribution becomes like the negative binomial distribution with k equal to p/ρ (Madden and Hughes, 1999). If ρ is small, this is true for moderate values of R, but as ρ increases, R must also increase for there to be much similarity between the two distributions.

4.12 On the Choice of Probability Distribution Model

We have intentionally restricted ourselves to four distributions: Poisson, negative binomial, binomial and beta-binomial. Other distributions have been suggested for modelling the frequency distributions of biological organisms, and routines for fitting data have been written (see, e.g. Gates *et al.*, 1987; Payne *et al.*, 1993). However, those that we have described here are the ones that have been found to fit most often, and therefore are the most commonly used.

Because each newly calculated frequency distribution is potentially unlike any other, it is normal practice to try fitting several probability distributions, in an attempt to find one that can be used in subsequent work. Frequently, more than one probability distribution fits adequately. There is no statistical procedure that can indicate which, if any, is the one to choose. Collecting more data in a wider range of environments should clarify the issue. What usually happens is that one distribution fits more often than the others, and the real question is whether the number of times it does not fit is a serious obstacle to using it as a foundation for decision-making (see, e.g. Nyrop *et al.*, 1989). Of course, there is a possibility that no theoretical distribution fits often enough to be convincing. We discuss how to deal with this problem in Chapter 9.

The fact that a probability distribution such as the negative binomial distribution fits a frequency distribution does not mean that the species represented by the data will always fit that probability distribution. Only if many independent data sets consistently follow a certain probability distribution can we begin to assume that at other times and places this probability distribution should be applicable. Once we have obtained enough data to convince ourselves, and others, that this is the case, we can calculate probability of decision (OC) functions for any sampling plan we wish. The data sets used to do this job of convincing do not need to have as detailed a coverage of individual fields as in the Colorado potato beetle example of Beall. Costs prohibit such extensive data collection. A reasonably large sample (say 100) is often sufficient. Of course, how large the data set should be depends on the spatial pattern itself. The greater the aggregation or variability is, the more sample units are needed. It is important to ensure that the 'application domain' that is, the pest densities around the critical threshold) is well covered by the sets of data that are used to fit the distributions. This advice is similar to that given in Section 3.4, for estimating TPL.

4.13 Summary

1. A frequency distribution is a summary of a spatial pattern, ignoring the location of sample units. It depends on the properties of the sample unit; in particular, its size and shape.
2. A probability distribution is a model for a frequency distribution, but only if it fits the data well.
3. The Poisson distribution is an appropriate model for a frequency distribution based on a random spatial pattern.

4. The negative binomial distribution, with two parameters (μ and k), is a useful model for a frequency distribution based on an aggregated spatial pattern. Both parameters depend on the size and shape of sample units. A variance–mean model can be used to estimate k for varying μ.
5. The binomial distribution is appropriate for modelling incidence when sample units are collected at random throughout the field.
6. The beta-binomial distribution is appropriate for modelling incidence when sampling is done by collecting random clusters of adjacent sample units throughout the field. Spatial aggregation is accommodated by a parameter that measures the intra-cluster correlation, ρ. The size of ρ depends on the shape and size of the cluster. A model relating the variance of incidence estimates under both types of sampling (random collection of sample units and random collection of clusters of sample units) can be used to estimate ρ.
7. The χ^2 test can be used to make an objective decision on whether a probability distribution is an adequate model for a frequency distribution. At least 100 sample units are needed to make an acceptable goodness-of-fit test.

References and Suggested Reading

Binns, M.R. (1986) Behavioural dynamics and the negative binomial distribution. *Oikos* 47, 315–318.

Campbell, C.L. and Madden, L.V. (1990) *Introduction to Plant Disease Epidemiology*. John Wiley, New York, 532 pp.

Chernoff, H. and Lehmann, E.L. (1954) The use of maximum likelihood estimates in χ^2 tests for goodness of fit. *Annals of Mathematical Statistics* 25, 579–586.

Diggle, P.J. (1983) *Statistical Analysis of Spatial Point Patterns*. Academic Press, London, 148 pp.

Faddy, M. (1997) Extended Poisson process modelling and analysis of count data. *Biometrical Journal* 39, 431–440.

Gates, C.E., Ethridge, F.G. and Geaghan, J.D. (1987) *Fitting Discrete Distributions: User's Documentation for the Fortran Computer Program DISCRETE*. Texas A&M University, College Station, Texas, 22 pp.

Hoel, P.G., Port, S.C. and Stone, C.J. (1971) *Introduction to Statistical Theory*. Houghton Mifflin, Boston, 237 pp.

Hughes, G., Madden, L.V. and Munkvold, G.P. (1996) Cluster sampling for disease incidence data. *Phytopathology* 86, 132–137.

Johnson, N.L. and Kotz, S. (1969) *Discrete Distributions*. Houghton Mifflin, New York, 328 pp.

Madden, L.V. and Hughes, G. (1999) Sampling for plant disease incidence. *Phytopathology* 89, 1088–1103.

Nyrop, J.P., Agnello, A.M., Kovach, J. and Reissig, W. (1989) Binomial sequential classification of sampling plans for European read mite (Acari: Tetranychidae) with special reference to performance criteria. *Journal of Economic Entomology* 82, 482–490.

Payne, R.W., Lane, P.W., Digby, P.G.N., Harding, S.A., Leech, P.K., Morgan, G.W., Todd, A.D., Thompson, R., Tunnicliffe Wilson, G., Welham, S.J. and White, R.P. (1993) *Genstat™ 5 Release 3 Reference Manual*. Clarendon Press, Oxford, 796 pp.

Pielou, E.C. (1977) *Mathematical Ecology*. John Wiley, New York, 385 pp.

Ripley, B.D. (1979) AS137, Simulating spatial patterns. *Applied Statistics* 28, 109–112.
Ripley, B.D. (1981) *Spatial Statistics*. John Wiley, New York, 252 pp.
Ross, G.J.S. (1970) The efficient use of function minimisation in non-linear maximum likelihood estimation. *Applied Statistics* 19, 205–221.
Ross, G.J.S. (1990) *Nonlinear Estimation*. Springer-Verlag, New York, 189 pp.
Taylor, L.R. (1984) Assessing and interpreting the spatial distributions of insect populations. *Annual Review of Entomology* 29, 321–359.

Appendix: Computer Generation of Aggregated Spatial Patterns

The computer-generated spatial patterns that we refer to in this book were derived using one or other of two methods described in the literature: the Poisson cluster processes (see, e.g. Diggle, 1983) and Gibbs processes (see, e.g. Ripley, 1981). We present here a brief outline of how we have implemented them.

Poisson cluster processes

For many years statisticians have developed a wide variety of probability distributions which are called compound Poisson distributions. The ones that are relevant here are those in which independent distributions of 'parents' and 'offspring' are specified. Our implementation uses the Poisson distribution for 'parents', and either a Poisson or a log-series (see, e.g. Pielou, 1977) distribution for the 'offspring'. This is put together in four stages to create a spatial pattern:

1. A number, N, of 'parents' is distributed randomly in the unit square.
2. Each 'parent' gives rise to M 'offspring', M being generated from a Poisson or log-series distribution with predefined mean.
3. The positions of these 'offspring' relative to the corresponding 'parent' are generated as bivariate normal with zero correlation.
4. Points generated outside the unit square are ignored.

Gibbs processes

Gibbs processes use an initial spatial pattern of points and derive, by iteration, a new spatial pattern. The goal is to adjust the initial pattern (by repeated rejection and replacement) of points, so that in the final pattern the distribution of the number of neighbouring points of each point conforms to a prespecified probability distribution. To implement this, a definition of 'neighbouring point' and rules for rejection and replacement are required. We start with a random spatial pattern of N points. Points are defined as neighbours of a given point if they are within a circle of radius r centred on that point. The rejection criterion is that new points are rejected with probability proportional to the number of their neighbours.

1. Start with a random spatial pattern of N points in the unit square.
2. Choose one of the N points at random to be replaced.
3. Generate a potential replacement point at random.
4. Calculate the number of neighbours, t, of the new point.
5. Generate a uniform random number, x, between 0 and 1.
6. If $x > t/N$ accept the new point; otherwise, return to step 3.
7. Repeat steps 2–6 a number of times. Using a slightly different rejection criterion, Ripley (1979) suggested that $4N$ times should be sufficient.

Faddy (1997) demonstrated how various probability distributions arise from inhomogeneous Poisson processes in the time domain. His definition of the negative binomial distribution as an inhomogeneous Poisson process in time is similar to how the above method generates spatial patterns (his definition was our motivation). When a grid is superimposed on a spatial pattern generated by this method to define sample units, then, in many instances, the frequency distribution of numbers of points in sample units is either Poisson or negative binomial. However, there are sufficient differences between a time dimension and two-dimensional space (in particular, time is an ordered dimension) that a negative binomial distribution is not guaranteed for these spatial patterns (Exhibit 4.2).

The generation of spatial patterns by the Gibbs process method tends to be much slower than by the Poisson cluster method. The computer-generated patterns referred to in this chapter were derived using the Gibbs method. Most of the computer-generated patterns referred to in subsequent chapters were derived by the Poisson method.

5 Sequential Sampling for Classification

5.1 Introduction

In this chapter we present three methods for sequential classification of pest abundance or pest incidence. With such methods, the need for further sample information is evaluated after each sample unit is examined by making one of three possible decisions: (i) the density is classified as greater than the critical density (*cd*), and sampling is stopped; (ii) the density is classified as less than or equal to *cd* and sampling is stopped; or (iii) an additional sample unit is taken and examined. A classification 'greater than the critical density' will probably trigger some kind of intervention. The decisions are reached by comparing total counts to stop boundaries. The three sequential methods differ in the shape of the stop boundaries produced and the parameters used to specify the boundaries. We describe each of the methods and illustrate how various parameters influence the operating characteristic (OC) and average sample number (ASN) functions for each. The OC and ASN functions are estimated using simulation. Finally, we compare the three procedures in reference with each other.

5.2 From Batch to Sequential Sampling

In Chapter 3 we introduced the notion of examining a batch of sample units and deciding whether, on the basis of the data so far, it was possible to come to a conclusion, or whether another batch of sample units was required. The idea of reducing the batch size to one sample unit was suggested, but there we ran up against the assumption of normality; there are few, if any, pest counts that can be described by a normal distribution, whatever the sample unit. If we had nevertheless proceeded to calculate OC and ASN functions (not difficult), these functions might have been misleading. The probability distributions described in Chapter 4 come much nearer to describing actual counts of pests, and, in any particular instance, one among them is usually good enough to give a reliable representation. We will make use of these probability distributions to model sample counts, and use these models to estimate OC and ASN functions for sequential sampling plans.

The expression *sequential sampling* was introduced in the early 1940s by Abraham Wald to describe the (then) new technique of taking sample units, one at a time, and (using what is called a likelihood ratio) deciding whether there is currently enough information to classify population density into one of two categories, or whether at least one more sample unit is required. In those days, before the widespread availability of electronic computers, Wald obtained analytical expressions for the corresponding OC and ASN functions. Unfortunately, although his sampling procedures might have been appropriate in the industrial work for which they were intended, they are not necessarily the most cost-effective elsewhere. In fact, a considerable literature arose with proposals for improving Wald's procedures, especially in the medical field. We shall discuss three methods of sequential sampling for classification, starting with batch sampling, where the batch size is reduced to one sample unit. This is often referred to as Iwao's procedure, in recognition of Syun'iti Iwao who first proposed it (Iwao, 1975).

5.3 Iwao's Sequential Procedure

Like the batch sequential sampling discussed in Chapter 3, the stop boundaries for Iwao's procedure are based on classification intervals (Equation 3.9). In fact, the boundaries for the cumulative sample total are the same as Equation 3.12 for batch sampling, but with $n_B = 1$ (we change the notation slightly so that it is more like what is found in the literature):

$$\left.\begin{aligned} L_n &= n\left(cd - z_{\alpha/2}\sqrt{\frac{V}{n}}\right) \\ U_n &= n\left(cd + z_{\alpha/2}\sqrt{\frac{V}{n}}\right) \\ &\text{for } n = 1, 2 \ldots (maxn - 1) \text{ and } L_{maxn} = U_{maxn} = maxn \times cd \end{aligned}\right\} \tag{5.1}$$

where n is the running total number of samples and $maxn$ is the maximum sample size (a decision is made at $maxn$, if not before). As in batch sampling, sampling is stopped if n reaches the maximum sample size, by making the two boundaries equal to each other (Equation 5.1). It is worth remembering that $\sqrt{V/n}$ is the standard error of the mean, and that V is the variance estimate when the true mean is equal to cd. Thus, four parameters are required to calculate Iwao stop boundaries; the critical density, a maximum sample size, the variance of a sample unit when the density is equal to cd and $z_{\alpha/2}$. All except $z_{\alpha/2}$ have readily understood interpretations. In Section 3.2.2, when we began to discuss classification sampling, $z_{\alpha/2}$ had a probability meaning, but this held only because the sample size, n, was fixed and we could assume a normal distribution for the sample mean m. Sequential sampling induces a probability distribution on the final sample size, n (when sampling stops), which in turn alters the probability distribution of the value of m when sampling stops. Therefore $z_{\alpha/2}$ is best thought of as a parameter that defines the distance between the boundary lines. In particular, α should not be interpreted as an 'error rate' or even as defining a single point on the OC function. However, for consistency with the literature, we retain the notation used by Iwao and others.

The variance used to calculate the stop boundaries, V, can be based either on the probability distribution used to describe the sample counts or on Taylor's Power Law (TPL) (Table 5.1). In estimating the properties of an Iwao plan, we must refer to a probability distribution (see below).

The stop boundaries calculated using Equation 5.1 are curvilinear lines that diverge with increasing n. To ensure a representative sample, the stopping rule is usually not applied until a minimum number of samples units, *minn*, is collected. Then, if the cumulative count after n samples, S_n, is greater than or equal to U_n, sampling is stopped and the density is classified as greater than *cd*; and if S_n is less than or equal to L_n, sampling is stopped and the density is classified as less than or

Table 5.1. Variances used for setting up Iwao sequential plans based on one of four probability distributions. Critical density equal to *cd*; critical proportion equal to *cp*.

Distribution	Variance when true mean = *cd* or *cp*
Poisson	Variance = *cd*
Negative binomial with constant k	Variance = $cd + \frac{cd^2}{k}$
Negative binomial with TPL	Variance = $a(cd)^b$ where a and b are parameters of TPL and *cd* is the critical density
Binomial	Variance = $cp\,(1 - cp)$ where *cp* is the critical proportion
Beta-binomial with constant ρ	Variance = $\left(\frac{cp\,(1-cp)}{R}\right)(1+\rho(R-1))$ where *cp* is the critical proportion, R is the number of sample units in a cluster and ρ is the intra-cluster correlation
Beta-binomial with variance-incidence power law	Variance = $A\left(\frac{cp\,(1-cp)}{R}\right)^b$ where A and b are parameters of the power law, *cp* is the critical proportion and R is the number of sample units in a cluster

equal to *cd*. The stop boundaries are adjusted so that sampling must terminate when *n* reaches a specified maximum value, *maxn*. If *n* = *maxn* is reached, the estimate of mean density, $S_{maxn}/maxn$, is compared with *cd* and the natural classification is made (Equation 5.1). Because at *maxn* the density is classified as greater than *cd* if $S_{maxn} > cd \times maxn$, this classification can be made if $S_n > cd \times maxn$, for any *n*. This results in a portion of the upper boundary being a horizontal line that ends at the point (*maxn*, *cd* × *maxn*). The general shape of the stop boundary is shown in Fig. 5.1.

To estimate the OC and ASN functions, simulation must be used. The general routine for all sequential procedures that use minimum and maximum sample sizes is as follows:

1. The probability distribution that will be used to simulate sampling is specified. For the negative binomial and beta-binomial distributions, an aggregation parameter must be provided (not necessary for the Poisson and binomial distributions). The value of the aggregation parameter can be either fixed or made to vary according to a variance–mean or variance–incidence model. For the negative binomial distribution, Taylor's variance–mean relationship is often used to calculate a *k* value that is appropriate for the mean. For the beta-binomial distribution, a similar type of model may be used to calculate a value for the intra-cluster correlation coefficient that is appropriate at a given level of incidence. Allowance can be made for biological variability around the fitted models. Details can be found in the Appendix.

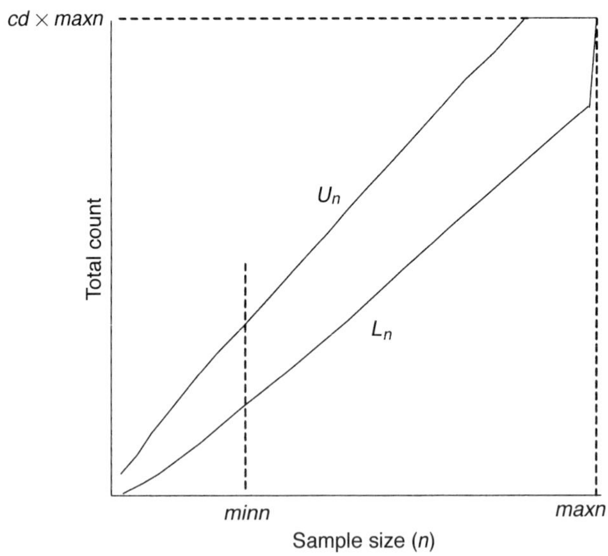

Fig. 5.1. Stop boundaries for a sequential classification sampling plan based on Iwao's method. U_n and L_n are the upper and lower stop boundaries respectively. Minimum and maximum sample sizes are *minn* and *maxn* and the critical density is *cd*.

2. A range of true means (densities or proportions: μ's or p's) is specified for which OC and ASN values will be estimated.
3. For each value of μ (or p), do steps 4–7.
4. Set up the distribution from which samples are to be generated. Note that, as in Section 3.5.1, the distribution used for simulation may differ from that used for setting up the plan.
5. Do steps 5a–5c *sr* times (*sr* is the number of simulation repetitions).
5a. To start each simulation, generate *minn* random variables from the specified distribution, set n equal to *minn*, and calculate S_n, the cumulative total.
5b. Compare S_n with the stop boundary (L_n, U_n) at n:
- if $S_n \leq L_n$, the density is classified as $\leq cd$; store this value of n, and go to step 5a for the next simulation,
- if $S_n \geq U_n$, the density is classified as $> cd$; store this value of n, and go to step 5a for the next simulation, otherwise,

5c. Increment n to $n + 1$, generate another random variable, recalculate S_n, and return to step 5b.
6. Calculate the number of times the density is classified as less than or equal to *cd*, and divide this by *sr*. This is the estimated probability of not recommending intervention; it can be plotted against the corresponding value of μ (or p) as part of the OC function.
7. Calculate the average value of n when sampling stopped; this can be plotted against the corresponding value of μ (or p) as part of the ASN function.
8. Repeat from step 3 until all values of μ (or p) have been done.

We use simulated OC and ASN functions to compare different sampling plans. Because a simulated OC function is closely related to the binomial distribution, we have a simple estimate of its standard error (Section 2.11): $\sqrt{p(1-p)/sr}$, where p is the OC value. We have found that $sr = 1000$ usually gives adequate accuracy, provided that the values of μ (or p) are not far apart: the estimated functions should be smooth over the chosen range.

Sequential procedures, including Iwao's, allow for classifications of the density to be made after very few sample units have been examined. Above, we stated that in practice it is desirable to take a minimum number of samples before comparing counts to the stop boundaries. The importance of this cannot be over-emphasized. In Chapter 2 we discussed at length four criteria for an adequate sampling plan: representativeness, reliability, relevance and practicality. The reliability and representativeness of the sample would almost certainly be violated if sequential sampling stopped at one, two, or even five sample units. If care is not taken in the implementation of sequential sampling, one or more of the necessary attributes of a good sample can be lost. Any sequential plan should include the specification of a minimum number of samples to be taken at the start and following the criteria of Chapter 2. Some experienced pest managers are so concerned with potential problems related to early stopping (or for other practical reasons) that they never use sequential sampling. We feel that this level of caution is unnecessary, but any sequential plan that we discuss from this point onwards will include a minimum number of samples.

Changes in the parameters of Iwao's sequential procedure and changes in the distribution of sample counts have similar general effects on the OC and ASN functions as in batch sampling (Section 3.3). They are summarized in Table 5.2.

Table 5.2. Main general effects of changing parameters of three sequential sampling plans: Iwao, Wald and Converging Lines (CL). The magnitude of any of these effects must be determined by simulation. For Wald, the critical density (*cd*) is taken to refer to $(\mu_0 + \mu_1)/2$. These effects hold when sampling for proportions: replace *cd* by *cp*.

Change		Main general effects
Iwao	Increase α (decrease z_α)	Decreases the interval between the boundaries, which in turn decreases the ASN function. The OC function is made flatter near *cd*
Wald CL	Increase α (β fixed) Increase aU (aL fixed)	Lowers the upper stop boundary.[a] Lowers the upper left part (< *cd*) of the OC, making it flatter, and lowers the right part (> *cd*) of the ASN
Wald CL	Increase β (α fixed) Increase αL (αU fixed)	Raises the lower stop boundary.[b] Raises the lower right part (> *cd*) of the OC, making it flatter, and lowers the left part (< *cd*) of the ASN
Iwao, CL	Increase *cd*	The OC and ASN functions are shifted to the right on the pest density axis
Wald	Increase μ_0 (μ_1 fixed, but $cd = (\mu_0 + \mu_1)/2$ increases)	Raises the upper stop boundary. Makes the upper left part (< *cd*) of the OC steeper, and increases the corresponding part of the ASN
Wald	Decrease μ_1 (μ_0 fixed, but $cd = (\mu_0 + \mu_1)/2$ decreases)	Decreases the lower stop boundary. Makes the lower right part (> *cd*) of the OC steeper, and increases the corresponding part of the ASN
Iwao, Wald	Increase *minn* (minimum sample size)	The sloping part of the OC becomes narrower, and the OC as a whole becomes steeper, especially the curved parts that converge into the horizontal asymptotes (at 0 and 1). The ASN is increased
CL	Increase *minn* (minimum sample size)	Increases the distance between the boundaries, which increases the ASN function. The OC function is made steeper
Increase *maxn* (maximum sample size)		Near *cd*, the OC function is made steeper and the ASN function is increased
Increase the true variance, σ^2, above the estimated variance, *V*		The ASN function is (usually) decreased near *cd*, but increased far from *cd*. The OC function is made flatter

[a] For Wald, the lower stop boundary is also affected: it is raised, but not by so much provided that α and β are small. The effects may appear counter-intuitive.

[b] For Wald, the upper stop boundary is also affected: it is lowered, but not by so much provided that α and β are small. The effects may appear counter-intuitive.

The effect of increasing σ^2 above V is worth more discussion than is found in Table 5.2. If the true distribution of sample counts is not the same as that used for calculating the stop boundaries, the effects are not immediately clear. Increasing the variance of the sample observations above the variance used to construct the stop boundary (this applies only to the negative binomial and beta-binomial distributions) flattens the OC function. The effect on the ASN function is not so straightforward: near *cd* (or *cp*) the ASN tends to be lower; otherwise, it tends to be higher. The basic reason for this is that the increased variability of S_n when the true mean is near *cd* (or *cp*) makes the sample path more erratic and it crosses the boundary earlier than it would otherwise; more incorrect classifications are also likely. However, if the true mean is far from *cd* (or *cp*), so that the sample path is expected to exit quite quickly, the increased variability of S_n can make the path remain within the boundary longer than it would otherwise. It should be noted that the variability would have to be very much larger before the path would be so erratic as to make an exit out of the 'wrong' side of the stop boundary, so the OC function is unlikely to be affected far from *cd*.

As noted for batch sampling, the magnitude of these effects on the OC and ASN functions is difficult to predict. The effects may be small and even ignorable; whether or not they are ignorable depends on practical considerations such as those discussed in Chapter 1. Some of these points are illustrated in Exhibit 5.1.

Exhibit 5.1. *Iwao's plan: the effect of altering parameters*

In this example, we demonstrate how the performance of Iwao's sequential procedure is influenced by different values for α and maximum sample size, and by variation in the model used to describe sample observations. We also demonstrate savings in sample resources that can be realized through the use of a sequential procedure. The example is based on sampling lucerne weevil (*Hypera postica*), a defoliator of lucerne with one generation per year throughout its range in North America. Critical densities for this pest range from 0.5 to 4.5 individuals per plant (Barney and Legg, 1988). Sampling procedures have been based either on removal of foliage from a specified area, or on stem sampling. Legg *et al.* (1985) proposed a sample unit of six stems taken from a 100 m^2 area and suggested that a minimum of five such samples be taken in each field. Barney and Legg (1988) determined the variance of six-stem sample counts taken from 100 m^2 as a function of the mean: $V = 1.0\mu^{1.2}$. Miller *et al.* (1972) had previously noted that the distribution of counts from six stems taken from an approximately 1 m^2 area can be described by a negative binomial distribution, so it is reasonable to suppose (although it has not been specifically shown) that counts from six stems taken from 100 m^2 are also distributed according to a negative binomial model. However, the variances among sample observations are less when stems are taken from 100 m^2 than when they are taken from 1 m^2, probably because the six stems are drawn from a larger sample universe, which would tend to reduce the effect of aggregation of the larvae on the sample variance. This would be in line with the simulations done in Section 4.6, where we showed that sample means from ever increasing areas had higher values of the parameter *k*.

Continued

Iwao stop boundaries were used to classify the density about $cd = 15$ (2.5 per plant × six plants in a sample unit) and OC and ASN functions were determined using simulation ($sr = 1000$ for each mean). In the first simulation comparisons were made among three values of α: 0.15, 0.05 and 0.0125. The minimum sample size was five units of six plants each, and the maximum sample size was 15. Taylor's Power Law with $a = 1.0$ and $b = 1.2$ was used to describe the variance as a function of the mean, both for setting up the stop boundaries and for generating the sample observations in the simulations. The stop boundaries and OC and ASN functions are shown in Fig. 5.2. The increased width between the stop boundaries

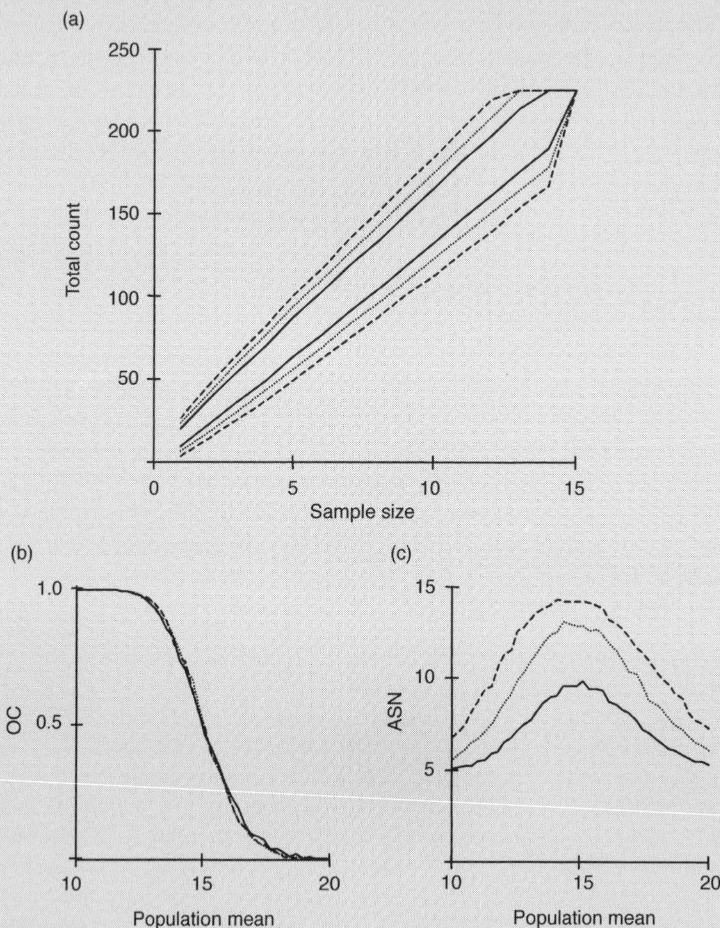

Fig. 5.2. Stop boundaries (a) and OC (b) and ASN (c) functions for an Iwao sequential classification sampling plan. The effect of changing the boundary parameter α: $\alpha = 0.15$ (——), 0.05 (···) and 0.0125 (- – -). Other boundary parameters: $cd = 15$, $minn = 5$, $maxn = 15$, TPL $a = 1.0$, TPL $b = 1.2$, k estimated by TPL. Simulation parameters: negative binomial with k estimated by TPL, 1000 simulations.

with decreasing values for α resulted in increased ASN values, but the corresponding changes in the OC function are nearly undetectable.

In the second simulation, comparisons were made among three maximum sample sizes; 15, 25 and 35, with $\alpha = 0.05$. All other parameters were as for the first simulation. The stop boundaries and OC and ASN functions are shown in Fig. 5.3. As when α was reduced, increasing the maximum sample size resulted in larger ASN values, but the effect on the OC function was easily detectable. Note that the greater steepness of OC was associated with an increase in the number of sample units, especially for those population means which are close to the critical density.

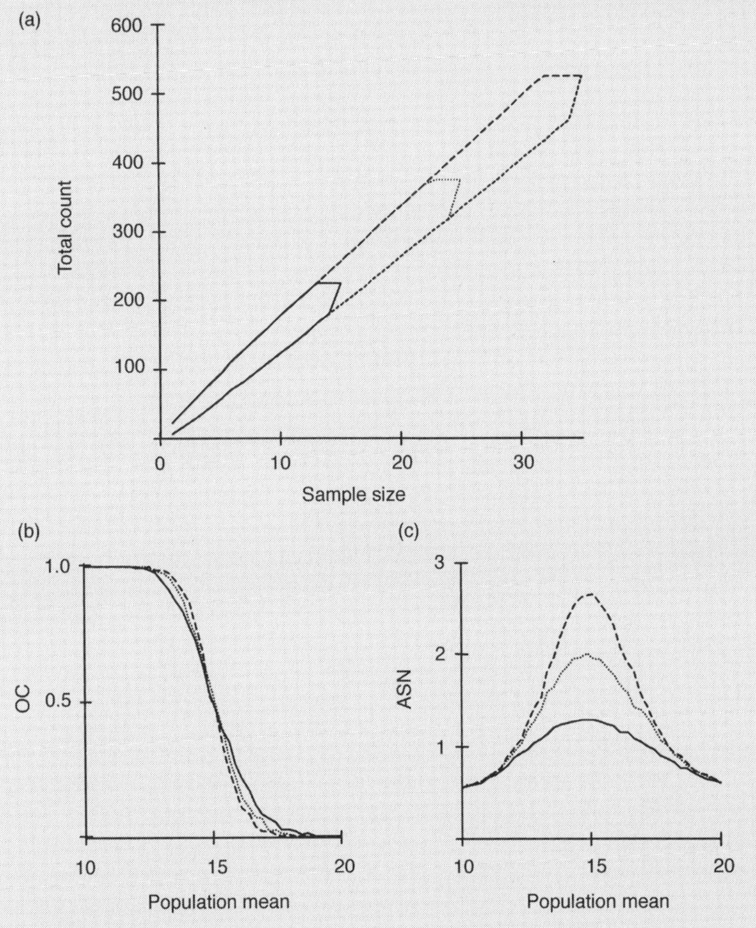

Fig. 5.3. Stop boundaries (a) and OC (b) and ASN (c) functions for an Iwao sequential classification sampling plan. The effect of changing the boundary parameter *maxn*: *maxn* = 15 (—), 25 (···) and 35 (- - -). Other boundary parameters: *cd* = 15, $\alpha = 0.05$, *minn* = 5, TPL *a* = 1.0, TPL *b* = 1.2, *k* estimated by TPL. Simulation parameters: negative binomial with *k* estimated by TPL, 1000 simulations.

Continued

In the third simulation, we took the same plan, constructed with TPL $a = 1$, $b = 1.2$, $\alpha = 0.05$, and minimum and maximum sample sizes equal to 5 and 25 (Fig. 5.3a, dotted line), and studied the effect of variation in the TPL parameters. The report on the TPL parameters (Barney and Legg, 1988) did not provide estimates of mean square error or estimates of parameter variances. We decided that a 30% adjustment to *a* would provide a good test, and used $a = 0.7$, 1.0 and 1.3 in turn. The OC and ASN functions computed using the three TPL models are shown in Fig. 5.4. When TPL *a* for the simulated data counts was greater than that used to construct the stop boundary, the OC function became flatter and ASN values were reduced, although, as noted above, for mean values far from *cd*, the ASN was increased (but only slightly). The effect of a 30% variation in the TPL parameter *a* on the OC and ASN functions is noticeable but, depending on other factors, may not be unduly large.

It is worth comparing the effects of widening the distance between the stop boundaries (Fig. 5.2) and of increasing the maximum sample size (Fig. 5.3). Both changes increased the average number of sample units (Figs 5.2 and 5.3), but only one of the changes, the increase in maximum sample size, paired the greater effort to an improved OC (Fig. 5.3). Widening the stop boundaries (Fig. 5.2) resulted in more work, but did not noticeably improve the accuracy of the classification. At first sight this appears perverse, but a more detailed examination of the variation in numbers of sample units required, *n*, provides some illumination. The variation can be quantified by quartiles of the distributions of the number of sample units required in the six sampling plans of Figs 5.2 and 5.3. Quartiles are defined as the 25th and 75th percentiles of the frequency distribution of sample sizes generated in the simulations, while the median is defined as the 50th percentile. The *p*th per-

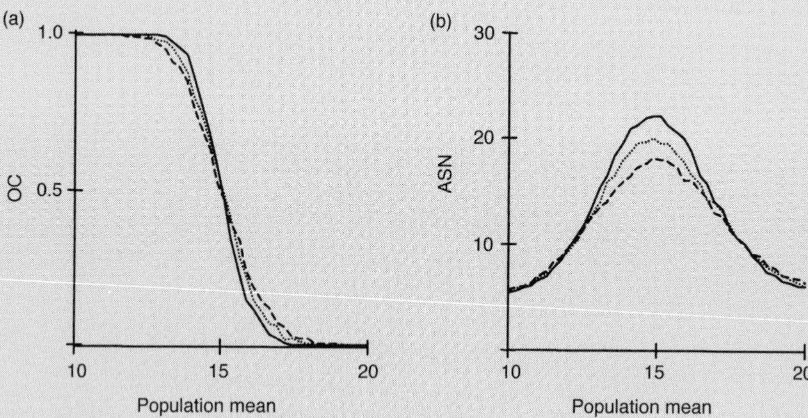

Fig. 5.4. OC (a) and ASN (b) functions for an Iwao sequential classification sampling plan. The effect of changing the simulation parameter TPL *a*: $a = 0.7$ (——), 1.0 (···) and 1.3 (- – -). Boundary parameters: $cd = 15$, $\alpha = 0.05$, $minn = 5$, $maxn = 15$, TPL $a = 1.0$, TPL $b = 1.2$, *k* estimated by TPL. Simulation parameters: negative binomial with *k* estimated by TPL (with different values of *a*), 1000 simulations.

centile is defined as the value below which *p*% of the population lies (e.g. Fig. 1.4). Figures 5.5a–c show quartiles for the three plans of Fig. 5.2, while Figs 5.5d–f show quartiles for the three plans of Fig. 5.3. Figures 5.5a–c illustrate that decreasing α in Fig. 5.2 meant only that more and more simulations stopped at the maximum sample size part of the stop boundary ($n = 15$): a relatively low *maxn* can nullify the expected effect (Table 5.2) of a wider stop boundary. On the other hand, increasing the maximum sample size in Fig. 5.3 allowed many more simulations to exit through the stop boundary before *maxn*, and the larger *maxn* also allowed higher precisions for estimates at *maxn*. If *maxn* is already reached quite frequently in the basic plan, adjusting parameters to widen the distance between the stop boundaries but not changing *maxn* will have little effect: all that happens is that the plan becomes more and more like a fixed sample size plan, with a sample size equal to *maxn*. Conversely, if *maxn* is decreased, the plan also becomes more and more like a fixed sample size plan, with a sample size equal to the new *maxn*.

Comparison of the ASN functions of Figs 5.2 and 5.3 with the solid line curves in Fig. 5.5 shows that presenting percentiles of the sample size distribution provides

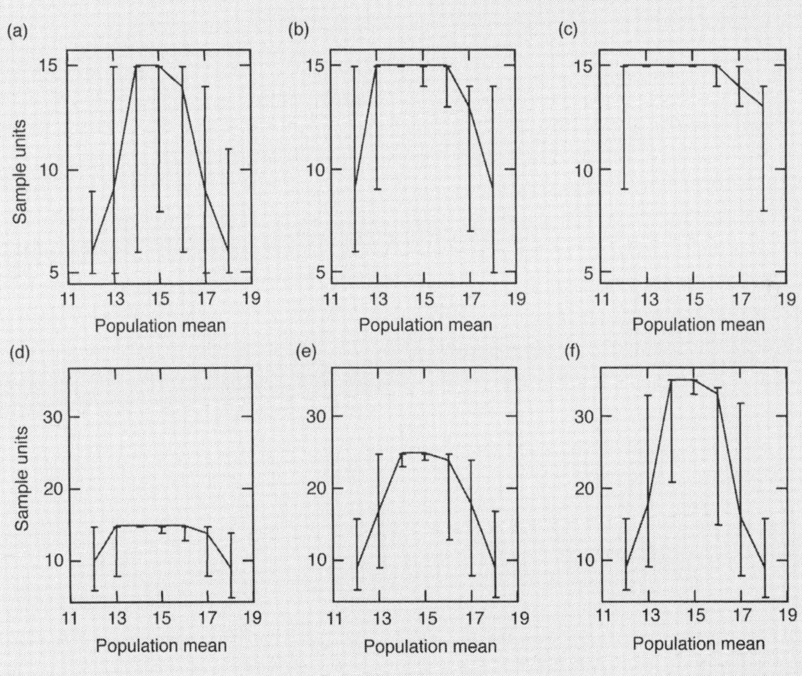

Fig. 5.5. Percentiles (25%, 50% and 75%) of the frequency distribution of sample size in the six sets of simulations presented in Figs 5.2 and 5.3. (a)–(c) refer to Fig. 5.2, *maxn* = 15 and $\alpha = 0.15$, 0.05, 0.0125, respectively; (d)–(f) refer to Fig. 5.3, $\alpha = 0.05$ and *maxn* = 15, 25, 35, respectively. Median number required (——), 25th and 75th percentiles (bars), based on 1000 simulations. Note that (b) and (d) are based on the same plan.

Continued

added insight into how a sampling plan is performing. When the median curves all reach the maximum sample size, this might warn an alert user that there is a problem. There may be a case for presenting the percentiles for the number of sample units instead of ASN or in addition to it, but we feel that this might often be more confusing than helpful. We shall continue to use ASN because it a useful summary statistic which is generally easy to interpret, but the reader should be aware that an average without an estimate of precision can be misleading.

5.4 Wald's Sequential Probability Ratio Test (SPRT)

Decision-making with batch sampling is based on estimating the mean pest density as each batch is sampled, and comparing the estimate with the critical density. Because Iwao's procedure can be envisaged as based on batch sampling with unit batch size, it can also be thought of as using estimation as an intermediate step to decision-making. Wald's procedure has no such intermediate step.

The SPRT is based directly on the classification problem: if the true mean density of a field is equal to μ_0 or less ($\mu_0 < cd$), classification must be correct at least $100(1 - \alpha)\%$ of the time, but if it is equal to μ_1 or greater ($\mu_1 > cd$) classification must be correct at least $100(1 - \beta)\%$ of the time. Concentrating on μ_0 and μ_1 alone, the goal of SPRT is to continue to collect sample units until there is enough evidence to be able stop and make a classification with predetermined probability of being correct if either of these values represents the true density. These requirements can be summarized in a table showing probabilities of correct and incorrect classification for $\mu = \mu_0$ and $\mu = \mu_1$ (Table 5.3). With two probabilities, α and β, specifying the probabilities of misclassification we can, if we want, allow relatively more correct 'no intervention' than 'intervention' actions (by letting α be less than β). In some pest management situations, this may be useful.

The stop boundaries for SPRT are based on the 'likelihood ratio', which is the ratio of likelihoods (Section 4.3):

$$\text{likelihood ratio, } LR = \frac{\text{probability of getting the sample data if } \mu = \mu_1}{\text{probability of getting the sample data if } \mu = \mu_0} \tag{5.2}$$

This can be represented as the product of ratios, each ratio referring to the data from one sample unit:

$$LR_n = \prod_{i=1}^{n} \frac{p(X_i \mid \mu_1)}{p(X_i \mid \mu_0)} \tag{5.3}$$

where X_i refers to the sample data in sample unit i. For example, if disease incidence were being tested based on the binomial distribution (Section 4.9), and r sample units were found to be infested among the first n units examined (i.e. $S_n = r$), the ratio would be

$$LR_n = \frac{\binom{n}{r} p_1^r (1-p_1)^{n-r}}{\binom{n}{r} p_0^r (1-p_0)^{n-r}} = \left(\frac{p_1}{p_0}\right)^r \left(\frac{1-p_1}{1-p_0}\right)^{n-r} \tag{5.4}$$

Wald defined the SPRT as the rule, starting at $n = 1$:

(i) continue collecting sample units while $\frac{\beta}{1-\alpha} < LR_n < \frac{1-\beta}{\alpha}$

(ii) stop and classify $\mu = \mu_0$ if $\quad LR_n \le \frac{\beta}{1-\alpha}$

(iii) stop and classify $\mu = \mu_1$ if $\quad LR_n \ge \frac{1-\beta}{\alpha}$

Wald showed that the SPRT resulted in probabilities of action as in Table 5.3, and that the SPRT was optimum in the sense that of all potential methods that are able to provide a decision between μ_0 or μ_1 at the specified error rates α and β, SPRT requires the smallest average number of sample units. Wald then gave formulae to calculate the rest of the OC function and the whole of the ASN function (see, e.g. Fowler and Lynch, 1987). These formulae were very useful when electronic computers were in their infancy, but nowadays it is better to use simulation to estimate both the OC and ASN functions, for two main reasons:

1. Wald's formulae are approximations.
2. Wald's formulae do not allow a minimum or maximum sample size.

The second of these disallows the formulae in practice, although they remain adequate approximations in general.

Classification by SPRT makes heuristic sense. Suppose, for example, that sampling stops with classification $\mu = \mu_1$. Based on (iii) above, this means that, when sampling stops, and whatever the value of n,

$$LR_n = \frac{\text{probability of getting the sample data if } \mu = \mu_1}{\text{probability of getting the sample data if } \mu = \mu_0} \doteq \frac{1-\beta}{\alpha} \tag{5.5}$$

Table 5.3. The probabilities, under SPRT, of classification for the pivotal mean densities μ_0 and μ_1 ($\mu_0 < cd < \mu_1$).

True value	Classify as 'acceptable' ($\mu = \mu_0$): action = no intervention	Classify as 'unacceptable' ($\mu = \mu_1$): action = intervene
$\mu = \mu_0$	$1 - \alpha$	α
$\mu = \mu_1$	β	$1 - \beta$

where $\doteq$ means that the equality is approximate, because the final ratio may be slightly greater than $(1 - \beta)/\alpha$. We proceed to classify μ as equal to μ_1, which we promised to do with probability $1 - \beta$ if $\mu = \mu_1$, and with probability α if $\mu = \mu_0$. This is precisely what Equation 5.5 states! The reader with some knowledge of statistics will by now recognize as two 'statistical hypotheses' the assumptions that the true mean density is either μ_0 or μ_1; the classification problem can be regarded as testing one hypothesis against the other.

For many distributions, the stop boundaries presented in (i), (ii) and (iii) above can be written out and reorganized into more convenient forms, involving cumulative totals, S_n, and the sample size, n. More specifically, the boundaries are parallel straight lines. This simplification works for the Poisson, negative binomial and binomial distributions. It does not work for the beta-binomial distribution; boundaries for it are so complicated that no one has calculated and published them. Equations for stop boundaries for the Poisson, negative binomial and binomial distributions are shown in Table 5.4.

Table 5.4. Formulae for computing stop boundaries for a SPRT based on binomial, Poisson, negative binomial and normal distributions. The normal distribution can be used when the boundaries for the desired distribution (e.g. beta-binomial) are not simple straight lines. $\mu_0 < \mu_1$ and $p_0 < p_1$.

Distribution and parameters	Low intercept	High intercept	Slope
Poisson; μ_0 and μ_1	$\dfrac{\ln\left(\dfrac{\beta}{1-\alpha}\right)}{\ln\left(\dfrac{\mu_1}{\mu_0}\right)}$	$\dfrac{\ln\left(\dfrac{1-\beta}{\alpha}\right)}{\ln\left(\dfrac{\mu_1}{\mu_0}\right)}$	$\dfrac{\mu_1-\mu_0}{\ln\left(\dfrac{\mu_1}{\mu_0}\right)}$
Binomial; p_0 and p_1, $q = 1 - p$	$\dfrac{\ln\left(\dfrac{\beta}{1-\alpha}\right)}{\ln\left(\dfrac{p_1 q_0}{p_0 q_1}\right)}$	$\dfrac{\ln\left(\dfrac{1-\beta}{\alpha}\right)}{\ln\left(\dfrac{p_1 q_0}{p_0 q_1}\right)}$	$\dfrac{\ln\left(\dfrac{q_0}{q_1}\right)}{\ln\left(\dfrac{p_1 q_0}{p_0 q_1}\right)}$
Negative binomial; μ_0 and μ_1, k	$\dfrac{\ln\left(\dfrac{\beta}{1-\alpha}\right)}{\ln\left(\dfrac{\mu_1(\mu_0+k)}{\mu_0(\mu_1+k)}\right)}$	$\dfrac{\ln\left(\dfrac{1-\beta}{\alpha}\right)}{\ln\left(\dfrac{\mu_1(\mu_0+k)}{\mu_0(\mu_1+k)}\right)}$	$\dfrac{k\ln\left(\dfrac{\mu_1+k}{\mu_0+k}\right)}{\ln\left(\dfrac{\mu_1(\mu_0+k)}{\mu_0(\mu_1+k)}\right)}$
Normal; μ_0 and μ_1, σ^2	$\dfrac{\sigma^2\ln\left(\dfrac{\beta}{1-\alpha}\right)}{(\mu_1-\mu_0)}$	$\dfrac{\sigma^2\ln\left(\dfrac{1-\beta}{\alpha}\right)}{(\mu_1-\mu_0)}$	$\dfrac{\mu_1+\mu_0}{2}$

As with Iwao's procedure, SPRT stop boundaries are adjusted so that sampling must terminate when n reaches a specified maximum value, $maxn$. If $n = maxn$ is reached, the estimate of mean density, $S_{maxn}/maxn$, is compared with cd, and the density is classified as greater than cd if $S_{maxn} > cd \times maxn$. This same classification can be made if $S_n > cd \times maxn$, for any n, which results in a portion of the upper boundary being a horizontal line ending at the point $(maxn, cd \times maxn)$.

The effect of SPRT parameters on the OC and ASN functions is similar to the effects noted with Iwao's procedure. With Iwao's procedure as originally formulated, there is only one parameter that controls the width of the stop boundaries ($z_{\alpha/2}$). While two parameters could be used (z_α for the lower boundary and z_β for the upper), we have retained the original formulation. With SPRT there are two parameters, α and β, one for each boundary. Increasing either α or β decreases the width between the stop boundaries and makes the OC flatter and reduces ASN values. The effects are summarized in Table 5.2, but because they are not altogether intuitive, it is worth describing how changing α affects the OC and ASN functions (the effect of changing β is similar).

Through its effect on the 'ln' functions in Table 5.4, increasing α perceptibly lowers the upper stop boundary and slightly raises the lower stop boundary (provided that α is small). Narrowing the distance between the parallel stop boundary lines reduces the number of sample units required to reach a decision, so the ASN is reduced for all values of μ; because the upper stop boundary is lowered more than the lower one is raised, the reduction is greater for $\mu > cd$. Therefore the ASN should be visibly reduced for $\mu > cd$.

The relatively greater lowering of the upper stop boundary increases the relative chance of exit through the upper stop boundary whatever the value of μ. In other words, the chance that true mean values less than cd are (incorrectly) classified as greater than cd is increased. Therefore the OC function to the left of cd is lowered. The effect on the OC function could alternatively be deduced from Table 5.3.

Some effects of changing the parameters are illustrated in Exhibit 5.2.

Exhibit 5.2. *Wald's SPRT plan: the effect of altering parameters and the distribution*

In this example we illustrate how α and β of Wald's SPRT and models for the distribution of sample counts influence the OC and ASN functions. The example is based on sampling nymphs of the three-cornered alfalfa hopper (*Spissistilus festinus* (Say)), which is a pest of soybean. The adult girdles stems and leaf petioles with its sucking mouthparts, which diverts plant sugars and may allow disease entry. Sparks and Boethel (1987) found that counts of first-generation hopper nymphs from 10 beat-net samples were distributed according to a negative binomial distribution, while counts from the second generation could be adequately modelled by the Poisson distribution. Taylor's Power Law fitted to all the data for both generations

Continued

together gave $a = 0.96$ and $b = 1.26$ (variance = $0.96\mu^{1.26}$). Sparks and Boethel estimated the relationship between nymph and adult numbers, and were able to determine a critical density for nymphs: 11.3 per 10 beat-net samples. The sample unit was defined as 10 beat-net samples. The basic plan was SPRT with $\mu_0 = 10.3$, $\mu_1 = 12.3$, $\alpha = 0.2$, $\beta = 0.2$, $minn = 10$, $maxn = 40$, TPL with $a = 0.96$ and $b = 1.26$, and negative binomial distribution with $k = 14$ (determined at the midpoint between μ_0 and μ_1 by Equation 5A.5 and TPL).

Changing α and β Three SPRT classification sampling plans were set up, using the above basic parameters, but with different values of α and β: α and β were both equal to 0.05, 0.10 or 0.2 for the three plans. The OC and ASN functions were determined using simulation with the sample counts distributed according to a negative binomial model and the variance modelled as a function of the mean using TPL. One thousand simulation replicates were made to estimate each OC and ASN value. Decreasing α and β caused the stop boundaries to lie further apart, the ASN values to be higher and the OC function to be slightly steeper (Fig. 5.6). However, the change in the OC function might be regarded as modest. Based on this finding, using larger values for α and β than those suggested by Sparks and Boethel (they used α and β equal to 0.05) might be justified because the average sample size would be reduced by almost one half over a wide range of pest densities without appreciable loss in classification precision.

Different simulation distributions Sparks and Boethel found that all the sample counts obtained with the beat-net method could be described using a negative binomial distribution, but also that 40% of these data sets could be described by the Poisson distribution. The authors designed a sequential classification sampling plan based on the negative binomial distribution: 'it permits conservative sampling with little added effort [when counts are randomly distributed] because of the low level of clumping as indicated by the value of k (approximately 14)'. By simulation we can determine just how conservative this sampling would be and how little the 'added effort' actually was. Stop boundaries were created using the above basic parameters. Sample counts were simulated in three ways; using a negative binomial distribution with $k = 14$, using a negative binomial distribution allowing TPL to determine the value of k based on Equation 5A.5, and using a negative binomial distribution with $k = 100$, which essentially means using the Poisson distribution (Section 4.9). The OC and ASN functions obtained for the three models for the sample counts are shown in Fig. 5.7. When the sample counts could be approximately described using the random distribution (Poisson), the ASN function was slightly larger and the OC function slightly steeper. There were essentially no differences between the OC and ASN functions obtained when the variance was described as a function of the mean using TPL or when a fixed value of k was used. The latter result is not surprising, because the variances among sample counts are nearly the same when $k = 14$ as when the variances are determined using TPL.

When the stop boundaries are based on the Poisson distribution but sample counts are more aggregated, the results are similar. Stop boundaries were created using the basic set of parameters, except that $k = 100$. The same three types of distribution were used to estimate OC and ASN functions (Fig. 5.8). As before, when the variance among sample counts was greater than that used to compute the stop boundaries, the OC functions were less steep and the ASN functions were reduced.

These results indicate that, for the three-cornered hopper that infests soybeans, what at first observation might be regarded as somewhat imprecise descriptions of

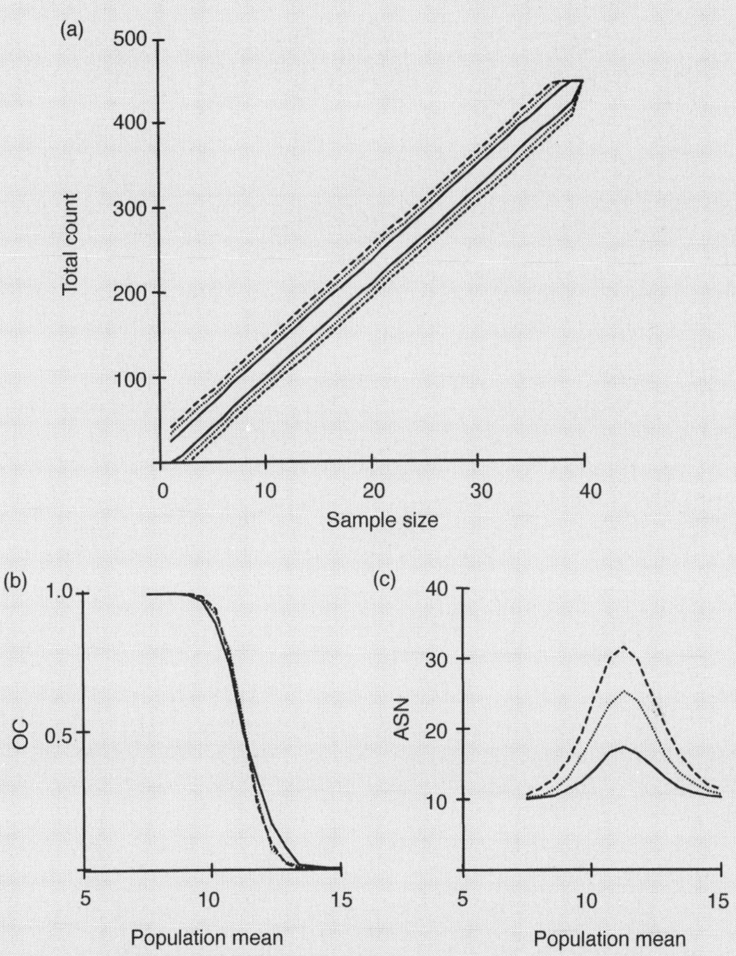

Fig. 5.6. Stop boundaries (a) and OC (b) and ASN (c) functions for an SPRT sequential classification sampling plan. The effect of changing the boundary parameters α and β: $\alpha,\beta = 0.2$ (——), 0.1 (···) and 0.05 (- – -). Other boundary parameters: $\mu_0 = 10.3$, $\mu_1 = 12.3$, *minn* = 10, *maxn* = 40, TPL *a* = 0.96, TPL *b* = 1.26, *k* estimated by TPL. Simulation parameters: negative binomial with *k* estimated by TPL, 1000 simulations.

sample counts are in fact quite adequate for developing a sequential classification plan. Sensitivity to different values of unknown parameters, such as *k*, can be examined by estimating OC and ASN functions for these different values. Conservative estimates of the OC and ASN functions can be obtained by using parameter values which make the sample variance greater (e.g. by using smaller values of *k* in the simulations than were used to set up the sampling plan).

Continued

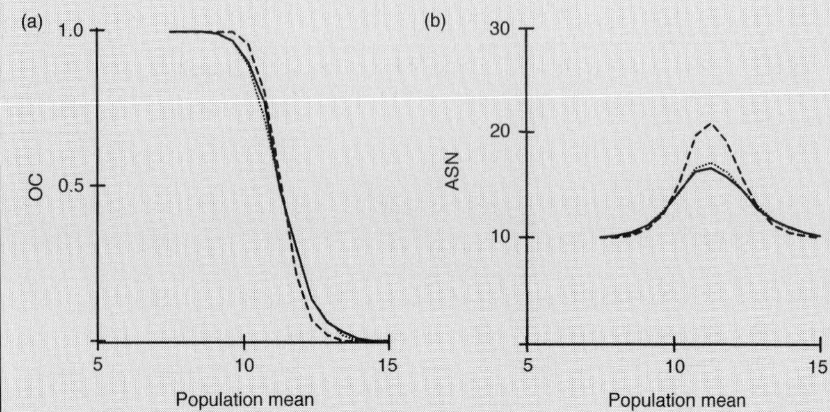

Fig. 5.7. OC (a) and ASN (b) functions for an SPRT sequential classification sampling plan. The effect of changing simulation parameters defining the negative binomial *k*: *k* = 14(——), *k* estimated by TPL (···), *k* = 100 (- – -). Boundary parameters: $\mu_0 = 10.3$, $\mu_1 = 12.3$, $\alpha = \beta = 0.2$, *minn* = 10, *maxn* = 40, TPL *a* = 0.96, TPL *b* = 1.26, *k* estimated by TPL. Simulation parameters: negative binomial with the above values of *k*, 1000 simulations.

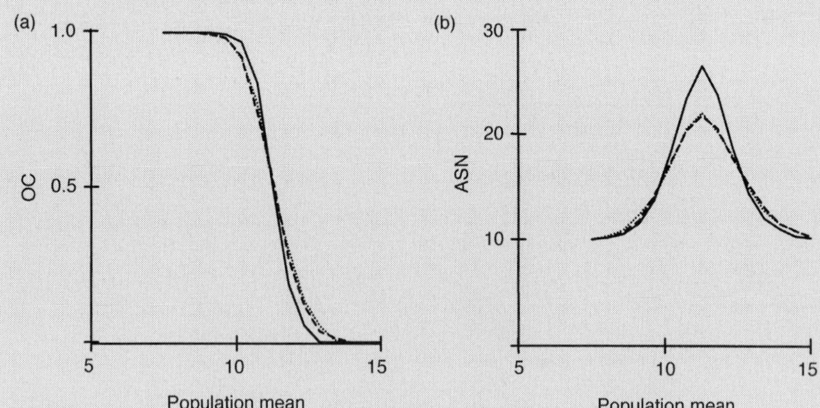

Fig. 5.8. OC (a) and ASN (b) functions for an SPRT sequential classification sampling plan. The effect of changing simulation parameters defining the negative binomial *k*: *k* = 14(——), *k* estimated by TPL (···), *k* = 100 (- – -). Boundary parameters: $\mu_0 = 10.3$, $\mu_1 = 12.3$, $\alpha = \beta = 0.2$, *minn* = 10, *maxn* = 40, *k* = 100. Simulation parameters: negative binomial with the above values of *k*, 1000 simulations.

Changing cd A parameter that has a large effect on the OC and ASN functions is the critical density. We illustrated this for batch sampling in Exhibit 3.1. We illustrate a similar effect with SPRT. Three SPRT plans were created using the basic parameters, but changing μ_0 and μ_1: we used the pairs (10.3, 12.3), (11.3, 13.3) and (9.3, 11.3). Stop boundaries and OC and ASN functions are shown in Fig. 5.9. These differences in μ_0 and μ_1 result in the OC and ASN functions being shifted either to the left or the right on the density axis. Of all the parameters studied in this example, an approximately 10% change in the critical density had the greatest effect on the OC and ASN. Often, the critical density is the least studied of all parameters used to formulate a decision guide. Although it may be unrealistic or impractical to obtain a very accurate estimate of the critical density, it is always worth using simulation to find out the effects on the OC and ASN functions of a slightly different value.

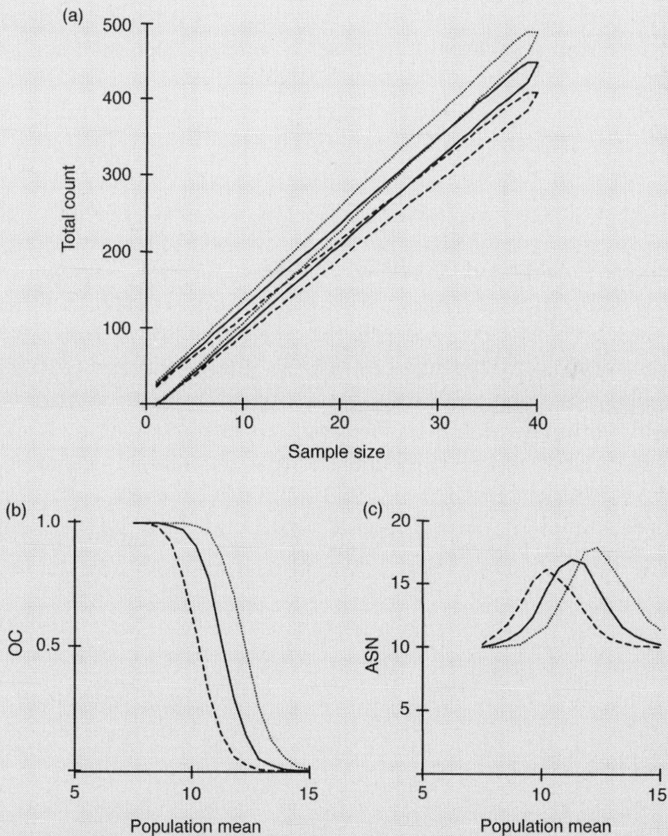

Fig. 5.9. Stop boundaries (a) and OC (b) and ASN (c) functions for an SPRT sequential classification sampling plan. The effect of changing boundary parameters μ_0 and μ_1: μ_0,μ_1 = 10.3,12.3 (—), 11.3,13.3 (···) and 9.3,11.3 (- – -). Other boundary parameters: $\alpha = \beta = 0.2$, *minn* = 10, *maxn* = 40, TPL *a* = 0.96, TPL *b* = 1.26, *k* estimated by TPL. Simulation parameters: negative binomial with *k* estimated by TPL, 1000 simulations.

5.5 Converging Lines

Both Iwao's and Wald's methods are based in some way or other on statistical theory. But it may be possible to improve the efficiency of classification by escaping from established theoretical frameworks and concentrating on practical goals.

Iwao's stop boundaries consist of divergent curvilinear lines and those for the SPRT are parallel lines (for most distributions of interest to pest managers). To force a decision at a maximum sample size, both sets of lines are abruptly brought together to a point. An alternative approach would be to bring the upper and lower boundaries together more gradually, to meet at the maximum sample size. This makes intuitive sense, because precision of a classification should improve as the sample size increases, so the stop boundaries should be allowed to converge. In fact, this type of boundary proved popular with the medical researchers, who found that Wald's SPRT was not entirely ideal for their purposes (Armitage, 1975). Converging Line stop boundaries have not to our knowledge been applied in pest management.

Converging Line stop boundaries could be developed in several ways. We use the following rationale: the boundaries consist of two straight lines, one of which meets the S_n-axis (the total count axis) above zero, and the other of which meets the n-axis (the sample size axis) above zero. The two lines converge and meet at the point ($maxn$, $cd \times maxn$) (see Fig. 5.10). The question to be answered when specifying one of these boundaries is: Where should the stop boundaries meet the horizontal (n) and vertical (S_n) axes? In general, the further apart the lines are, the greater the sample size and consequently the greater the precision in classification. An intuitive way to quantify the spread in the stop boundaries is to specify the minimum sample size, *minn*, and calculate classification intervals there about the critical density (as in Equation 3.11, with *minn* replacing n_B). This allows for symmetric stop boundaries as well as boundaries that differentially guard against the two types of misclassification.

The upper and lower boundary points corresponding to the minimum sample size are calculated exactly as for Iwao's method:

$$\begin{aligned} L_{minn} &= minn\left(cd - z_{\alpha L}\sqrt{\frac{V}{minn}}\right) \\ U_{minn} &= minn\left(cd + z_{\alpha U}\sqrt{\frac{V}{minn}}\right) \end{aligned} \tag{5.6}$$

where $z_{\alpha L}$ and $z_{\alpha U}$ are standard normal deviates, and V is the variance when the mean is equal to cd.

The complete stop boundary consists of straight lines joining the point ($maxn$, $cd \times maxn$) to each of the points ($minn$, L_{minn}) and ($minn$, U_{minn}). These lines can be extended to meet the axes, but of course the boundary is not strictly relevant below *minn* (Fig. 5.10).

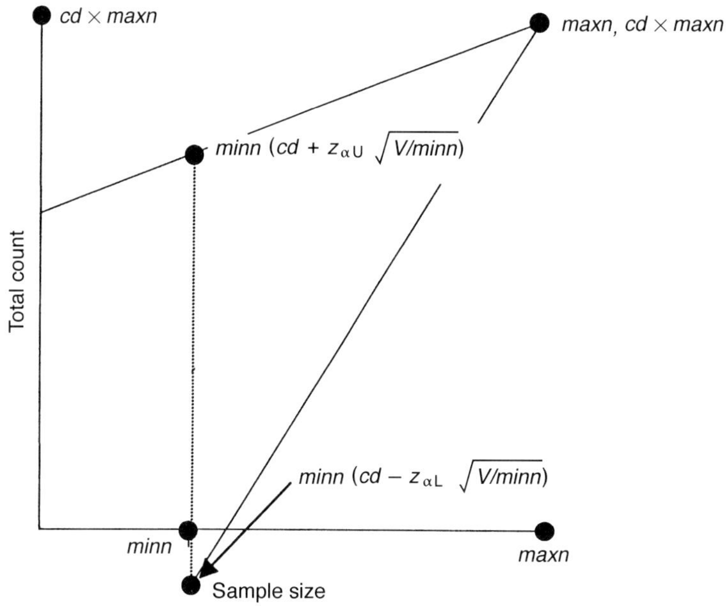

Fig. 5.10. Converging Line sequential classification stop boundaries. The boundaries consist of Converging Lines that meet at the point *maxn, cd* × *maxn,* where *maxn* is the maximum sample size and *cd* is the critical density. The upper and lower boundary points corresponding to the minimum sample size (*minn*) are determined as

$$U_{minn} = minn\left(cd + z_{\alpha U}\sqrt{\frac{V}{minn}}\right) \text{ and } L_{minn} = minn\left(cd - z_{\alpha L}\sqrt{\frac{V}{minn}}\right),$$

where $z_{\alpha U}$ is a standard normal deviate such that $P(Z > z_{\alpha U}) = \alpha U$, $z_{\alpha L}$ is a standard normal deviate such that $P(Z > z_{\alpha L}) = aL$, and V is the variance when the mean = *cd.* The complete stop boundary consists of straight lines joining the boundary points at *minn* to the point (*maxn, cd* × *nmax*) and extending to the 'Total count' and 'Sample size' axes.

The quantities used to specify the width of the stop boundaries can be interpreted as classification intervals (Section 3.2.2), but this is not particularly meaningful. It is best to do as we have suggested with other sequential plan parameters: αU and αL should be viewed as tools for manipulating the stop boundaries so that desirable OC and ASN functions are achieved (Table 5.2). Additional parameters that can be used for this purpose are *minn*, *maxn* and *cd* (or *cp*). In general, decreasing αU and/or αL will increase the width of the boundaries, make the OC more steep and increase ASN. Decreasing αU but keeping αL fixed raises the upper arm of the stop boundary and makes the upper part of the OC steeper, and similarly for αL. Increasing *minn* and *maxn* will increase the width of the boundaries, make the OC more steep and increase ASN. As for Iwao's and Wald's plans, increasing the variance of the sample observations above the variance used to construct the stop boundary (which applies only to the negative binomial and beta-binomial distributions) flattens the OC function, while near *cd* (or *cp*) the ASN tends to be lower; otherwise, it tends to be higher. These patterns are illustrated in Exhibit 5.3. The OC and ASN functions can only be obtained by simulation.

Exhibit 5.3. *Converging Lines plan: the effect of altering parameters*

In this example we illustrate how some of the parameters for Converging Line stop boundaries influence the OC and ASN functions. The example is based on sampling lepidopteran pests of fresh market sweetcorn in New York State. Hoffman *et al.* (1996) showed that the sampling distribution of plants infested with caterpillars could be described by a beta-binomial distribution when plants were sampled in groups of 5 ('clusters of five', to use the terminology of Chapter 4). These authors parameterized the beta-binomial distribution using θ, which is a measure of aggregation that ranges from 1 to infinity, rather than ρ. However, $\rho = \theta/(1 + \theta)$. These authors reported the median value for ρ to be 0.1, and the 90th and 95th percentiles to be 0.23 and 0.44. The critical proportion of plants infested with larvae was 0.15 prior to silking of the maize and 0.05 after silking. In this example, we use $cp = 0.15$.

The influence of αU and αL was first studied by setting both of these parameters equal to one of three values: 0.05, 0.1 or 0.2. The remaining parameters for these plans were the same: $cp = 0.15$, the number of plants examined at each sample location in the field (R) = 5, $\rho = 0.1$, $minn = 15$ and $maxn = 50$. One thousand simulation replicates were made to estimate each OC and ASN value. The stop boundaries and OC and ASN functions are shown in Fig. 5.11. Increasing αU and αL caused the stop boundaries to move closer together. This resulted in decreasing ASN functions, but only a small reduction in the precision of the classifications.

The influence of the maximum sample size was examined by setting αU and αL to 0.05 and allowing *maxn* to have one of three values: 25, 50 or 75. All other parameters were as described above. The stop boundaries and OC and ASN functions are shown in Fig. 5.12. Increasing *maxn* resulted in steeper OC functions (increased precision of classification) and increased ASN functions. Note that the maximum ASN value is much less than *maxn* when $maxn = 75$. This is because the stop boundaries are very narrow for this value of *maxn*.

The influence of the minimum sample size was examined by setting $maxn = 50$ and allowing *minn* to have one of three values: 10, 20 or 30. All other parameters were as described above. The stop boundaries and OC and ASN functions are shown in Fig. 5.13. Increasing the minimum sample size over the range 10, 20, 30 caused the stop boundaries to lie further apart and resulted in an increase in the ASN functions by nearly 50% (from 10 to 20) and by nearly 25% (from 20 to 30). The corresponding changes in the OC functions were small and might not be justified by the extra sampling costs.

The parameter ρ of the beta-binomial distribution measures the correlation in incidence rate among plants within a cluster of plants. As noted in Section 4.8, as ρ increases, the variability of p, the proportion of plants infested, also increases. This in turn increases the variance in the estimated proportion of infested plants throughout the field. Shown in Fig. 5.14 are OC and ASN functions when stop

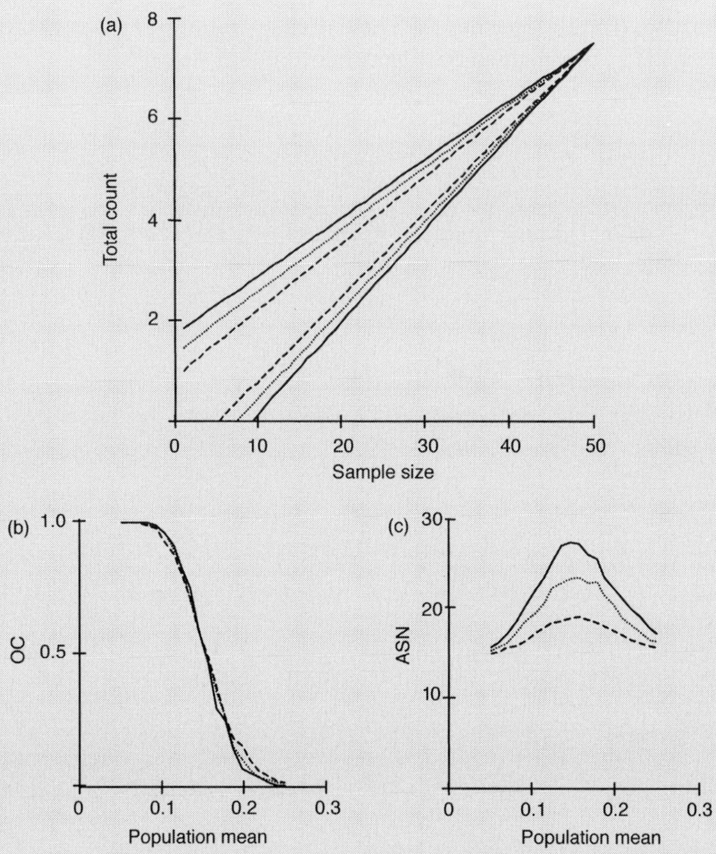

Fig. 5.11. Stop boundaries (a) and OC (b) and ASN (c) functions for a Converging Lines sequential classification sampling plan and the beta-binomial distribution. The effect of changing the boundary parameters αU and αL: $\alpha U = \alpha L = 0.05$ (——), 0.1 (···) and 0.2 (- – -). Other boundary parameters: $cp = 0.15$, $R = 5$, $\rho = 0.1$, $minn = 15$ and $maxn = 50$. Simulation parameters: beta-binomial with $\rho = 0.1$, 1000 simulations.

boundaries were calculated using $\rho = 0.1$ and sampling was simulated using $\rho = 0.1$, 0.23 and 0.44. As might be expected, increasing the ρ used in the simulations above the value used to calculate the stop boundaries caused the ASN function to decrease and the OC function to become more shallow (less precision). This is the same type of effect as we found in Exhibits 5.1 (Iwao) and 5.2 (SPRT). Whether or not these changes are important depends on the frequency with which different values of ρ might occur in practice.

Continued

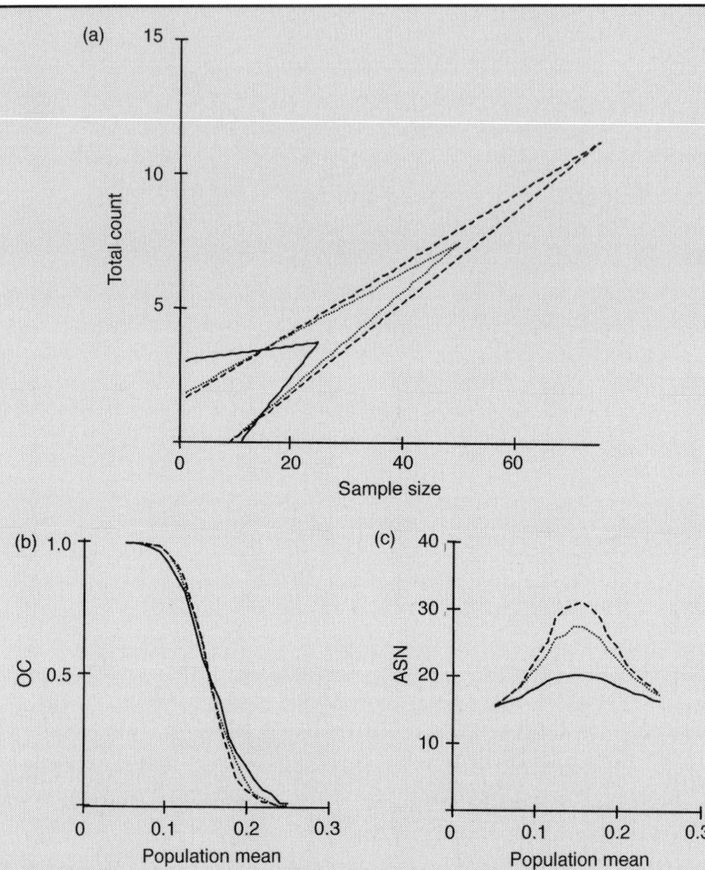

Fig. 5.12. Stop boundaries (a) and OC (b) and ASN (c) functions for a Converging Lines sequential classification sampling plan and the beta-binomial distribution. The effect of changing the boundary parameter *maxn*: *maxn* = 25 (——), 50 (···) and 75 (- – -). Other boundary parameters: $cp = 0.15$, $\alpha U = \alpha L = 0.05$, $R = 5$, $\rho = 0.1$ and $minn = 15$. Simulation parameters: beta-binomial with $\rho = 0.1$, 1000 simulations.

Continued

5.6 Green's Stop Boundary for Estimation

In this chapter, we have not (so far) mentioned estimation. As noted in Chapter 3, the goals of estimation and classification are different, so stop criteria for sequential plans for estimation should probably be different from those for classification. Sequential sampling plans for estimation do exist. Some discussions are highly mathematical (see, e.g. Siegmund, 1985) and others are less so (see, e.g. Green, 1970). In contrast to stop boundaries for classification, stop boundaries for estimation have no critical density and they tend to bend in towards the starting point (*n*

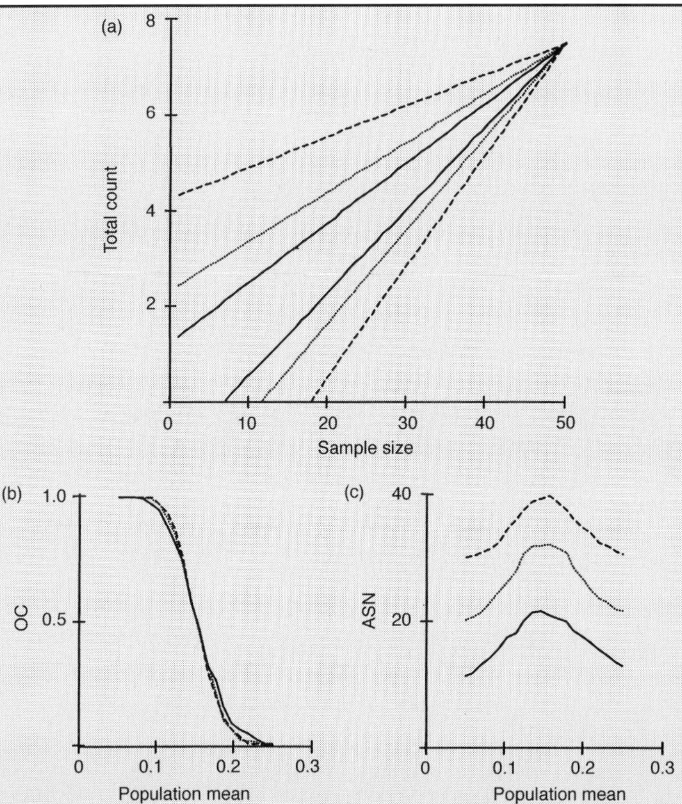

Fig. 5.13. Stop boundaries (a) and OC (b) and ASN (c) functions for a Converging Lines sequential classification sampling plan and the beta-binomial distribution. The effect of changing the boundary parameter *minn*: *minn* = 10 (——), 20 (···) and 30 (- – -). Other boundary parameters: $cp = 0.15$, $\alpha U = \alpha L = 0.05$, $R = 5$, $\rho = 0.1$ and $maxn = 50$. Simulation parameters: beta-binomial with $\rho = 0.1$, 1000 simulations.

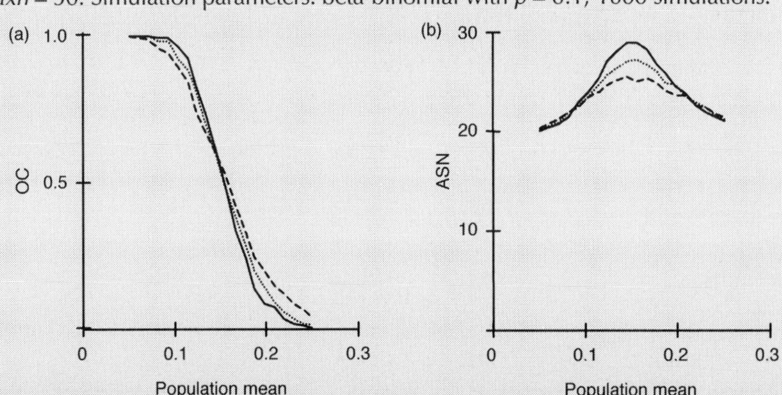

Fig. 5.14. OC (a) and ASN (b) functions for a Converging Lines sequential classification sampling plan and the beta-binomial distribution. The effect of changing simulation parameters defining the beta-binomial ρ: $\rho = 0.1$ (——), $\rho = 0.23$ (···) and $\rho = 0.44$ (- – -). Boundary parameters: $cp = 0.15$, $\alpha U = \alpha L = 0.05$, $R = 5$, $\rho = 0.1$, *minn* = 20 and *maxn* = 50. Simulation parameters: beta-binomial with the specified value of ρ, 1000 simulations.

and total count, S_n, both equal to zero). Green's boundary is based on TPL and a precision requirement for the CV. From Equation 3.7,

$$n = \frac{V}{\mu^2 CV^2} = \frac{a\mu^b}{\mu^2 CV^2} = \frac{a\mu^{b-2}}{CV^2} \tag{5.7}$$

Replacing μ by S_n/n,

$$n = \frac{a\,(S_n/n)^{b-2}}{CV^2} \tag{5.8}$$

which can be turned around to give S_n in terms of n:

$$S_n = \left(\frac{CV^2}{a}\right)^{\frac{1}{b-2}} \times n^{\frac{b-1}{b-2}} \tag{5.9}$$

This defines the boundary. As soon as the point representing (n, S_n) crosses the boundary, sampling stops and μ is estimated by S_n/n. The boundary is exemplified in Fig. 5.15. As noted for the simple case in Section 3.2, such a boundary is not likely to be useful for classification; for example, because sample size is large for large μ, whatever the value of *cd*. We now return to classification.

5.7 Which Stop Boundary?

We have described three types of stop boundary which can be used to construct a sequential classification sampling plan. Which stop boundary should be used? In most cases, the choice of boundary type (Iwao, SPRT or Converging Lines) is much less important than the choice of parameters for whatever boundary type is chosen. Having said this, there is some evidence that Converging Line boundaries provide OC functions that are nearly identical to those obtained using either Iwao or SPRT

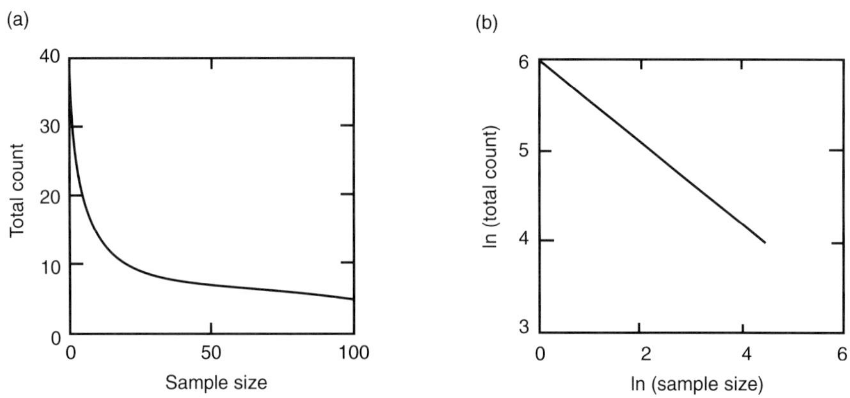

Fig. 5.15. The stop boundary for estimation by Green's method. The required *CV* is 25%; TPL parameters are $a = 4$ and $b = 1.3$. (a) Untransformed numbers; (b) logarithms.

boundaries, but with smaller ASN functions. Thus, Converging Line boundaries seem to be more efficient than either SPRT or Iwao boundaries. This evidence is not extensive and consists only of a few sets of simulations that we have completed. One such example is shown in Exhibit 5.4.

5.8 Summary

1. Sequential sampling requires fewer sample units than fixed sample size sampling to obtain the same OC function.
2. Three types of sequential classification sampling plan have been presented: Iwao, Wald's SPRT and Converging Lines. Of these, Iwao and Wald's SPRT are commonly used in pest management sampling. There is some evidence that Converging Lines may be at least as efficient as the other two.
3. Each of the types has a number of parameters – minimum and maximum sample sizes, plus:

	Iwao	SPRT	Converging Lines
Critical densities	*cd* or *cp*	μ_0, μ_1	*cd* or *cp*
Misclassification probabilities	α	α, β	αL, αU
Distribution parameters	*V* or variance–mean relationship (e.g. TPL)	Negative binomial *k*	*V* or variance–mean relationship (e.g. TPL)

4. The general effects of changing plan parameters are summarized in Table 5.2.
5. The only requirement for setting up Iwao or Converging Lines stop boundaries is that an estimate of the variance at *cd* (or *cp*) is available.
6. When the distribution parameters are not known precisely, the effect of using incorrect values in setting up the stop boundaries must be examined by simulation.

Exhibit 5.4. *A comparison among the three sequential plans: Iwao, SPRT and Converging Lines*

Vincelli and Lorbeer (1987) described a sequential sampling plan for use in determining when fungicide applications should be started for control of botrytis leaf blight of onion, caused by *Botrytis squamosa*. The sample unit was the three oldest leaves on a plant and the critical density was three disease lesions per sample unit. When this density is reached, fungicide applications are to be initiated. Counts of lesions per plant can be described using a negative binomial distribution. The authors modelled the variance as a function of the mean using a model different from TPL. For demonstration purposes, we estimated TPL parameters by computing the variance as a function of the mean using the model reported by the authors and then fitting TPL to these values using regression. The estimates for *a* and *b* were 3.05 and 1.02 respectively.

Continued

Sequential sampling plans were constructed using Iwao, SPRT and Converging Line boundaries. The minimum and maximum sample sizes were the same for all three boundary types; 15 and 50 respectively. Other parameters for each sampling plan were adjusted until OC functions for plans based on the three boundary types were approximately the same. This allowed for comparison of sampling costs required to achieve comparable levels of classification precision.

Shown in Fig. 5.16 are the stop boundaries and OC and ASN functions for the Iwao and Converging Line sampling plans. Recall that the minimum sample size was set to 15, so the big difference in stop boundaries is that those for Converging Lines become narrower with increasing sample size, while those for Iwao become wider until the maximum sample size is reached. The OC functions for the two plans are nearly identical, but the ASN function of the Converging Line plan is either close to or less than the ASN function for the Iwao plan.

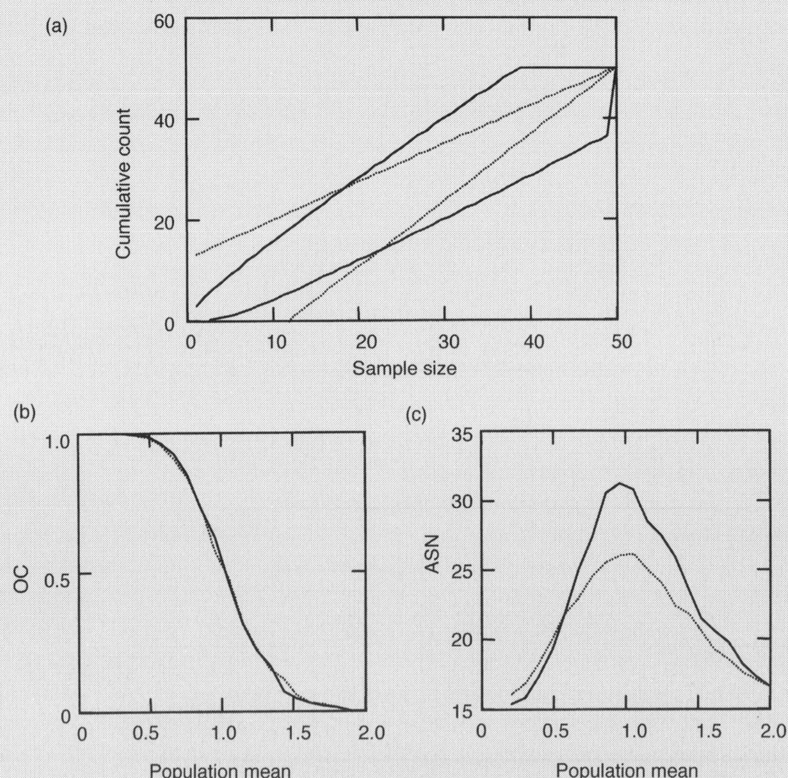

Fig. 5.16. Stop boundaries (a) and OC (b) and ASN (c) functions for an Iwao (——) and a Converging Lines (···) sequential classification sampling plan and the negative binomial distribution. Boundary parameters: *cd* = 1, *minn* = 15, *maxn* = 50, TPL *a* = 3.05 and TPL *b* = 1.02. Iwao plan, $\alpha = 0.15$. Converging Lines plan, $\alpha L = 0.05$ and $\alpha U = 0.1$. Simulation parameters: negative binomial with *k* estimated by TPL, 1000 simulations.

A similar, although less extreme, pattern is evident when the Converging Line and SPRT plans are compared (Fig. 5.17). Here also, the OC functions are nearly identical, but the ASN function for the Converging Line is again either close to or less than the ASN function for the SPRT.

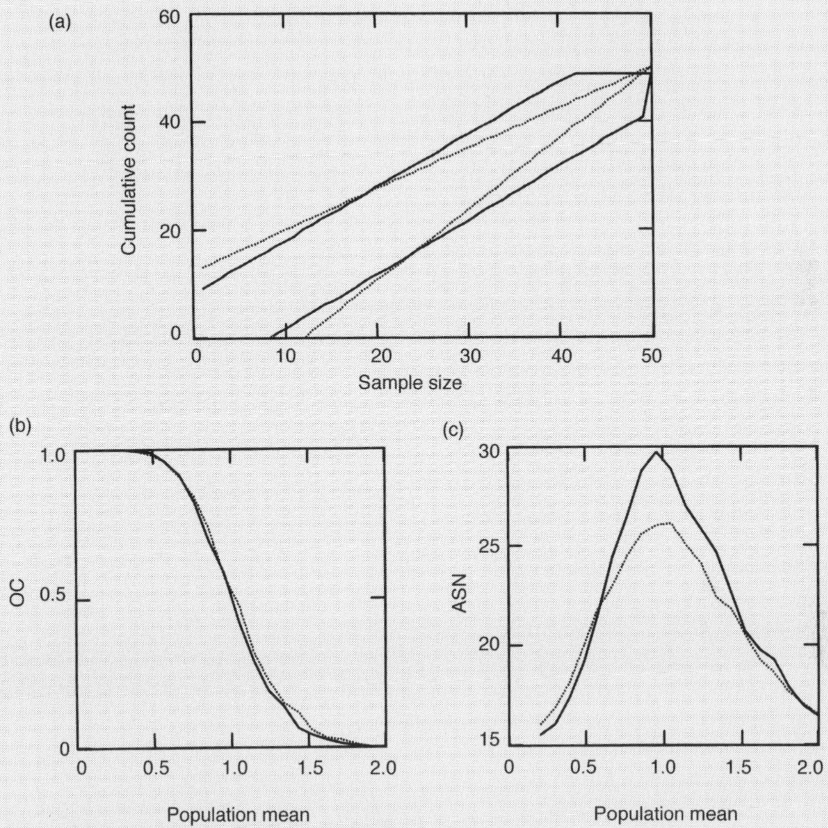

Fig. 5.17. Stop boundaries (a) and OC (b) and ASN (c) functions for an SPRT (——) and a Converging Lines (···) sequential classification sampling plan and the negative binomial distribution. Boundary parameters: *cd* = 1, *minn* = 15, *maxn* = 50, TPL *a* = 3.05 and TPL *b* = 1.02. SPRT plan, $\mu_0 = 0.8$, $\mu_1 = 1.2$ and $\alpha = \beta = 0.25$. Converging Lines plan $\alpha L = 0.05$ and $\alpha U = 0.1$. Simulation parameters: negative binomial with *k* estimated by TPL, 1000 simulations.

References and Suggested Reading

Anscombe, F.J. (1949) The statistical analysis of insect counts based on the negative binomial distribution. *Biometrika* 5, 165–173.

Armitage, P. (1975) *Sequential Medical Trials*, 2nd edn. John Wiley, New York, 194 pp.

Barney, R.J. and Legg, D.E. (1988) Accuracy of a single 30-stem sample for detecting alfalfa weevil larvae (Coleoptera: Curculionidae) and making management decisions. *Journal of Economic Entomology* 80, 512–516.

Binns, M.R. and Nyrop, J.P. (1992) Sampling insect populations for the purpose of IPM decision making. *Annual Review of Entomology* 37, 427–453.

Fowler, G.W. and Lynch, A.M. (1987) Sampling plans in insect pest management based on Wald's sequential probability ratio test. *Environmental Entomology* 16, 345–354.

Green, R.H. (1970) On fixed precision level sequential sampling. *Researches on Population Ecology* 12, 249–251.

Hoffman, M.P., Nyrop, J.P., Kirkwyland, J.J., Riggs, D.M., Gilrein, D.O. and Moyer, D.D. (1996) Sequential sampling plan for scheduling control of lepidopteran pests of fresh market sweet corn. *Journal of Economic Entomology* 89, 386–395.

Hughes, G., Madden, L.V. and Munkvold, G.P. (1996) Cluster sampling for disease incidence data. *Phytopathology* 86, 132–137.

Iwao, S. (1975) A new method of sequential sampling to classify populations relative to a critical density. *Researches on Population Ecology* 16, 281–288.

Legg, D.E., Shufran, K.A. and Yeargan, K.V. (1985) Evaluation of two sampling methods for their influence on the population statistics of alfalfa weevil (Coleoptera: Curculionidae) larva infestations in alfalfa. *Journal of Economic Entomology* 78, 1468–1474.

Miller, C.D.F., Mukerji, M.K. and Guppy, J.C. (1972) Notes on the spatial pattern of *Hypera postica* (Coleoptera: Curculionidae) on alfalfa. *Canadian Entomologist* 104, 1995–1999.

Nyrop, J.N. and Binns, M.R. (1991) Quantitative methods for designing and analyzing sampling programs for use in pest management. In: Pimental, D. (ed.) *Handbook of Pest Management in Agriculture*, 2nd edn, vol. II. CRC Press, Boca Raton, Florida, pp. 67–132.

Siegmund, D. (1985) *Sequential Analysis*. Springer-Verlag, New York, 272 pp.

Sparks, A.N. Jr., and Boethel, D.J. (1987) Evaluation of sampling techniques and development of sampling plans for threecornered alfalfa hoppers (Homoptera: Membracidae) on soybeans. *Journal of Economic Entomology* 80, 369–375.

Vincelli, P.C. and Lorbeer, J.W. (1987) Sequential sampling plan for timing initial fungicide application to control Botrytis leaf blight of onion. *Phytopathology* 77, 1301–1303.

Wald, A. (1973) *Sequential Analysis*. Dover, New York, 212 pp. (Wald's original paper, 'On cumulative sums of random variables', was published in 1944.)

Appendix: Calculating Aggregation Parameters for Use in the Simulations

This appendix provides details on how to calculate the aggregation-related parameters k of the negative binomial distribution and ρ of the beta-binomial distribution. For both distributions, three cases may be distinguished:

1. The aggregation parameter is constant, and a single value can be used for all densities (negative binomial) or incidences (beta-binomial). In this situation, the methods in this appendix are not needed. In practice, however, k and ρ are rarely constant.
2. The parameter varies in value with the level of density or incidence, and a model is needed to capture the relationship. For both the negative binomial and the beta-binomial distributions, good descriptions of the aggregation parameter are obtained on the basis of a relationship between the variance and the mean density or incidence.
3. The parameter varies in value with the level of density or incidence, but in addition to this, there are significant differences in variance between fields with similar densities. This variability affects expected sampling performance.

This appendix deals with situations 2 and 3.

Negative binomial k

The value of k can be modelled as a function of the mean, using TPL:

$$\sigma^2 = a\mu^b \tag{5A.1}$$

Some variability always exists in variance–mean models, so actual values of σ^2 will vary about the value predicted by the model (Equation 5A.1). In the simulations where OC and ASN functions are estimated, allowance can be made for variability around the model as follows. If TPL is fitted by linear regression, we generate a value for σ^2 by

$$\left.\begin{array}{l} \ln(\sigma^2) = \ln(a) + b\ln(\mu) + z(0,\sigma_\varepsilon) \\ or \quad \sigma^2 = a\mu^b e^z \end{array}\right\} \tag{5A.2}$$

where z is normally distributed with mean 0 and standard deviation σ_ε (the square root of the mean square error for the regression used to estimate TPL can be used as an estimate of σ_ε). By equating sample and theoretical variances for the negative binomial distribution,

$$a\mu^b e^z = \mu + \frac{\mu^2}{k} \tag{5A.3}$$

we obtain an estimate of k:

$$k = \frac{\mu^2}{a\mu^b e^z - \mu} \tag{5A.4}$$

Each time a sampling plan is used to classify a particular density during a simulation, Equation 5A.4 is used to generate k for the population being sampled. If an OC determined using simulation is to be based on 500 simulation runs, then 500 different values of k would be determined, one for each simulation run at each value of μ. If variability is not to be included in the simulations, the value used for k is

$$k = \frac{\mu^2}{a\mu^b - \mu} \tag{5A.5}$$

Equating theoretical and sample variances is an example of the 'Method of Moments', because according to statistical terminology the variance is one of the 'moments' of a probability distribution (it is the second moment, the first being the mean).

Beta-binomial

The derivation for ρ is very similar. Its value can be modelled as a function of the incidence, p, using the model (Hughes *et al.*, 1996)

$$\sigma^2 = A\left(\frac{p(1-p)}{R}\right)^b \tag{5A.6}$$

where R is the cluster size, and A and b are the parameters of the variance–incidence model. Again, there is always some variability in incidence–mean models, so actual values of σ^2 will vary about the value predicted by the model (Equation 5A.6). In the simulations where OC and ASN functions are estimated, allowance can be made for variability around the model as follows. If the model is fitted by linear regression, we generate a value for σ^2 by

$$\left.\begin{aligned} \ln(\sigma^2) &= \ln(A) + b\ln\left(\frac{p(1-p)}{R}\right) + z(0,\sigma_\varepsilon) \\ \text{or}\quad \sigma^2 &= A\left(\frac{p(1-p)}{R}\right)^b e^z \end{aligned}\right\} \tag{5A.7}$$

where z is normally distributed with mean 0 and standard deviation σ_ε. By equating sample and theoretical variances,

$$A\left(\frac{p(1-p)}{R}\right)^{b} e^{z} = \left(\frac{p(1-p)}{R}\right)(1+\rho[R-1]) \tag{5A.8}$$

we obtain an estimate of ρ:

$$\rho = \frac{1}{(R-1)}\left(\frac{AR^{1-b}e^{z}}{[p(1-p)]^{1-b}} - 1\right) \tag{5A.9}$$

This formula is used exactly as described above for Equation (5A.4). If variability is not to be included in the simulations, the value used for ρ is

$$\rho = \frac{1}{(R-1)}\left(\frac{AR^{1-b}}{[p(1-p)]^{1-b}} - 1\right) \tag{5A.10}$$

6 Enhancing and Evaluating the Usefulness of Sampling Plans

6.1 Introduction

In this chapter we review how the usefulness of sampling plans for pest management decision-making may be maximized by proper choice of design ingredients, by suitable presentation of the decision tool to users, and by use of simulation tools and field methods, along with user interaction, for evaluation and target-oriented design of plans. Evaluation of a sampling plan involves more than just the operating characteristic and average sample number functions, which were discussed in the preceding chapters. Qualitative aspects, such as the practicality and simplicity of the decision guide, and the reliability, representativeness and relevance are equally important. These aspects, although not easily expressed quantitatively, have a considerable influence on adoption in practice. It is therefore necessary to evaluate qualitative aspects of performance in direct consultation and interaction with end-users. Quantitative indicators of performance, such as the expected outcome from sampling and time spent sampling, can be readily defined and analysed using computer simulation. While field evaluations of sampling plans can generally not be used to estimate operating characteristic (OC) and average sample number (ASN) functions, they are useful in other contexts and have the advantage that there are no model-based artefacts in the evaluation results. When evaluating the results of a field or computer evaluation, interaction with end-users is again indispensable. The economics of sampling are an important aspect of sampling plan evaluation, and we need a decision-theoretic framework that allows a quantitative evaluation of the value of sampling.

The text covers the full range of components which go into creating and implementing a good decision guide. We look first at the nature of the design process: the process between the recognition that a decision guide based on sampling is needed and the implementation of a decision guide in practice. We follow this with an examination of the necessary design ingredients of a decision guide, expanding the introductory discussion in Chapter 2. Then we discuss the various ways decision guides can be evaluated. Finally, we show how to estimate the overall economic value of decision guide sampling under a variety of crop loss formulations and expected likelihoods of high pest densities.

6.2 The Cyclical Process of Learning, Improvement and Design

Decision guides must satisfy the objectives and constraints of users. If they do not, the usefulness of a proposed tool may be compromised and the decision guide may never be adopted. The specific objectives of crop and pest managers evolve as product prices vary and pressures from society change, but certain basic objectives are always present (see Chapter 1):

Minimize
- pest damage
- control costs
- the use of environmentally harmful control measures
- the potential for resistance to chemicals

Maximize
- natural controls
- the use of techniques that promote a natural and healthy image of the farm business and the farm product
- economical and ecological sustainability (robustness) of the farming system

Compared with pest management approaches that rely on calendar- or phenology-based pesticide applications, the use of a sample-based decision guide has potential benefits: a reduced risk of severe crop loss, reduced pest control costs, reduced environmental contamination, a reduced risk of pests developing resistance to chemicals, conservation of natural enemy populations, a better public image and increased sustainability of the production system. However, growers or pest management advisors will use a sample-based decision guide only if it fits their overall objectives and management approach (or if it is compulsory to do so). Assuming free will, they will adopt a proposed procedure for collecting sample information only if it provides key information in a cost-effective manner.

Unfortunately, sampling is often framed in the narrow perspective of economic optimization, where all the other factors that make life and farming worthwhile seem to play no role. This is demeaning to the real values and priorities of farmers and other stakeholders in the agricultural landscape, as well as to researchers and extension specialists who assist in the development of sampling plans. We need a richer approach in evaluation, where all the quantitative and qualitative aspects of management approaches are evaluated more comprehensively. However, we cannot escape the need to make economic evaluations, because one of the primary goals of farming is to make a profit and earn a living.

The design and evaluation of sampling plans as crop protection decision guides follows a cyclical process that consists of three phases:

1. Proposing or modifying a sampling method.
2. Determining and evaluating the performance of the proposed method.
3. Implementing the decision guide.

Phases 1 and 2 cycle together until the evaluation indicates that the sampling plan is acceptable; only then is phase 3 (implementation) entered. In the process of cycling through these phases, a learning process unfolds which should result in a successful design being implemented – or the realization that there is no viable sampling approach for the problem.

6.2.1 Proposing or modifying a sampling method (phase 1)

Proposing a sampling method entails:

- Determining the basic design ingredients of sampling, such as the size of the management unit to which sample results will be applied, the sample unit, the timing of sampling and the sample path in the crop (Section 6.3).
- Determining a critical density (Chapter 1), with the understanding that this parameter can be changed in order to modify sampling plan performance (e.g. the OC function, as in Exhibits 3.2 and 5.2).
- Choosing the design parameters of the sampling plan (Chapter 5): a template such as the SPRT, and all its parameters (e.g., for the SPRT, the lower mean, μ_0, and associated risk, α, the upper mean, μ_1, and associated risk, β, and minimum and maximum sample sizes).

Modifying a sample method entails using theory and simulation results to suggest changes in sampling plan parameters which are preferable in their own right or which may improve performance indicators. Possibly, another template for the sampling plan may be used.

6.2.2 Determining and evaluating the performance of the proposed method (phase 2)

Determining the performance of a method entails:

- Choosing performance indicators (these may include non-economic measures)
- Conducting field tests to place values on performance indicators (Section 6.4.1)
- Conducting simulation experiments to place values on performance indicators (Section 6.4.2)

Evaluating the performance of a method entails:

- Consulting, individually or together, with people who represent the knowledge base capable of judging the usefulness of the proposed decision guide
- Discussing the pros and cons of alternative sampling plans in view of their design ingredients and performance measures

6.2.3 Implementing the decision guide (phase 3)

Following successful completion of phase 2, implementation can proceed. This may include describing the procedure in bulletins, or by speaking to groups of farmers or crop consultants. Often, an intermediary step of demonstrating the utility of the proposed procedure on a practical scale is needed (similar to a field evaluation).

To complete the cycle successfully, a variety of expertise is needed:

- for phase 1 – crop protection experience; theoretical and practical knowledge on sampling
- for phase 2 – mathematical and computer skills; practical experience in the field
- for phase 3 – communication and innovation skills with respect to crop protection

It should be clear that there are many potential collaboration and linkage problems in this cycle. In particular, there always tends to be a gap between the two more research-oriented phases 1 and 2, and the extension to practice, phase 3 (Pedigo, 1995).

This book concentrates on phases 1 and 2. The purpose of all the theory and simulation tools presented here is to create opportunities and provide simulation and analysis tools for developing sample methods that are as useful as possible for pest management.

6.3 Design Ingredients of Sampling Plans that Affect Usefulness

A useful sampling plan is one in which sample information is obtained quickly and economically: the plan should be easy to follow, and it should fit within normal management practice. The sample information should answer questions that the grower is actually asking and the procedure must be clearly defined in terms of: (i) the size of the management unit; (ii) the sample unit; (iii) when and what to sample; (iv) the sample path in the crop; and (v) the procedure for recording data.

6.3.1 The size of the management unit

In principle, the size of the management unit should be based on the uniformity of the required management: the more homogeneous the crop, the larger the management unit may be. However, in practice, labour deployment and practicality have the greatest influence. Risk also plays a role, as do external pressures from society. These aspects are illustrated in the following exhibit.

Exhibit 6.1. *Modernizing the management of fire blight disease in Dutch flower bulbs*

Until recently, Dutch flower bulb growers used routine spraying of fungicides as the main strategy for combatting fire blight disease, caused by *Botrytis* spp., in flower bulb crops. These crops (e.g. lily and tulip) are grown for market and can be worth more than US$100,000 per hectare. The bulbs are thus very valuable and worth protecting at almost any cost. Another important threat to the crop is the spread of viruses by aphids. Regular spraying with either pesticides or mineral oils (which affect virus transmission), or a combination of these, is practised to limit virus spread. Because growers need to control virus spread, and botrytis is a continuous threat to the crop, tank mixing of aphicides and fungicides has become standard practice. Spraying has tended to be done on a calendar basis with large management units. From a labour standpoint, this is economical, but from the perspective of trying to limit pesticide usage it is not beneficial. What options are there for limiting fungicide usage?

It was recently rediscovered that some cultivars are hardly or not at all susceptible to botrytis and require very few, if any, treatments. Farmers who had never grown flower bulbs without an umbrella of pesticides in the past 20 or more years were unaware of this fact. The situation is currently being improved as disease warning systems are being developed and farmers are becoming more aware of differences between varieties in disease susceptibility (van den Ende *et al.*, 2000) and more sensitive to environmental issues (Rossing *et al.*, 1997). A consequence of this may be a reduction in the size of the management unit. In order to reduce their input of unneeded pesticides, growers will have to do more fine-tuning. In principle, this would save costs, but it would also make their farming operations more complex. Moreover, growers may perceive a reduction in sprays as risky, especially in view of the high value of flower bulb crops. Therefore, an economic incentive may not be enough to pull them over to a less pesticide-reliant management approach. Regulation by the government may help here, and indeed, in the framework of the Dutch Multi-year Crop Protection Plan (Ministry of Agriculture, Nature Conservation and Fisheries, 1991), the bulb growers and the government have reached an agreement on reducing the pesticide input in this sector by about 60% from the year 1987 to the year 2000. Many European countries have implemented such policy plans in the late 1980s and early 1990s in order to make agriculture in the year 2000 more sustainable and socially acceptable (Jansma *et al.*, 1993). Such policy developments help the adoption of more environment-friendly approaches which make use of decision tools. A benefit of reducing pesticide inputs may be the restoration of populations of natural enemies and microbial antagonists, reducing the need for pesticides in the long term. But it remains to be seen how beneficial such restoration will be.

There are several take-home messages here. One is that the growers' evaluation of the usefulness of sampling plans and other decision tools depends heavily on their perceived risks and management constraints, and on the pressures and demands from the world around. The size of the management unit varies according to these internal and external objectives and constraints. Another message is that multiple pests are important.

6.3.2 The definition of the sample unit

Statistical and practical considerations both play a role in the definition of the sample unit, and there may be a trade-off between them. The precision of sample estimates is generally greater if many small sample units are collected, rather than a few large units. This is because, within larger units, counts on smaller sub-units are likely to be correlated. For instance, the probability of finding mildew on a haulm of wheat is higher when the disease has already been observed on another haulm of the same plant than when the disease was not observed on that other haulm. This implies that inspecting another haulm on a plant does not provide quite as much information as inspecting a haulm on another, distant, plant. The principle of correlation between sub-units within units applies to many different levels: leaves on stems, stems on plants, plants in clusters and so on. Practicality, however, may suggest that sample units be taken in clusters (i.e. a few plants at a sample location, a few branches on a tree and so on), because it reduces the travel time between sample units and may speed sample collection. However, if correlation within clusters is large, the statistical usefulness of extra subsamples within clusters is minimal. Hence, a compromise is often necessary between the ease of taking only a few large samples versus obtaining a more representative sample by examining a greater number of smaller sample units. We deal with this subject more formally in Chapter 8.

Practical considerations may dictate using a smaller sample unit when a larger one would provide better information. For example, the larvae of some scarab beetles feed on the roots of turf grass, and sampling these organisms requires digging soil and sod to count the larvae. On turf plantings such as golf courses it is desirable to minimize damage to the turf from the actual sampling. Thus, while a better picture of scarab larval densities might be obtained by examining a 30 cm × 30 cm area of turf, this would result in excessive damage. Instead, a circle of turf 12 cm in diameter is examined (Dalthorp *et al.*, 1999).

6.3.3 When and what to sample

Sampling should be done when pests are readily observed and controlled, when observable indicators of injury are good predictors of damage and when damage is still (largely) preventable. Often, the time of sampling insects can be based on the development of the crop, as herbivorous insects are frequently synchronized with their host plants. It is important to choose the moment of sampling carefully. If sampling is done too early, the presence of damaging pest populations may not be noticed or might be underestimated, whereas if sampling is too late, the target stage may no longer be present or damage may no longer be preventable. In addition to crop stage, temperature sums, regional warnings or pheromone catches of adult flying insects can be used to trigger the need for sampling.

It is important that the proposed decision guide should be robust to errors in the timing of sampling, because it may not be possible to predict precisely when the sample should be taken. For example, the spotted tentiform leafminer (*Phyllonorycter blancardella*) is a pest in apple orchards whose phenology cannot be precisely predicted using temperature summations. To deal with this problem, a decision guide was proposed which called for sampling on more than one occasion (Schmaedick and Nyrop, 1995), making use of a procedure called 'cascaded sequential sampling' (described in Chapter 11). The robustness of sampling plans to variations in the timing of the sample may also be enhanced by clever choice of the measurement variable. For instance, signs of feeding damage or 'frass' are sometimes more reliable indicators of pest presence than the pest themselves, which may be hiding, or in a stage that is difficult to find.

When developing a decision guide, it is obvious that the pest will be sampled. An important question is whether natural enemies should also be sampled. There are good reasons for including natural enemies: if we were to use sample-based knowledge about natural enemy abundance, we might be able to predict the future dynamics of pest populations. This might save pesticide applications by raising thresholds if natural enemies are present in sufficient densities (Brown, 1997). Refraining from pesticides if natural enemies are abundant might also conserve natural enemy populations, with a view to the evolution of a cropping system that would not be reliant on using corrective tactics as much.

Very few practical plans have been worked out on the principle of sampling pests and their natural enemies (Croft and Nelson, 1972; Nyrop, 1988; Baillod *et al.*, 1989) and adoption in practice has been low. The principal reason for this must be the multiplication of work that goes with expanding sampling plans to include natural enemies in addition to pests, and at the same time requiring a great deal of biological knowledge on the part of the grower or pest management practitioner. It might, however, be argued that the latter is an advantage rather than a disadvantage, as sociologists have found that learning is one of the prime 'qualities' of decision support systems (Leeuwis, 1993).

6.3.4 The sample path in the crop

In theory, samples have to be taken at random, because only then can one be assured that the mathematical expectation (Section 2.2) of the sample estimate is equal to the true value: in other words, that the estimate is truly representative of the whole management unit. In practice, sample units are always collected in a pseudo-random way, usually by specifying an approximate V- or W-shaped path to be followed through the crop. Samples are taken at (ir)regular intervals along the path in such a way that broad coverage of the management unit is obtained. This approach is generally deemed to be adequate; possible errors have been investigated in some instances and found to be acceptable or insignificant (see, e.g. Legg *et al.*, 1985). Care must still be exercised to ensure that bias does not creep into the selection of sample units while traversing the sample path. Ways to avoid such bias have been discussed in Chapter 2.

6.3.5 Procedures for recording data

Data may be recorded into charts, forms or computer software. Charts and forms, on which users can log and evaluate their observations, are universally used. Ideally, the advised management action (e.g. a stop boundary) should appear as a natural consequence of the sample information as it is recorded. It is also helpful to present the decision guide as a graphic and to include numerical values with the graphic. This is illustrated in Fig. 6.1. The benefit of a graphic is that it helps users to understand the logic of what they are doing during sampling. The benefit of including numerical values for stop boundaries is that interpolation is simplified.

6.4 Evaluation by Users, Field Tests and Simulation

In the previous sections a lot of emphasis has been put on meeting practitioners' needs when developing decision guides. This might suggest that development of decision guides is predominantly a customer satisfaction enterprise, in which developers of decision tools should develop what practitioners like to use. We do not intend to give this impression. There is a large body of knowledge about pest sampling that can be put to good use by smart designers who take account of the management objectives of practitioners, and who have the insight and tools to create decision guides suited to specific needs. The simulation tools presented in previous chapters allow a smart designer to develop sampling plans that may be unfamiliar

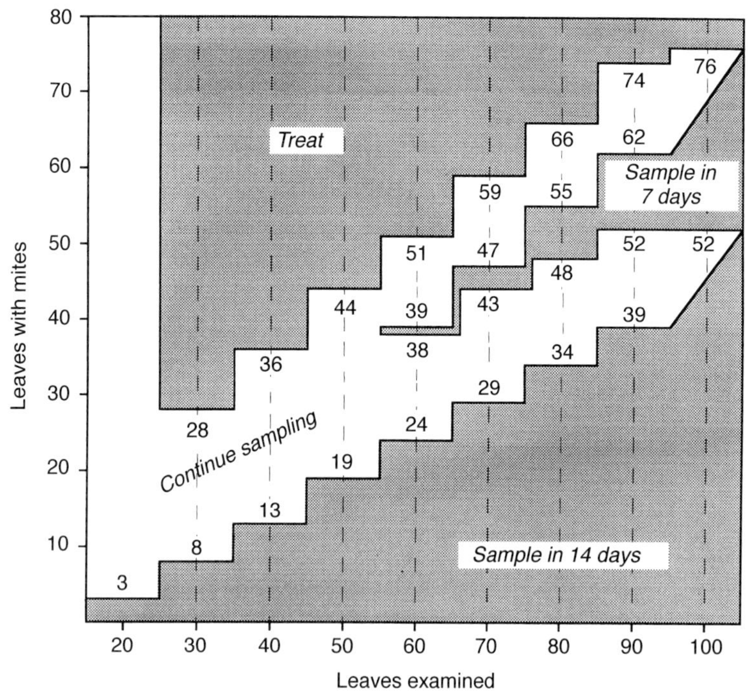

Fig. 6.1. Stop boundaries for classifying European red mite densities into one of three categories: greater than the action threshold (treat), less than the action threshold and requiring resampling in 7–10 days (sample in 7 days), and much less than the action threshold, with sampling not needed again for 14–20 days (sample in 14 days). The numbers are the actual stop boundary values (from Breth and Nyrop, 1998).

to the client, but which provide satisfactory management advice (as described in the OC function), and yet keep the sampling effort at as low a level as possible (as described by the ASN function). In a fruitful collaboration, practitioners will ask questions and developers will provide answers, as well as the other way round. This produces a continuous cooperative evaluation process in which all participants learn and teach.

6.4.1 Field evaluation

What, if any, is the role of field experiments in the evaluation process? One reason why field tests can be viewed as indispensable is that the models used to represent sample observations may be inadequate and may not cover the full variability of real systems. If this were the case, estimates of OC and ASN functions calculated by the tools described in previous chapters would not be representative of what happens in the real world. However, this should not be a critical concern, because OC and ASN functions are relatively robust to departures from the models used to represent sample observations (Nyrop *et al.*, 1999). Therefore, field experiments are not essential for estimating OC and ASN functions. This is indeed fortunate, because field experiments cannot realistically be used to estimate OC and ASN functions directly. This is because in order to estimate a point on the OC and ASN functions, the sampling plan must be applied many times to populations with the same or similar means, and this process must be repeated over a range of means. This would be extremely difficult, if not impossible, to accomplish in the field. As an aside, we note that field data can be used in a simulation context to estimate OC and ASN functions; we will return to this in Chapter 9.

While field experiments can generally not be used to estimate OC and ASN functions, they do serve three important roles in the development and evaluation of crop protection decision guides. First, they can provide real-world estimates of the actual time and resources required to obtain sample information. Second, they can alert developers to possible erroneous assumptions or factors that they may have overlooked in the development process. These assumptions or factors are not related to the distribution of the sample observations, but to the timing of sampling during the day, the selection of sample units and other types of bias that may influence the sample counts. Finally, field experiments can serve to demonstrate to practitioners the usefulness of the decision guide.

The simplest way to test a guide using a field experiment is to apply the sample protocol at a site and follow it immediately with validation sampling, to provide a precise and accurate estimate of pest density or incidence. If the influence of pest phenology on sample outcome is being evaluated, the timing of the validation sampling must be carefully considered. By comparing the results from the decision guide with the validation estimate, one can determine whether the guide provided the correct advice. The decision guide might also be used by several testers at the same site to provide information on possible observer bias. Another possibility is to test the sampling plan over a short period of time during which the actual pest density does not change, but the conditions (e.g. light, wind and temperature) under which the sampling plan is used do change.

6.4.2 Simulation, sensitivity and robustness

We have stated that OC and ASN functions are relatively robust to variability in the models used to represent sample observations. Given this robustness, we believe that a viable strategy for developing sampling methods for a new crop–pest system is to guess the parameters used to describe the distribution of sample counts on the basis of similar crop–pest systems, and perform a sensitivity analysis (see, e.g. Exhibits 5.1–5.3) to see whether the performance of a pest-control decision guide would be sensitive to any differences in parameter values in the plausible range (Nyrop *et al.*, 1999). This is not a novel insight. Green (1970) and Jones (1990) came to similar conclusions, although from the perspective of estimating abundance rather than from the view of sampling for decision-making. Elliott *et al.* (1994) also suggested that generic models of sampling distributions might suffice for developing sampling plans for pests of wheat and rice. From the perspective of sampling for decision-making, we think an even stronger case can be made for the use of generic models of sampling distributions.

Based on work described in the literature and the examples provided thus far in this book, we make three recommendations for developing sampling plans for use in pest management decision-making.

6.4.2.1 *Critical density*

Ideally, the first step is to obtain a good estimate of the critical density or critical proportion (*cd* or *cp*). Of course, this may not be possible, and it is often the case that using some value for this parameter, even though not the theoretically correct value, would markedly improve pest management decision-making, compared to calendar- or phenology-based tactics. It is then important to recognize the vagueness of *cd* or *cp*, and to understand that precise information about pest abundance provides a false sense of security. In such cases, reductions in decision errors will come from improved knowledge of *cp* or *cd*, rather than from increasing the precision of the sample information.

6.4.2.2 *Sensitivity with respect to sampling distribution*

The second recommendation is that when developing sampling plans, one should start by determining what is known about the sampling distribution, use this information to develop a sampling plan, and then perform a sensitivity analysis to determine if further refinement of this information is warranted. When nothing is known about the sampling distribution for a particular pest, sampling distribution models might be grouped for similar taxa on similar crops (e.g. aphids on small grains, or disease incidence on perennial fruit crops). This hypothesis is easily tested using simulation.

6.4.2.3 Size and representativeness of sample

The third recommendation is that sample size should first be based on the requirement that the sample information should be representative. This means that the sample information must be a good indicator of pest abundance or incidence in the management unit. Sample information may not be representative if sample observations are not collected from throughout the sample universe, which can happen when very few samples are taken. Therefore, when using sequential sampling, a reasonable minimum sample size and sample path need to be defined.

A final important evaluation question to answer is whether collection of the sample information leads to reduced pest control costs. This is addressed in the next section.

6.5 The Economic Value of Sample Information

What is the value of using decision guide sampling to recommend a management action, as against always intervening, or as against always leaving nature to take its course? At one extreme, farmers whose crops rarely have pest problems might feel that spending time and money collecting samples to tell them mostly what they already know (don't intervene) would be wasteful. At the other extreme, farmers whose crops nearly always require treatment might feel the same – and continue to treat. For these two extremes, and between them, we can calculate the long-term economic value of decision guide sampling. The results provide a comprehensive economic scale for comparing decision guides.

The economic value of the sample information used to guide pest management decision-making is the reduction in pest control costs that can be attributed to basing management decisions on the sample data. While it is desirable that pest control costs be reduced as a result of using the sample information, a sampling plan with a negative economic gain may still be worthwhile if it carries other benefits, such as reduced pesticide inputs (Nyrop *et al.*, 1989). We use the approach for quantifying the economic benefit of sampling proposed by Nyrop *et al.* (1986), which is rooted in economic decision theory.

To quantify the economic benefit, we require more detailed information about the context in which decisions are made than is needed to estimate OC and ASN functions, although these functions are still used in the analysis. First, we need a quantitative relationship between pest density, μ, and monetary loss when no control is used, $L(\mu)$. In theory, such a relationship is used to provide an initial estimate of the critical density (see Chapter 1), but a provisional critical density can be formulated when the pest–loss relationship is rather poorly defined. Second, we need to know the likelihood of encountering each pest density within the range of possible densities. A probability distribution function is used to model this information. Because it provides the likelihood of any pest density before a sample is taken, it is called the prior distribution, $P(\mu)$, of the pest density, μ. Another way of looking at this prior distribution is to think of it as a long-term frequency of occurrence for each pest density.

The value of the sample information is determined as the difference between the pest control costs incurred when using the best control strategy not based on sample information, and the pest control costs when decision strategies are based on sample information.

6.5.1 Best control strategy with no sampling

The first step is to determine the best control strategy when decision guide sampling is not used. We compare two strategies (more than two could be considered, but the essentials are the same). The first strategy is a *laissez faire* approach, which is to always do nothing (Decision 0), and the second is an *always control* approach (Decision 1).

The best strategy to use when no sample information is collected is the one with the lowest expected costs, where the expectation is taken with respect to the prior pest density. For each pest density, the cost of damage is multiplied by the likelihood of the density. These products are then summed to produce the expected cost.

The total cost for doing nothing (not treating, Decision 0) is calculated as follows:

$$Cost(0) = \Sigma P(\mu)L(\mu) \tag{6.1}$$

where $P(\mu)$ is the prior probability that models the likelihood of pest density μ and $L(\mu)$ is the loss (e.g. the reduction in crop value) due to pest density μ. Note that $L(\mu)$ is expressed on a per management unit basis, such as per hectare or per acre. The total cost for always treating (Decision 1) is as follows:

$$Cost(1) = \Sigma P(\mu)L(\mu(1 - ControlEffect)) + ControlCost \tag{6.2}$$

where *ControlEffect* is the effectiveness of the control treatment as measured by the proportion of the pest population removed by the treatment (*ControlEffect* necessarily lies between 0 and 1), and *ControlCost* is the cost of treatment on a per unit basis. The smaller of these two (Equations 6.1 and 6.2) is the minimum total cost due to pest damage and pest control (without sampling). The best strategy to adopt when not sampling is that which minimizes total cost:

$$MinimumTotalCostWithoutSampling = \text{Min}(Cost(0), Cost(1)) \tag{6.3}$$

If $Cost(0) = Cost(1)$, we choose $Cost(0)$.

These equations confirm what we would intuitively reason, when we have estimates for $P(\mu)$ and $L(\mu)$. If the pest density is regularly less than the critical density, then the best strategy without sampling is to do nothing. This pattern of pest abundance would be reflected by a prior distribution in which the likelihoods for densities less than the critical density are greater than the likelihoods for densities greater than the critical density. Conversely, when the pest density is regularly greater than the critical density, the best strategy without sampling is always to treat. It follows that sampling can be advantageous only in situations in which it is not certain *a priori* whether the density is above or below the critical density.

6.5.2 Calculating the value of sample information

When sampling is used, intervene or not-intervene decisions depend on the pest density, μ, as specified by the operating characteristic function, $OC(\mu)$ (the probability of deciding on action 0: do nothing). The costs which are incurred for non-intervention and intervention when a decision is based on sampling are as follows:

Classification	Cost
No intervention	$C_{ni} = \Sigma P(\mu) OC(\mu) L(\mu)$
Intervention	$C_i = \Sigma P(\mu)(1 - OC(\mu))\{L(\mu(1 - ControlEffect)) + ControlCost\}$

Therefore

$$TotalCostWithSampling = C_{ni} + C_i \tag{6.4}$$

The (economic) value of sampling is the difference between *TotalCostWithSampling* (Equation 6.4) and *MinimumTotalCostWithoutSampling* (Equation 6.3):

$$ValueOfSampling = TotalCostWithSampling - MinimumTotalCostWithoutSampling \tag{6.5}$$

But sampling is not free. The cost of sampling includes a fixed cost and a variable cost per sample unit. The variable cost can be estimated from the ASN function of the sampling plan, so

$$TotalCostOfSampling = FixedCostOfSampling + \Sigma P(\mu) ASN(\mu) CostPerSampleUnit \tag{6.6}$$

The net value of sampling is the difference between the *ValueOfSampling* (Equation 6.5) and the *TotalCostOfSampling* (Equation 6.6):

$$NetValueOfSampling = ValueOfSampling - TotalCostOfSampling \tag{6.7}$$

A final quantification which can provide insight about the decision guide is the expected frequency of intervention:

$$ExpectedFrequencyOfIntervention = \Sigma P(\mu)(1 - OC(\mu)) \tag{6.8}$$

This can be related to the frequency of intervention for the best strategy without sampling, namely 1 if $Cost(0) > Cost(1)$ and 0 otherwise, as a change in treatment frequency (from 0 or 1, as appropriate):

$$\begin{aligned} &\text{if } Cost(0) \leq Cost(1) \\ &\quad ChangeInTreatmentFrequency = ExpectedFrequencyOfIntervention \\ &\text{if } Cost(0) > Cost(1) \\ &\quad ChangeInTreatmentFrequency = ExpectedFrequencyOfIntervention - 1 \end{aligned} \tag{6.9}$$

We have presented all of these computations as discrete summations even though pest density is continuous. We use discrete equations because it simplifies computations on a computer and, for those not currently familiar with calculus, they are easier to visualize. Use of discrete representations in lieu of continuous ones will not introduce large errors provided that the interval between discrete values is not too large.

6.5.3 The value of sample information in practice

How should these concepts be used to satisfy the needs of farmers? Farmers show a variety of styles characterized by values, interests, objectives, preferences, inclinations and constraints (Van der Ploeg, 1990). Some farmers may make a purely economic evaluation and choose the management tactic that gives the best net gain. However, it is strongly in the farmers' interests to use integrated management and reduce pesticide inputs. Hence, they may be willing to accept a negative net gain (in monetary terms) because of benefits that may be attributed to reduced pesticide inputs (Section 6.2). It is also possible that the best strategy to choose when no sample information is collected is never to treat, which leads to the value of the sample information being zero or even negative. Even then, farmers may still choose to sample because, while the expected net value of the sample information is on average zero or negative, sample information may alert them to relatively rare but catastrophic events where the pest density is unexpectedly high. We illustrate these concepts with an example.

Exhibit 6.2. *The value of sampling for lepidoptera larvae infesting sweetcorn*

Three lepidopterous insects infest sweetcorn grown in New York State: European corn borer (*Ostrinia nubilalis*), fall armyworm (*Spodoptera frugiperda*) and corn earworm (*Helicoverpa zea*). These pests may feed throughout the plant and render the ears unmarketable. The first of these pests is indigenous to the northeastern United States, while the latter two are migratory. Hoffman *et al.* (1996) determined that the sampling distribution of plants infested with these caterpillars could be described by a beta-binomial distribution (see Exhibit 5.3). Prior to the whorl stage of crop growth, a *cp* of 0.3 is recommended to schedule control; after this stage (the tassel and silk stages) the critical proportion should be decreased to 0.15. Here we scrutinize the value of sample information collected after the whorl stage. Because in this exhibit we are dealing with incidences and critical proportions, we replace μ in the above formulae by a more generic symbol, θ.

Loss, $L(\theta)$ A pest–loss relationship was determined from data provided by Riggs *et al.* (1998), in which several treatment thresholds were tested. On the basis of their data, we estimated a linear relationship through zero, between the last estimate of the proportion of infested plants (θ) and the proportion of infested ears at harvest. The relationship has a slope equal to 0.43 (intercept = 0). In using this relationship, we assume that the incidence rate estimated at any time during the silk stage of crop growth will remain constant until the end of the sampling period. This is unlikely, but we lack the information needed to describe changes in the incidence rate over time. The remaining parameters for the cost functions were obtained from Nyrop and Binns (1991), with some adjustments for inflation. We used 17,500 ears of maize per acre as a yield estimate and each ear was valued at US$0.02 (so the value, V, of the crop per acre is 17,500 × 0.02 dollars per acre). The damage per infested plant is derived from the above regression: $D = 0.43$. The cost of insecticide treatment, C, was set to US$15 per acre with an effectiveness, K, equal to 0.75 (or 75%). Based on the above, the loss without control for plant infestation equal to θ is defined by the equation:

$$L(\theta) = V\,D\,\theta = (17{,}500 \times 0.02)0.43\theta = 150.5\theta \text{ US\$/acre}$$

and the loss with control (i.e. with θ reduced to $\theta(1-K)$ and the addition of C) is

$$L(\theta(1-K)) = C + V\,D\,\theta\,(1-K) = 15 + 37.60\theta \text{ US\$/acre}$$

The economic injury level, *EIL*, is where these two losses intersect; that is at $\theta = 0.13$ (Fig. 6.2). Because the loss function $L(\theta)$ is linear, *EIL* can also be calculated using Equation 1.1. The value ($\theta = 0.13$) is close to the treatment threshold recommended by Riggs *et al.* (1998), although their treatment threshold was based on a lack of measurable crop damage rather than economic criteria.

Prior distributions, $P(\theta)$ Three prior distributions were used to model the likelihood of an incidence rate (Fig. 6.3). We assumed initially that pest incidences would never be greater than 0.4. The first prior distribution (uniform prior) modelled each incidence rate within the range 0.0–0.4 as equally likely. This range was based on the range of incidences after the whorl stage as reported by Riggs *et al.* (1998) (the actually estimated range was from 0.01 to 0.42). The remaining two priors were based on the frequency with which classes of incidence rates occurred in the Riggs *et al.* (1998) study. Two groups of incidence rates were used: the first applied to fields that had been treated with an insecticide before the silk stage and

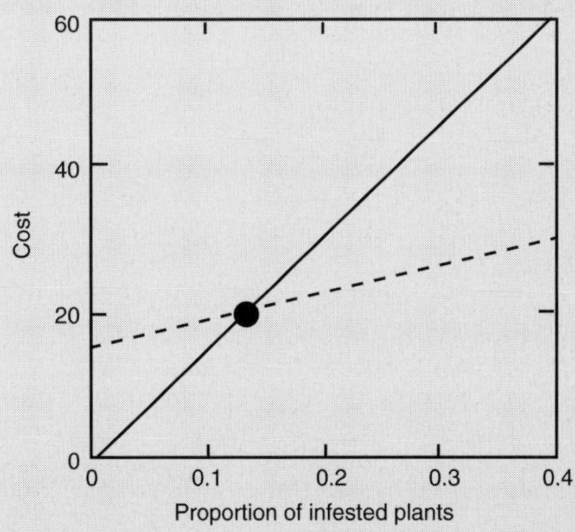

Fig. 6.2. Costs per acre (in US$) with (- – -) and without (——) control of lepidoptera infesting maize. The filled circle corresponds to the economic injury level.

Continued

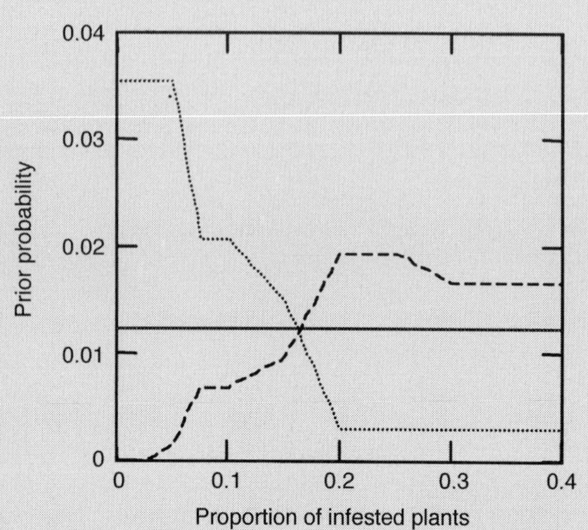

Fig. 6.3. Prior distributions used to model the likelihood of the proportion of maize plants infested by lepidoptera larvae; uniform infestation (——), low infestation (···) and high infestation (- – -).

the second to fields that had not been treated. In the first group incidences were much lower (the *low-incidence prior*) than in the second (the *high-incidence prior*). To quantify the prior distributions, we created five incidence classes (0–0.05, 0.05–0.1, 0.1–0.15, 0.15–0.2 and 0.2–0.4) and calculated the frequency with which incidence rates fell into them.

For the uniform prior distribution (all incidences equally likely), the best strategy in the absence of sampling is always to treat, because this minimizes total costs (Equations 6.1 and 6.2). The same is true when higher incidence rates are more likely (high-incidence prior). When low incidence rates are most likely (low-incidence prior), the best strategy is never to treat.

We used these values for $L(\theta)$ and $P(\theta)$ to study three questions:

1. Is the net value of sampling positive and, if so, how much sampling should be done?
2. What is the effect of shortening the range of incidences for the prior distributions?
3. What is the effect of different pest–loss models?

The net value of sampling The value of sample information was determined for fixed sample sizes of 25, 50 and 100 plants, sampling groups of five plants at each location in a field. The density was classified with respect to cp = 0.13. The OC functions were estimated using a beta-binomial distribution (Section 4.10), and the results are shown in Fig. 6.4. The value of the sample information (Equation 6.5) and its net value (Equation 6.7) for each sampling plan and prior distribution are displayed in Fig. 6.5. The summations in these equations used an increment of 0.01 between levels of incidence. The value of sample information is positive for each prior distribution and each sample size, although it is much lower for the high-incidence prior.

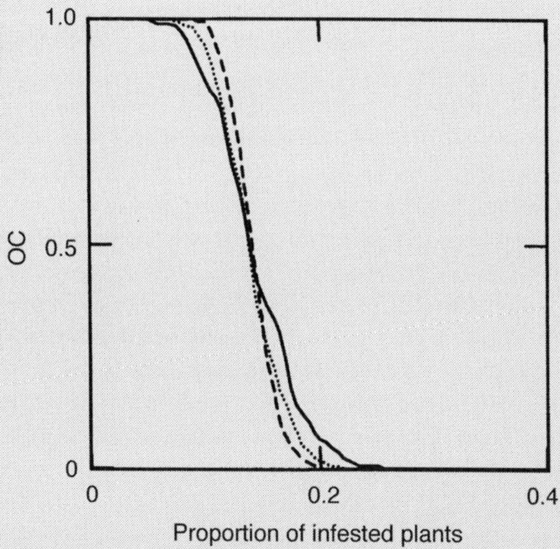

Fig. 6.4. OC functions for sampling plans with fixed sample sizes of 25 (——), 50 (···) and 100 (- – -), used to classify the proportion of infested plants with respect to a critical proportion of 0.13. The OC functions were estimated using a beta-binomial distribution with $R = 5$ and $\rho = 0.1$.

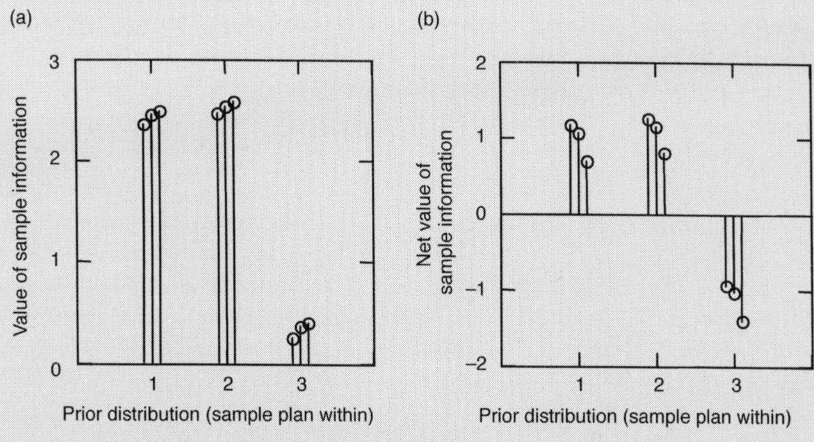

Fig. 6.5. The value (a) and net value (b) in US$ per acre of sample information used to classify the proportion of infested maize plants with respect to a critical proportion of 0.13. Prior distribution 1 is the uniform model, distribution 2 is the low-infestation model and distribution 3 is the high-infestation model. The sampling plans consisted of 25, 50 or 100 samples and are nested within the prior distributions.

Continued

For the high-incidence prior distribution, the value of the sample information is increased by reducing pesticide applications, because without sampling, the best strategy is always to treat. However, with the high-incidence prior, densities less than the critical proportion occur infrequently, so there is little chance to realize savings by not applying a pesticide. For the uniform and low-incidence prior distributions, the value of the sample information is about the same. This similarity occurs by chance alone. With the uniform prior, the value of the sample information accrues through savings in pesticide applications, whereas with the low-incidence prior, the value in the sample information accrues because sampling prevents loss when there are high rates of incidence. For each prior distribution the value of the sample information increased (slightly) with increasing sample size, but the net value of the sample information decreased with increasing sample size, signalling that the increased precision obtained when more samples were taken was not economically justified (Fig. 6.5b). The net value of the sample information was positive for the uniform and low-incidence priors, but was always negative for the high-incidence prior.

An obvious question is whether sequential sampling can be used to improve the net value of the sample information. To study this, we constructed a sequential sampling plan using Iwao stop boundaries with cp = 0.13, α = 0.1, minimum sample size equal to 25, maximum sample size equal to 60, cluster size R = 5 and intra-cluster correlation coefficient ρ = 0.1. The maximum sample size was chosen to make the maximum ASN value approximately 50. The value of the sample information obtained using this plan was compared to those for fixed sample sizes of 25 and 50. OC and ASN functions for the three sampling plans are shown in Fig. 6.6. The original three prior distributions were used in the calculations and the resulting value and net value of the sample information for each prior and sampling plan are shown in Fig. 6.7. These graphs show that the sequential plan has both the greatest

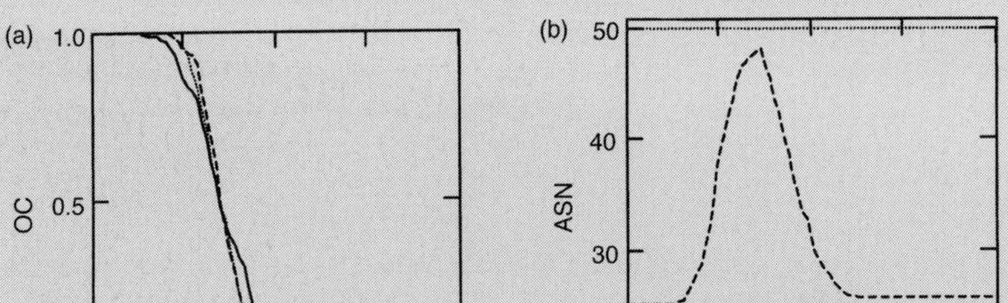

Fig. 6.6. OC (a) and ASN (b) functions for a fixed sample size plan (n = 25[——], n = 50 [···]), and an Iwao sequential plan (- – -) used to classify the proportion of infested plants with respect to a critical proportion of 0.13. The OC functions were estimated using a beta-binomial distribution with R = 5 and ρ = 0.1. For the sequential plan, α = 0.1, the minimum sample size was 25 and the maximum sample size was 60.

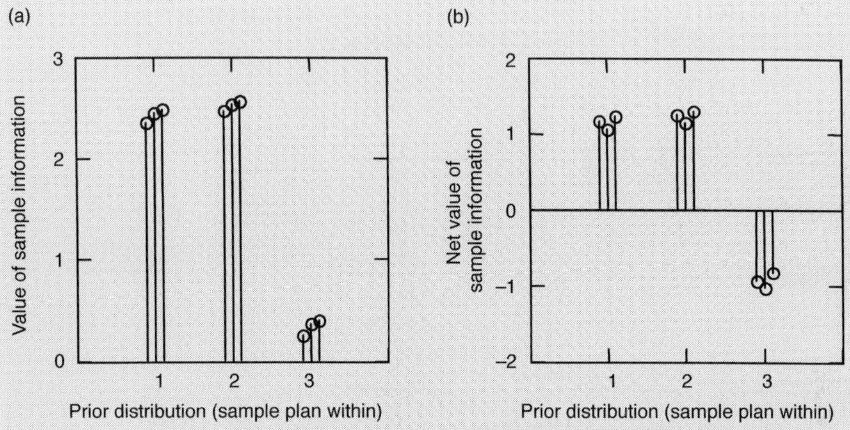

Fig. 6.7. The value (a) and net value (b) in US$ per acre of sample information used to classify the proportion of infested maize plants with respect to a critical proportion of 0.13. Prior distribution 1 is the uniform model, distribution 2 is the low-infestation model and distribution 3 is the high-infestation model. The sampling plans consisted of 25 (1) or 50 (2) samples, or a sequential stop boundary (3), as for Fig. 6.6.

value and the greatest net value of all plans, although differences among sampling plans are small.

Shortening the range of the prior distributions It is already apparent that the prior distribution strongly influences the value of the sample information. We have demonstrated this by using priors with different shapes. Does the possible range of pest incidences have an equally large effect? The data that we have used to estimate the prior distributions suggest that very low (< 0.05) and very high (> 0.3) incidence rates are uncommon. Therefore, we made calculations using prior distributions incorporating these restrictions and using the sequential plan described above. The adjusted prior distributions restrict incidences to between 0.05 and 0.3 (Fig. 6.8). Restricting the range of pest abundance reduced the value and net value of the sample information (Table 6.1).

At the original limits of pest abundance, the sample information almost certainly leads to a correct decision, and this decision contributes to the value of the sample information. However, this contribution is lessened when these extreme incidences are eliminated. If the best strategy without sampling is always to treat (uniform and high-incidence priors), then at the lower limit of pest abundance the sample information leads to reduced costs by eliminating unnecessary pesticide applications. With the reduced lower range of possible pest incidences, the magnitudes of these cost reductions are lessened. On the other hand, if the best pest control strategy without sampling is never to treat (low-incidence prior), at the upper limit of pest incidence the sample information leads to reduced costs by calling for a pesticide application and thereby reducing pest damage. With the reduced upper range of pest abundance, the magnitude of these savings is also reduced. Note that

Continued

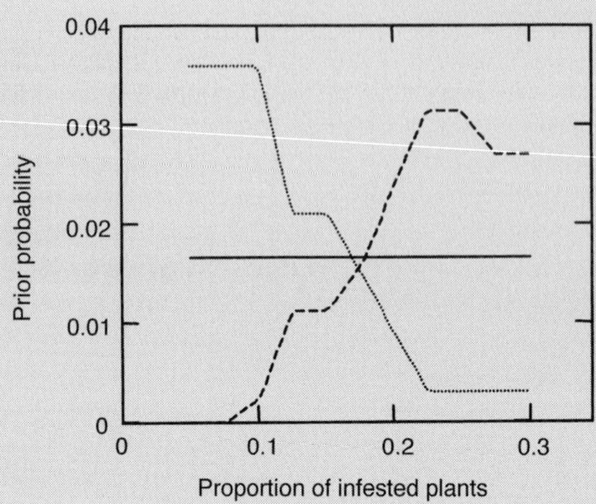

Fig. 6.8. Prior distributions with restricted ranges used to model the likelihood of the proportion of maize plants infested by lepidoptera larvae; uniform infestation (——), low infestation (···) and high infestation (- – -).

Table 6.1. The influence of the range of prior distributions[a] on the value of sample information for lepidoptera larvae infesting sweetcorn. Sampling: Iwao sequential classification with $cp = 0.13$, $\alpha = 0.1$, minimum sample size = 25, maximum sample size = 60. Simulation: beta-binomial distribution, R (cluster size) = 5 and $\rho = 0.1$.

	Uniform infestation		Low infestation		High infestation	
	Full range[b]	Restricted range[c]	Full range	Restricted range	Full range	Restricted range
Best action without sampling[d]	Always treat	Always treat	Never treat	Never treat	Always treat	Always treat
Value of sampling[e]	2.47	1.47	2.57	1.77	0.38	0.01
Net value of sampling[f]	1.23	0.20	1.32	0.50	−0.86	−1.23
Change in treatment frequency[g]	−0.34	−0.37	0.25	0.32	−0.12	−0.10

[a] Refers to prior distributions in Fig. 6.2.
[b] 0.0–0.4.
[c] 0.05–0.3.
[d] Equations 6.1–6.3.
[e] Equation 6.5.
[f] Equation 6.7.
[g] Equation 6.9. A negative value represents a reduction in pesticide applications and a positive value an increase.

Table 6.2. The influence of the slope of the pest–loss model on the value of sample information for lepidoptera infesting sweetcorn. Sampling: Iwao sequential classification with $cp = 0.13$, $\alpha = 0.1$, minimum sample size = 25, maximum sample size = 60. Simulation: beta-binomial distribution, R (cluster size) = 5 and $\rho = 0.1$.

	Prior distribution[a]								
	Uniform infestation			Low infestation			High infestation		
	Slope of pest–loss model[b]								
	200	150.5	100	200	150.5	100	200	150.5	100
Critical proportion[c]	0.1	0.13	0.2	0.1	0.13	0.2	0.1	0.13	0.2
Best action without sampling[d]	AT	AT	AT	NT	NT	NT	AT	AT	AT
Value of sampling[e]	1.87	2.49	3.73	5.00	2.57	0.88	0.16	0.40	1.15
Net value of sampling[f]	0.63	1.24	2.48	3.78	1.32	−0.35	−1.07	−0.84	−0.13
Change in treatment frequency[g]	−0.26	−0.34	−0.50	0.38	0.25	0.125	−0.06	−0.12	−0.30

[a] Refers to prior distributions in Fig. 6.2.
[b] Refers to the slope of the solid line in Fig. 6.2.
[c] Equal to the *EIL*.
[d] AT = always treat, NT = never treat. Equations 6.1–6.3.
[e] Equation 6.5.
[f] Equation 6.7.
[g] Equation 6.9. A negative value represents a reduction in pesticide applications and a positive value an increase.

in both cases the change in frequency of treatment is not greatly influenced by the reduced range in the prior distributions. This is because both very low and very high incidence rates are eliminated from the priors. These are the reasons why, in this example, the range of pest abundance used in the prior distributions plays a relatively large role in determining the value of sample information.

Different pest–loss models The third question posed was what is the effect of different pest–loss models on the value of the sample information? This is an important question, because pest–loss relationships are almost never known precisely. To study this, we set up three loss functions with slopes equal to 100, 150.5 (the calculated value) and 200, and recalculated the *EIL* for each (0.2, 0.13 and 0.1, respectively). Using the original prior distributions and sequential sampling (Iwao, as before) with critical proportions equal to the *EIL*s, we obtained the results shown in Table 6.2.

Continued

The best action to take when no sample information is collected did not change with the different loss models, although the expected costs for the two strategies (always treat, never treat) did change (data not shown). For the uniform and high-incidence prior distributions, the lower slope (100) for the pest–loss model (increased economic injury level, 0.2) resulted in an increased value of the sample information, while the converse was true for the higher slope. This occurred because with the increased economic injury level (and critical proportion for the sampling plan), there was a greater likelihood that densities less than the critical proportion would occur and, concordantly, an increased savings in pesticide applications. These reductions in control treatments, which are also quantified as change in frequency of treatment applications (Table 6.2), led to an increased value of the sample information. For the low-incidence prior, the opposite pattern occurred: an increased economic injury level led to a decrease in the value of the sample information. This occurred because, for this prior distribution, the value of the sample information was based on making pesticide applications when needed, and the best action to take when no sample information was collected was never to treat. However, with a higher economic injury level, treatment becomes less necessary, so there is reduced value in the sample information. This is again reflected in the change in frequency of treatment applications.

Summary Overall, sampling to determine the need for control is probably economically justified when the uniform or low-incidence prior, or any prior that spans the range between these two, applies. In contrast, sampling is probably not economically justified when the high-incidence prior is applicable. This means that when a corn field has not been treated before the tassel stage, it should probably always be treated in the silk stage. These recommendations are valid for data following the pattern of incidence as reported by Riggs *et al.* (1998). Under different conditions, the recommendations could change.

The bottom line of the calculations in this exhibit is that, when the value of sampling is to be evaluated, one needs to have information about the pest densities that may occur. Without an insight into this, it is not possible to make an evaluation. Given the large number of sampling plans and decision guides that have been developed over the past 30 years, the number of studies that have looked at the value of sample information is very small. This is unfortunate, because important insights might be gained from doing these analyses, both in successful and in unsuccessful systems. We can suggest reasons why there are so few studies: the required information, such as crop loss and expected pest abundance, is not easy to obtain; the value of sample information is likely to turn out to be small, and there is an assumption that such a result is difficult to publish; studies at higher integration levels are less attractive, because the publication climate favours reductionist rather than integrative studies.

Although the relationships that are required to do the analysis may not be well defined, the ability to do the computations quickly on a computer allows a quick sensitivity analysis of the major assumptions made for decision sampling. These assumptions are made regardless of whether decision theory analysis (as outlined here) is conducted. It therefore seems worthwhile to study the possible consequences of such assumptions, given the possible economic and ecological consequences.

6.6 Summary

Thus far in this book, we have emphasized the quantitative evaluation of sampling plans used for pest management decision-making and, especially, using the OC and ASN functions for this purpose. These quantitative evaluations are essential, and in this chapter we have extended these evaluations to include the economic value of the sample information. Qualitative evaluation of the sampling plans is equally important. Practicality, simplicity, reliability, representativeness and relevance are key considerations when developing and evaluating a sampling plan. In the developmental process, it is important that users and developers engage in a collaborative effort. While it will often be the case that insufficient information is available to estimate the value of sample information with great confidence, these valuations are still useful, because they force assumptions to be explicitly stated.

References and Suggested Reading

Baillod, M., Antonin, Ph., Guignard, E. and Jermini, M. (1989) Vers une généralisation de la lutte biologique contre les acariens phytophages en verger de pommiers. *Revue Suisse de Viticulture, Arboriculture et Horticulture* 21, 279–284.

Breth, D. and Nyrop, J.P. (1998) *A Guide for Integrated Mite Control in Apples in the Northeast.* Cornell IPM Publ. No. 215.

Brown, G.C. (1997) Simple models of natural enemy action and economic thresholds. *American Entomologist* 43, 117–124.

Croft, B.A. and Nelson, E.E. (1972) An index to predict efficient interaction of *Typhlodromus occidentalis* in control of *Tetranychus macdanielli* in Southern California apple trees. *Journal of Economic Entomology* 64, 310–312.

Dalthorp, D., Nyrop, J.P. and Villani, M.J. (1999) Estimation of local mean population densities of Japanese beetle grubs (Scarabeiidae: Coleoptera). *Environmental Entomology* 28, 255–265.

Elliott, N.C., Hein, G.L. and Shepard, B.M. (1994) Sampling arthropod pests of wheat and rice. In: Pedigo, L.P. and Buntin, G.D. (eds) *Handbook of Sampling Methods for Arthropods in Agriculture*. CRC Press, Boca Raton, Florida, pp. 627–666.

Green, R.H. (1970) On fixed precision level sequential sampling. *Researches in Population Ecology* 12, 249–251.

Hoffman, M.P., Nyrop, J.P., Kirkwyland, J.J., Riggs, D.M., Gilrein, D.O. and Moyer, D.D. (1996) Development and validation of a sequential sampling plan for lepidopterous pests of fresh market sweet corn in New York. *Journal of Economic Entomology* 89, 386–395.

Jansma, J.E., van Keulen, H. and Zadoks, J.C. (1993) Crop protection in the year 2000: a comparison of current policies toward agrochemical usage in four West European countries. *Crop Protection* 12, 483–489.

Jones, V.P. (1990) Developing sampling plans for spider mites (Acari: Tetranychidae): those who don't remember the past may have to repeat it. *Journal of Economic Entomology* 83, 1656–1664.

Leeuwis, C. (1993) *Of Computers, Myths and Modelling. The Social Construction of Diversity, Knowledge, Information and Communication Technologies in Dutch Horticulture and Agricultural Extension.* Wageningen Studies in Sociology 36. Pudoc, Wageningen, 468 pp.

Legg, D.E., Shufran, K.A. and Yeargan, K.V. (1985) Evaluation of two sampling methods for their influence on the population statistics of Alfalfa Weevil (Coleoptera: Curculinidae) larva infestation in alfalfa. *Journal of Economic Entomology* 78, 1468–1474.

Ministry of Agriculture, Nature Conservation and Fisheries (1991) *Meerjarenplan Gewasbescherming Regeringsbeslissing* [*Multi-year Crop Protection Plan, Governmental Decision*]. Sdu Publishers, The Hague, 298 pp.

Nyrop, J.P. (1988) Sequential classification of predator-prey ratios with application to European red mite (Acari: Tetranychidae) and *Typhlodromus pyri* in New York apple orchards. *Journal of Economic Entomology* 80, 14–21.

Nyrop, J.P. and Binns, M.R. (1991) Quantitative methods for designing and analyzing sampling programs for use in pest management. In: Pimental, D. (ed.) *Handbook of Pest Management in Agriculture*, 2nd edn, Volume II. CRC Press, Boca Raton, Florida, pp. 67–132.

Nyrop, J.P. and van der Werf, W. (1994) Sampling to predict or monitor biological control. In: Pedigo, L.P. and Buntin, G.D. (eds) *Handbook of Sampling Methods for Arthropods in Agriculture*. CRC Press, Boca Raton, Florida, pp. 245–336.

Nyrop, J.P., Binns, M.R. and van der Werf, W. (1999) Sampling for IPM decision making: Where should we invest time and resources? *Phytopathology* 89, 1104–1111.

Nyrop, J.P., Foster R.E. and Onstad, D.W. (1986) The value of sample information in pest control decision making. *Journal of Economic Entomology* 79, 1421–1429.

Nyrop, J.P., Shelton, A.M. and Theunissen, J. (1989) Value of a control decision rule for leek moth infestations in leek. *Entomologia Experimentalis et Applicata* 53, 167–176.

Pedigo, L.P. (1995) Closing the gap between IPM theory and practice. *Journal of Agricultural Entomology* 12, 171–181.

Riggs, D.I.M., Hoffmann, M.P. and Thetford, L.C. (1998) Treatment thresholds for European corn borer (*Lepidoptera: Pyralidae*) infestations in early-season fresh market sweet corn. *Journal of Entomological Science* 33, 393–399.

Rossing, W.A.H., Jansma, J.E., de Ruijter, F.J. and Schans, J. (1997) Operationalizing sustainability: exploring options for environmentally friendly flower bulb production systems. *European Journal of Plant Pathology* 103, 217–234.

Schmaedick, M.A. and Nyrop, J.P. (1995) Methods for sampling arthropod pests with uncertain phenology with application to spotted tentiform leafminer (Lepidoptera: Gracillariiidae). *Journal of Economic Entomology* 88, 875–889.

Southwood, T.R.E. (1978) *Ecological Methods*, 2nd edn. Chapman & Hall, London, 524 pp.

van den Ende, J.E., Pennock-Vos, M.G., Bastiaansen, C., Koster, A.Th.J. and van der Meer, L.J. (2000) BoWaS; a weather based warning system for the control of Botrytis blight in lily. In: Challa, H. and Monteiro, A.A. (eds) *Proceedings of the XXVth International Horticultural Congress. Computers and Automation. Electronic Information in Horticulture.* Acta Horticulturae 519, pp. 215–220.

Van der Ploeg, J.D. (1990) *Labor, Markets, and Agricultural Production.* Westview Press, Boulder, Colorado, 313 pp.

7 Binomial Counts

7.1 Introduction

In Chapter 4, we noted that sometimes, with disease assessment, the only information that could readily be obtained from a sample unit was whether the unit did or did not display disease symptoms. We recognized that trying to get more information from the sample unit, such as the severity of the infestation, might not be worth the effort. Under these circumstances, it is natural to use a critical proportion (*cp*) as the criterion for decision-making, instead of a critical density (*cd*), and to use the binomial (or beta-binomial) distribution for setting up sample plans, as in Chapter 5.

An infestation by whatever kind of pest can be recognized in binomial terms. A sample unit can be regarded as 'infested' if it contains any pests at all, and the severity of the infestation (i.e. how many individual pests are present) can be ignored. The only sample information collected is the mean proportion of infested sample units. If the critical density, estimated following the principles noted in Chapter 1, can be related to a critical proportion, then such a 'presence–absence' binomial sample plan can be used as a decision guide. Because it takes more effort to count all the pests on a sample unit than to note merely whether any pests are present, a binomial count plan incurs fewer sample costs than does a traditional full count plan. However, less sample work equals less sample information. Before recommending a binomial sample plan, it is necessary to discover whether the information loss is more than balanced by the savings in sample costs.

In this chapter, we show how to construct presence–absence binomial count sample plans and other more general binomial plans as replacements for full count plans. Intuition suggests that some information is lost. We show how to assess the loss of information, and how to ensure that the loss is not critical to pest management decision-making. We present types of sample plans which have been found useful and cost-effective in practice.

7.2 What is Binomial Sampling?

The basic principles of field sampling remain those noted in Section 2.5: representativeness (no unaccountable bias), reliability (no effect of uncontrolled variables), relevance (relationship with crop loss) and practicality. Suppose that we have developed a sample plan in which all individual pests are counted on each sample unit (what we refer to as a *full count sample plan*) and that we are satisfied that it follows the first three of these basic principles, but the fourth presents a problem. We realize that practitioners are reluctant to use the plan, because the pests are difficult or tedious to count. Potential users of the full-count sample plan have decided that the value of the information obtained from this sampling procedure is not worth the time expenditure. Binomial methods can sometimes be used to make sampling less tedious and less costly.

The foundation of a binomial sample plan is the mathematical link between counts and proportions. This link is essential, because it is used to translate sample information on the proportion of infested sample units into the pest density per sample unit, the criterion for decision-making. Two types of link exist: one where a theoretical probability distribution (Chapter 4) can be assumed, and one where no theoretical distribution can be used to describe sample counts. Both types of link are described in this chapter, beginning with the link based on a theoretical probability distribution.

7.3 Presence–Absence Binomial Sampling with the Poisson Distribution

With the Poisson distribution (Section 4.3), the probability of finding at least one individual on a sample unit when the mean density is μ can be written as

$$p = \text{Prob}(x > 0 \mid \mu) = 1 - p(0 \mid \mu) = 1 - e^{-\mu} \tag{7.1}$$

If, out of n sample units, r contain at least one pest, then the ratio, r/n, is an estimate of p in the above equation. Rearranging the equation provides an estimate m_{bin} of μ itself:

$$\begin{aligned} r/n &= 1 - e^{-m_{bin}} \\ m_{bin} &= -\ln\left(\frac{n-r}{n}\right) \end{aligned} \tag{7.2}$$

What this means is that, without actually counting how many individuals there are in each sample unit, we still have an estimate of the mean density per sample unit. The potential savings in sampling time can be very great, depending on how hard it is to count the pests or assess pest severity on a sample unit. However, these savings come with two penalties: an increase in variance and the introduction of bias.

The variance of m_{bin}, var(m_{bin}), can be shown to be approximately (see the Appendix)

$$\mathrm{var}(m_{bin}) = \frac{1}{n}\frac{p}{1-p} = \frac{e^{\mu}-1}{n} \tag{7.3}$$

which is large when μ is large, or p is small, $\mathrm{var}(m_{bin})$ is always greater than the variance for a full count sample from a Poisson distribution (μ/n). This can be shown by using the series expansion of e^{μ}:

$$e^{\mu} = 1 + \frac{\mu}{1!} + \frac{\mu^2}{2!} + \frac{\mu^3}{3!} + \ldots \tag{7.4}$$

We have already noticed that, with full count sampling, the higher the variance, the flatter is the operating characteristic (OC) curve (see, e.g. Exhibit 2.2). The increased variance that occurs with binomial sampling has the same effect: the OC function of a binomial sample plan is flatter than the OC function of the corresponding full count plan.

As an estimate of μ, the mean density per sample unit, m_{bin}, is not only comparatively imprecise, but it is also biased. This bias is a result of the underlying mathematics of the curvilinear relationship between the proportion of infested sample units and the mean density, and is not due to the way in which sample units are collected. It can be shown that the bias in m_{bin} is positive and approximately equal to (see the Appendix)

$$\mathrm{bias}(m_{bin}) \doteq \frac{1}{2n}\frac{p}{1-p} = \frac{e^{\mu}-1}{2n} \tag{7.5}$$

which, like the variance, is large when μ is large or p is small. Like the variance, the bias is reduced if sample size, n, is increased. Explicit removal of the bias is possible, but turns out to be unnecessary for classification sampling, given the tools at our disposal (see below).

In presence–absence binomial sampling, we use an estimate of Prob($x > 0$) to give us information on the mean, μ. If μ is large, most of the distribution is far from 0 and Prob($x > 0$) is close to 1. There is less scope for precision when the only reasonable choice for r/n is 1, or possibly $(n - 1)/n$. It is not surprising that there are problems when μ is large (or when p is small). We will discuss this later, in connection with what we call 'tally numbers'. Note also that a sample proportion equal to 0 or 1 can raise arithmetic problems. For example, if $r = n$ in Equation (7.2), the estimate of μ does not exist. In practice, this is avoided by adjusting such data:

Sample proportion	r	Adjusted r
0	0	δ
1	n	$n - \delta$

where δ is a small number between 0 and 1. Naturally, the estimate depends on the value of δ. However, because we are interested in classification rather than estimation, we do not need to worry about the exact value of δ, provided that small changes in its value do not change the classification.

The following table summarizes how to set up a presence–absence binomial plan and simulate its properties:

	Full count plan	Poisson presence–absence plan
Criterion	cd	$cp = 1 - e^{-cd}$
Range of means for simulation	μ_i	$p_i = 1 - e^{-\mu_i}$
Distribution for simulation	Poisson	Binomial

Setting up a sample plan based on the second column above was discussed in Chapter 5. Setting up a binomial plan with a given value of cp was also discussed in Chapter 5. The extra feature here is that cp and the range of values p_i depend on cd and μ_i. As far as the mechanics of setting up a presence–absence sample plan are concerned, the only novelty is this relationship between probabilities and means.

7.4 Tally Numbers Other than Zero

When the decision criterion is the level of infestation, or incidence, of a pest, the only information required from a sample unit is whether or not it is infested. However, when the decision criterion is a pest density, but we want to take advantage of the simplicity of binomial sampling, we are not tied to a 'presence–absence' definition of incidence. Other definitions depend on what we call a tally number, T. A sample unit is defined as infested if the number of organisms in the sample unit exceeds T. The presence–absence definition corresponds to $T = 0$.

There are two reasons why we may wish to increase T: variance and bias. We noted that the variance and bias of the Poisson presence–absence sample estimate are large when μ is large. We suggested that this had something to do with an estimate of p giving little information about μ if μ is large. It is obvious from Equations 7.3 and 7.5 that increasing the sample size would reduce both variance and bias, but it turns out that moving T closer to μ would also reduce them both. Whether variance or bias is a problem and, if so, which of n and T should be adjusted, must be assessed on a practical basis by cost comparisons of OC and average sample number (ASN) functions.

When $T > 0$, the equation corresponding to Equation 7.1 is

$$p = \text{Prob}(x > T \mid \mu) = 1 - \left(\text{p}(0 \mid \mu) + \text{p}(1 \mid \mu) + \ldots + \text{p}(T \mid \mu)\right)$$
$$= 1 - e^{-\mu} \sum_{j=0}^{T} \frac{\mu^j}{j!} \tag{7.6}$$

Setting up sample plans for $T = 0$ and $T > 0$ is similar:

	Full count plan	Poisson binomial plan with T
Criterion	cd	$cp_T = 1 - e^{-cd} \sum_{j=0}^{T} \frac{cd^j}{j!}$
Range of means for simulation	μ_i	$p_i = 1 - e^{-\mu_i} \sum_{j=0}^{T} \frac{\mu_i^j}{j!}$
Distribution for simulation	Poisson	Binomial

Note that even though *cd* does not change, a change in T forces a change in *cp*. Because of this, we put a suffix on *cp* to avoid confusion, to give cp_T.

Unfortunately, it is now impossible to give a direct formula corresponding to Equation 7.2 for the binomial estimate of μ, but the calculation is easy with a computer. However, because we are interested in classifying density rather than estimating it, this is not of particular concern. All that is required is that cp_T be calculated and the rest is as for $T = 0$: the decision made in the field is based on comparing the sample proportion with cp_T. The concepts that we have presented thus far are illustrated in Exhibit 7.1.

Exhibit 7.1. *Binomial classification sampling with the Poisson distribution*

In this example, we demonstrate how the tally number, *T*, and sample size, *n*, influence the OC functions for binomial count sample plans that classify density with respect to a critical density, *cd*. The example is based on sampling the potato leafhopper, *Empoasca fabae*, which is considered to be one of the most consistently damaging pests of lucerne in north-central and northeastern United States. The potato leafhopper feeds by inserting its mouthparts into cells of lucerne stems. This results in the clogging of phloem tissue, which may lead to reduced plant growth and plant quality. Leafhopper density is assessed by sweeping a net through the plants and counting the number of hoppers. Various numbers of sweeps of a net have been proposed as sample units; an acceptable one is a ten-sweep unit (Shields and Specker, 1989). Counts of leafhoppers using a ten-sweep sample unit can be described using a Poisson distribution (Shields and Specker, 1989). Critical densities for potato leafhopper vary depending on the height of the plants. For plants greater than 30 cm in height, a proposed *cd* is two hoppers per sweep, or *cd* = 20 hoppers per 10 sweeps (Cuperus *et al.*, 1983).

It might be tedious and even difficult to count all the leafhoppers in a net following 10 sweeps, because the net can harbour insects other than the leafhopper along with plant material. Because the leafhoppers can fly, there is the additional danger of missing some of the leafhoppers captured in the net. For these reasons, binomial sampling is an attractive option. Because of the relatively high *cd*, using *T* = 0 is not practical: with a true mean of 10, the proportion of sweep samples infested with leafhoppers would be nearly 1. Therefore, development of a binomial count sample plan was begun using *T* = 10.

Continued

Initially, three fixed sample size plans with $cd = 20$ were studied. The first plan used complete counts with $n = 30$, the second used binomial counts ($T = 10$) with $n = 30$ and the third used binomial counts ($T = 10$) with $n = 60$. The relationship between p and μ, and the OC functions, are shown in Fig. 7.1. The p–μ relationship is clearly not ideal, because p is close to one when $\mu = cd$. Due to the positive bias (estimates are greater than they 'should' be), the OC function with $n = 30$ is noticeably shifted to the left – that is, to lower densities – in comparison to the chosen critical density of 20 hoppers per ten sweeps. However, classification using the binomial count plan with $n = 60$ is less biased than when $n = 30$. It is difficult to see much effect of the increased sample size on the steepness of the OC function.

Increasing T beyond 10 towards cd should decrease the bias and increase the precision of the classification. The magnitude of these effects was studied by formulating three further sample plans, each with $n = 30$: complete counts, binomial with $T = 10$ and binomial with $T = 20$. The p–μ relationships and the OC functions are shown in Fig. 7.2. With $T = 20$ the OC functions for the binomial and complete count sample plans are nearly identical. Note that p is close to 0.5 for $\mu = 20$ when $T = 20$. Even though such a high tally number probably requires considerably more work than $T = 10$ or $T = 0$, using $T = 20$ may still be worthwhile, because it should be easy to determine if the number of leafhoppers in the net greatly exceeds 20 or is much less than 20, and more effort might have to be made only when hopper numbers in the net were in the range 10–30.

Sequential methods can be used with binomial sampling. Sequential binomial sampling was studied by constructing three SPRT plans: full counts based on the Poisson distribution (α and β both equal to 0.1), binomial with $T = 20$ (α and β both equal to 0.1) and binomial with $T = 20$ (α and β both equal to 0.05). The OC and ASN functions for these plans are shown in Fig. 7.3. Both binomial plans required more sample units than the full count plan, and (as shown in Chapter 5) the use of

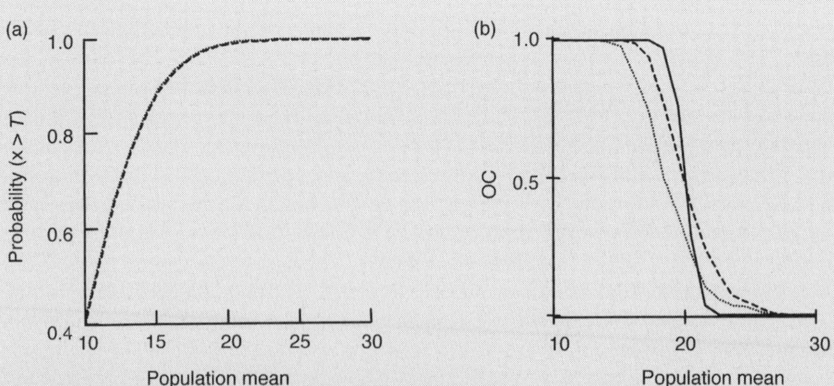

Fig. 7.1. The relationship between p and μ for the Poisson distribution (a) when $T = 10$, and OC functions (b) for three sampling plans that classify the density about $cd = 20$; complete enumeration and $n = 30$ (—), binomial counts with $T = 10$ and $n = 30$ (···), and binomial counts with $T = 10$ and $n = 60$ (- – -).

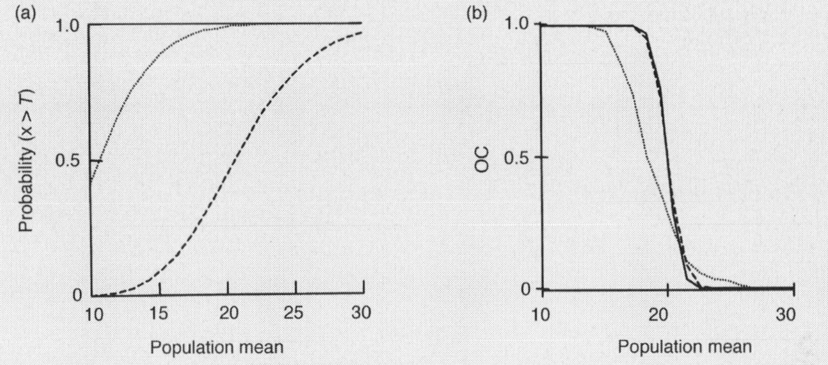

Fig. 7.2. The relationship between p and μ for the Poisson distribution (a) when $T = 20$ (- – -) and $T = 10$ (···), and OC functions (b) for three sampling plans with $n = 30$ that classify the density about $cd = 20$; complete enumeration (—), binomial counts with $T = 10$ (···), and binomial counts with $T = 20$ (- – -).

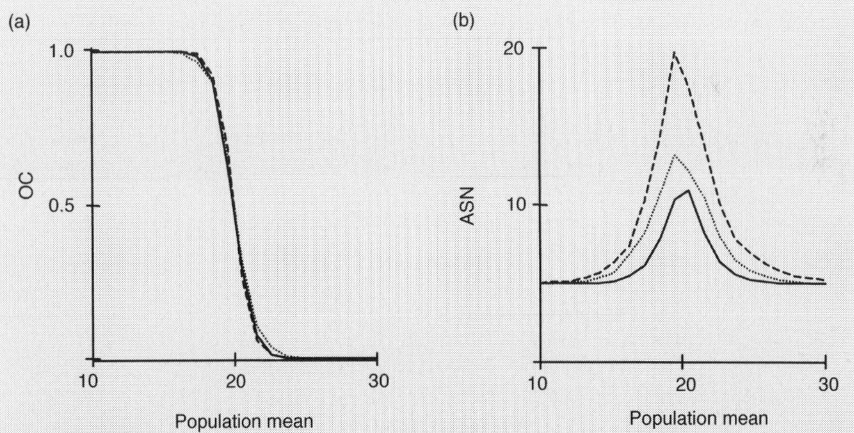

Fig. 7.3. OC (a) and ASN functions (b) for three sequential sampling plans used to classify the density with respect to $cd = 20$. Stop boundaries constructed using the SPRT. Parameters for each plan are as follows: full count, Poisson distribution, $\mu_1 = 22$, $\mu_0 = 18$, $\alpha = \beta = 0.1$, $minn = 5$, $maxn = 50$ (—); all parameters the same except binomial counts used with $T = 20$ (···); all parameters the same except binomial counts used with $T = 20$ and $\alpha = \beta = 0.05$ (- – -).

lower values of α and β results in a higher ASN function. However, these increases in sampling effort had almost no effect on the OC function: there was little difference among all three OC functions.

Continued

As with a fixed sample size plan, T affects the ASN. It does this by determining the critical proportion cp_T and the upper and lower critical proportions used to construct the stop boundaries. These patterns are illustrated in Fig. 7.4, which presents OC and ASN functions for three binomial count sample plans. The plans differ only in T ($T = 20$, 15 and 12). The OC functions for all three plans are approximately the same, but with $T = 12$, the ASN is greatly increased. This is because, on the binomial scale, the upper and lower proportions used to construct the stop boundaries are determined by T. With $T = 20$ they are 0.61 and 0.27, with $T = 15$ they are 0.92 and 0.71, and with $T = 12$ they are 0.98 and 0.91. When sampling with $T = 12$, one has to decide between two very similar proportions (0.98 and 0.91), which naturally requires many sample units. Because of the high ASN, the plan with $T = 12$ seems the least attractive. Of the other two plans, $T = 15$ is preferable to $T = 20$, for the practical reason that $T = 20$ may be so tedious that enumerative bias could creep in and nullify any theoretical advantage that it might have.

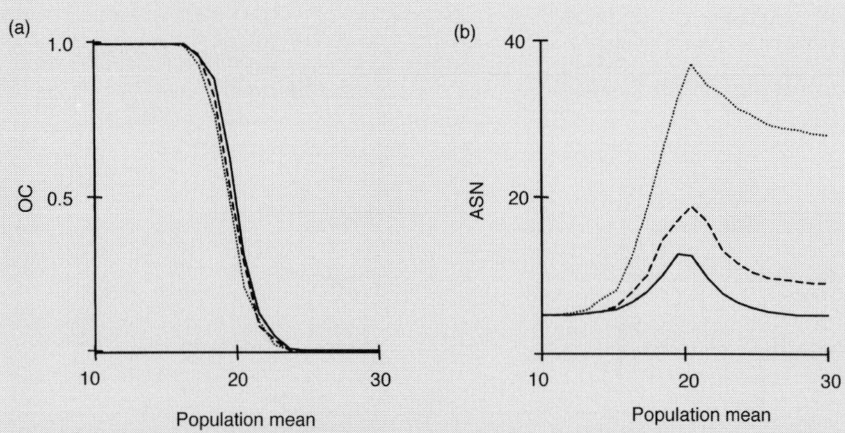

Fig. 7.4. OC functions (a) and ASN functions (b) for three binomial count sequential sampling plans used to classify the density with respect to $cd = 20$. Stop boundaries constructed using a Poisson SPRT. Parameters for each plan are as follows: $T = 20$, $\mu_1 = 22$, $\mu_0 = 18$, $\alpha = \beta = 0.1$, $minn = 5$, $maxn = 50$ (——); all parameters the same except $T = 12$ (···); all parameters the same except $T = 15$ (- – -).

7.4.1 The tally number and precision

We have shown how the tally number, T, influences the OC function for a binomial count sample plan, especially for large *cd*. Now we provide further insight into why this occurs. When $T = 0$, we use the sample data to estimate $p = \text{Prob}(x > 0)$. This estimate is then used to obtain an estimate of the mean, μ, which determines the management recommendation. If μ is very large and p is close to 1, all or almost all of the sample units will contain at least one individual, so the sample proportion will usually be equal to 1, and only occasionally equal to $1 - 1/n$, or possibly $1 - 2/n$. This, and its effect on the estimate of μ, is illustrated in Fig. 7.5. With only 25 sample units, sample proportions less than 24/25 are unlikely (Fig. 7.5a), so, in practice, there are only two possible 'estimates' of μ, and both of these are of inferior quality (Fig. 7.5b). (Note that, as described above, a sample proportion equal to 1 is adjusted before obtaining the estimate of μ; we used $\delta = 0.0005$: values as large as 0.02 gave the same type of unacceptable result, but with all estimates below the true value.) What this means is that presence–absence sampling may provide only scanty information about μ. If we use a larger T, the problem can be reduced considerably. For example, if $T = 4$, many more distinct sample proportions are likely to occur (Fig. 7.5c) and the distribution of the sample estimate of μ (Fig. 7.5d) shows that the estimate is more useful for discriminating among values close to μ.

Such explanations are part of the reason why binomial estimates can be biased, and can have large variances. The biases and variances for these examples, and

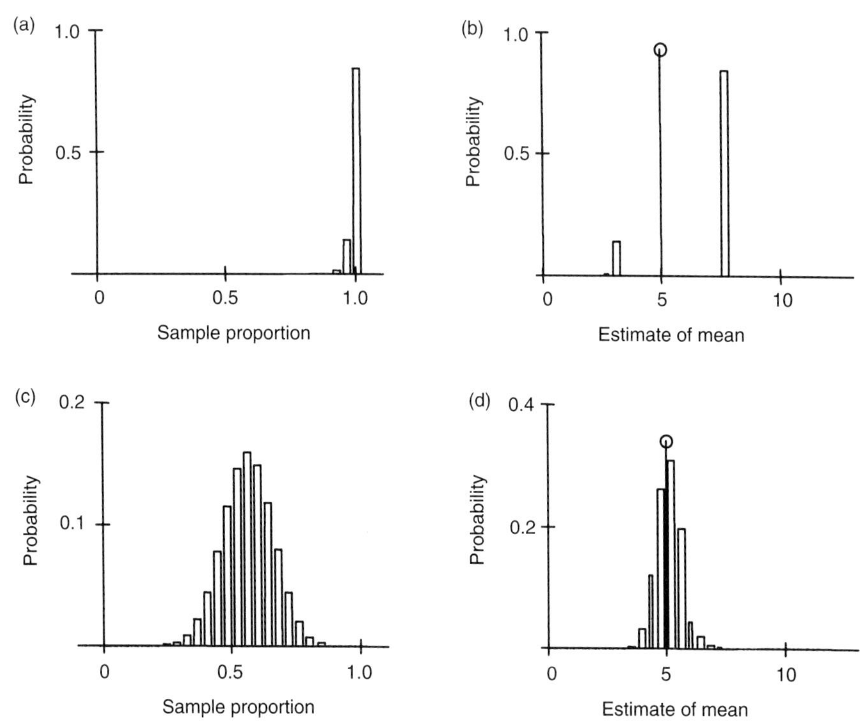

Fig. 7.5. The effect of T on m_{bin}, the estimate of μ ($\mu = 5$, $n = 25$). (a) Distribution of sample proportion with $T = 0$; (b) distribution of m_{bin} with $T = 0$ (the tethered balloon represents the true mean, μ); (c) distribution of sample proportion with $T = 4$; (d) distribution of m_{bin} with $T = 4$ (the tethered balloon represents the true mean, μ).

Table 7.1 The effect of the tally number, *T*, on the bias and variance of the sample mean for binomial count sampling from the Poisson distribution with mean equal to 5. Results are for sample sizes $n = 25$ and $n = 50$. Note that the slightly anomalous values for $T = 0$, especially for $n = 25$, arise because of the adjustment that has to be made when the sample proportion is equal to 1 or 0; in this table, 1 is replaced by $1 - 0.1/n$, and 0 is replaced by $0.1/n$.

	T					
	0	2	4	5	6	8
$n = 25$						
Bias	0.16	0.28	0.03	0.00	−0.04	−0.24
Variance of sample mean	0.74	1.28	0.35	0.33	0.38	0.90
$n = 50$						
Bias	0.52	0.11	0.02	0.00	−0.02	−0.12
Variance of sample mean	1.22	0.42	0.17	0.16	0.18	0.42

others, were calculated from the distributions of the estimates of μ and are presented in Table 7.1. These illustrate the fact that the bias and variance decrease as T increases, until around the value of μ, after which they begin to increase. There are usually several values of T near μ for which there are no practical differences among the results: a choice can be made based on sample costs. They also illustrate the reduction in bias and variance when n is increased.

Another way of investigating the effect of T on the precision of sampling is to study the mathematical relationship between p and μ. This relationship depends on T (Fig. 7.6). In the Appendix to this chapter, we show how to derive an approximation to the variance of the sample estimate (m_{bin}) of μ based on the mathematical relationship, $\mu = f(p)$:

$$\mathrm{var}(m_{bin}) = \left(\frac{\mathrm{d}f}{\mathrm{d}p}\right)^2 \frac{p(1-p)}{n} \tag{7.7}$$

Unfortunately, the relationship $\mu = f(p)$ is often very complicated when $T > 0$, while the inverse relationship $p = g(\mu)$ (see, e.g. Equation 7.6) is simpler. We can use the fact (which can be proved by elementary calculus) that

$$\left(\frac{\mathrm{d}f}{\mathrm{d}p}\right) = 1 \bigg/ \frac{\mathrm{d}g}{\mathrm{d}\mu} \tag{7.8}$$

to obtain the approximation

$$\mathrm{var}(m_{bin}) = \frac{p(1-p)}{n} \bigg/ \left(\frac{\mathrm{d}g}{\mathrm{d}\mu}\right)^2 \tag{7.9}$$

In practice, μ is replaced by m_{bin} in $\mathrm{d}g/\mathrm{d}\mu$.

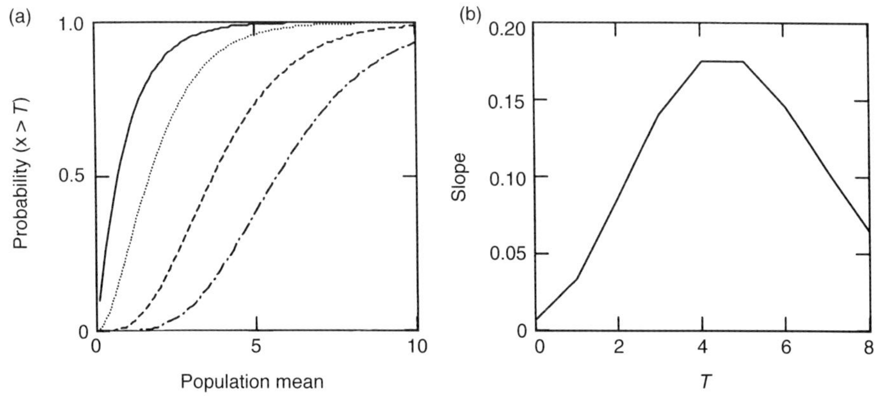

Fig. 7.6. The relationship between p and μ (mean) for the Poisson distribution (a) when $T = 0$ (——), 1 (···), 3 (- – -) and 5 (– · – · –) and the slope (b) of the p–μ model for $\mu = 5$ as a function of T.

We can make formula (Equation 7.9) a little more palatable by noting that $dg/d\mu$ is the slope of the p–μ relationship, $p = g(\mu)$. Equation 7.9 states that the variance of m_{bin} decreases as this slope increases. Because changing T changes this slope, it also changes the variance. These relationships are illustrated in Fig. 7.6a. When $\mu = 5$, the slope, $dg/d\mu$, is maximum when $T = 4$ or $T = 5$ (Fig. 7.6b). The slope is often greatest when T is near the true mean.

7.5 Binomial Sampling Based on the Negative Binomial Distribution

As we noted in Chapter 4, the Poisson distribution is rarely satisfactory for counts of arthropods or diseases, because the spatial pattern of counts is usually aggregated. One distribution that can often be used to describe sample counts is the negative binomial. Of course, there are instances in which no probability distribution is satisfactory, and we discuss later in this chapter how to deal with that. Now, we focus on the situation in which the negative binomial distribution provides a good description of the distribution of sample counts.

Unlike the Poisson distribution, the negative binomial distribution has two parameters, μ and k, and this extra parameter, k, adds further difficulties for binomial sampling. It is not practical in sampling for decision-making to estimate both μ and k, so we must choose a value of k independently of the sample data. In full count sampling, not knowing a precise value for k influences the precision of the estimate of μ but not the bias: the bias is zero whatever value is chosen for k. We observed this in Chapter 5, where the OC functions were all centred on cd.

With binomial sampling, however, incomplete knowledge of k can result in bias. This occurs because the negative binomial distribution can take many forms, depending on k. In particular, the probability of zero incidence, Prob(count = 0) can correspond to a large number of mean pest densities, μ, each of which corresponds to a different value of k. If Prob(count = 0) is moderately large, μ must be small, and does not change much for different values of k. However, if Prob(count = 0) is small, μ must be large, and its estimated value depends greatly on the assumed value for k (Table 7.2). This is presented graphically in Fig. 7.7: for Prob(count = 0) = 0.6, the possible negative binomial distributions that have the same p, but different μ and k are all close together, but for Prob(count = 0) = 0.2, they are greatly different.

Independently of the effect of k, there remain similar effects of T and n on the variance and bias as with the Poisson distribution. Formulae derived from Equations 7A.4 and 7A.6 in the Appendix summarize these effects. Increasing n decreases the bias and the variance, making the OC steeper. Increasing T to around the critical density also reduces bias, including the effect of uncertainty in

Table 7.2. Mean values, μ, for negative binomial distributions corresponding to fixed probabilities of zero incidence and a range of values of k.

	$k = 0.5$	$k = 1$	$k = 2$
Prob(count = 0) = 0.2	12	4	2.5
Prob(count = 0) = 0.4	2.6	1.5	1.2
Prob(count = 0) = 0.6	0.89	0.67	0.58
Prob(count = 0) = 0.8	0.28	0.25	0.24

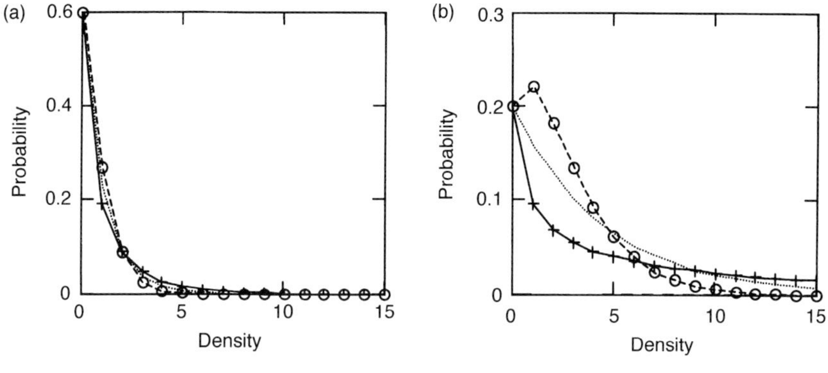

Fig. 7.7. Negative binomial distributions with common values for Prob(count = 0) but unequal values of k ($k = 0.5$ (+–+), $k = 1$ (···), $k = 2$(o-o-o)). (a) Prob(count = 0) = 0.6 ($\mu = 0.9, 0.7, 0.6$); (b) Prob(count = 0) = 0.2 ($\mu = 12, 4, 2.5$).

k, and makes the OC steeper (Binns and Bostanian, 1990a). An OC obtained when sampling a population with k less than that used to formulate the stop boundaries will lie to the right of the nominal OC and will move towards the nominal OC (to the left) as T is increased. However, with continued increases in T to values greater than the critical density, the OC for the populations having k less than the nominal value will eventually cross the nominal OC and lie to the left of it (Fig. 7.8).

When formulating binomial count classification sample plans based on the negative binomial distribution, it is essential that the effect of uncertainty in k be evaluated and perhaps ameliorated by manipulating T and the sample size. These points are illustrated in Exhibit 7.2.

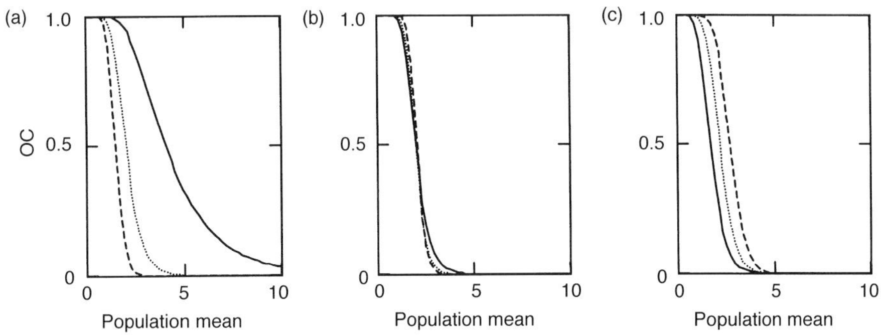

Fig. 7.8. Binomial sampling based on the negative binomial distribution with $cd = 2$ and $n = 30$: the effect of T on OC when k is not known precisely. The assumed value for k is 1; true k is equal to 0.5 (——), 1 (···), 2 (- – -). (a) T = 0; (b) T = 3; (c) T = 8.

Exhibit 7.2. *Binomial classification sampling with the negative binomial distribution*

In this example, we illustrate the influence of k on the OC functions for binomial count classification sample plans based on the negative binomial distribution, and how an appropriate choice of T can reduce bias due to not knowing the true value for k. The example is based on sampling European red mite (*Panonychus ulmi*), a small plant-feeding mite which is a common pest in commercial apple orchards. These arthropods damage leaves by inserting their mouthparts into cells to remove fluids. In doing so, they reduce the leaf's ability to photosynthesize. Severe red mite injury can lead to reduced crop yield and quality.

Nyrop and Binns (1992) found that the sampling distribution of counts of European red mite on leaves could frequently be described by a negative binomial distribution. Although k tended to increase with increasing mite density, there was considerable variability in k at any particular mite density. European red mites are small and difficult to count. As such, they are good candidates for binomial count sample plans. The potential for damage by the European red mite depends on the time of year, the crop load and the geographical location, so the critical density is not constant. One common working critical density is five mites per leaf.

Continued

Nyrop and Binns (1992) reasoned that because k increased with μ, variability in k should be assessed for a restricted range of densities around each critical density and not over the entire range of densities observed. For $cd = 5$, they found that the median value for k was 0.8, with 10th and 90th percentiles equal to 0.4 and 1.5 respectively.

Initial comparisons were made among three sample plans: full count fixed sample size ($n = 50$), binomial count fixed sample size ($T = 0$, $k = 0.8$, $n = 50$), binomial count SPRT ($T = 0$, $k = 0.8$, $\mu_1 = 6$, $\mu_0 = 4$, $\alpha = \beta = 0.1$, $minn = 5$ and $maxn = 50$). The results are shown in Fig. 7.9. The sequential and fixed sample size binomial count plans had nearly identical OC functions that were flatter than the OC for the complete enumeration plan. This reflects the greater variability inherent in binomial counts. The sequential plan resulted in some savings over the fixed sample size plan for small mean densities. However, the p–μ relationship for $T = 0$ is quite flat for densities greater than 5 (Fig. 7.6), so the variance is high (Equation 7.9). In turn, this means that the sampling plan requires close to the maximum number of sample units – the ASN function for means greater than 5 is relatively flat.

The OC functions for these binomial count sample plans can be made to look more like the OC for the full count plan by increasing the sample size. However, the effect of imperfect knowledge about k is more important. A binomial count Sequential Probability Ratio Test (SPRT) plan ($T = 0$, $k = 0.8$, $\mu_1 = 6$, $\mu_0 = 4$, $\alpha = \beta = 0.1$, $minn = 5$ and $maxn = 50$) was set up and tested on negative binomial distribu-

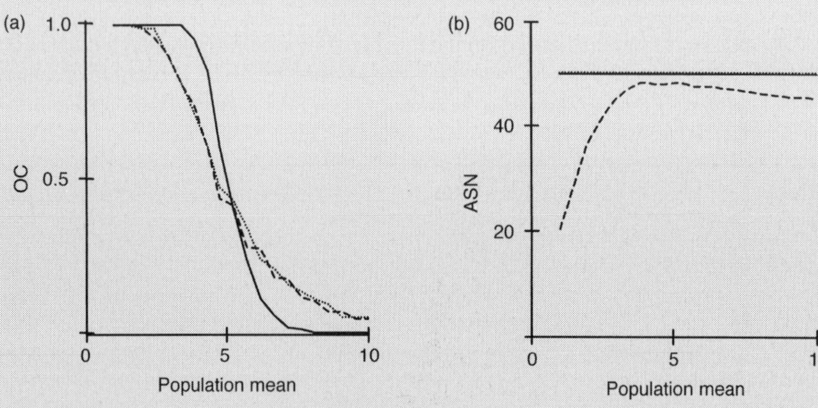

Fig. 7.9. OC (a) and ASN (b) functions for three sampling plans used to classify the density with respect to $cd = 5$. Sample counts were described by a negative binomial distribution ($k = 0.8$). Plan 1 (——) used a fixed sample size ($n = 50$) and complete counts. Plan 2 (···) used binomial counts with $T = 0$ and $n = 50$. The third plan (- – -) used binomial counts, was sequential and based on the SPRT, with the parameters $T = 0$, $\mu_0 = 4$, $\mu_1 = 6$, $\alpha = \beta = 0.1$, $minn = 5$ and $maxn = 50$.

tions with different values of k: $k = 0.8$ (corresponding to the parameters of the sample plan), $k = 0.4$ and $k = 1.5$ (representing possible lower and upper bounds for k). The effect of these values on the OC functions is extreme: the OC for $k = 1.5$ is shifted far to the left of the nominal OC and the OC for $k = 0.4$ is shifted far to the right (Fig. 7.10).

By increasing T, the differences among the three OC functions depicted in Fig. 7.10 can be greatly reduced, and in many instances effectively eliminated. This is shown in Fig. 7.11, where the sample plan parameters are identical to those described in the above paragraph, but with T equal to 3, 5, 8 and 11. The use of $T = 7$ would probably minimize bias due to imperfect knowledge about k. However, it would also make scoring samples much more time-consuming than using $T = 0$. It is necessary to estimate OC and ASN functions for different scenarios, so that the properties of any given plan can be assessed against user needs and expectations. For example, plans with $T = 7$ have been used successfully by growers and extension personnel in Quebec for several years.

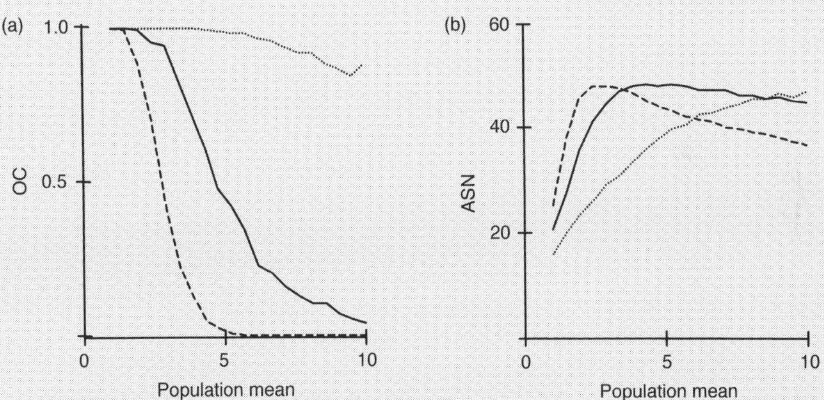

Fig. 7.10. The OC (a) and ASN (b) functions for a binomial count SPRT plan used to classify the density with respect to $cd = 5$. Sample counts were described by a negative binomial distribution. Parameters for the sampling plan was $T = 0$, $k = 0.8$, $\mu_0 = 4$, $\mu_1 = 6$, $\alpha = \beta = 0.1$, $minn = 5$ and $maxn = 50$. The parameter k of the negative binomial distribution used to describe the populations sampled was 0.8 (——), 0.4 (···), and 1.5 (- – -).

Continued

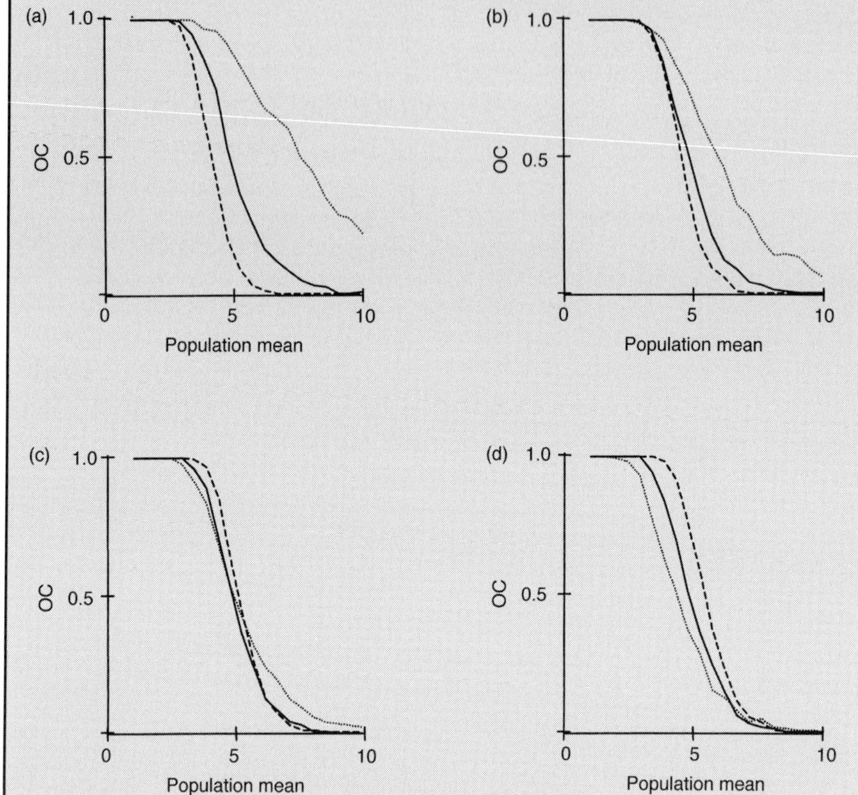

Fig. 7.11. The OC functions for binomial count SPRT plans with $T = 3$(a), 5(b), 8(c) and 11(d) used to classify the density with respect to $cd = 5$. Sample counts were described by a negative binomial distribution. Parameters for all sampling plans were $k = 0.8$, $\mu_0 = 4$, $\mu_1 = 6$, $\alpha = \beta = 0.1$, $minn = 5$ and $maxn = 50$. The parameter k of the negative binomial distribution used to describe the populations sampled was 0.8 (——), 0.4 (···), and 1.5 (- – -).

7.6 Incorporating a Variance–Mean Relationship

Up to this point, we have assumed that while k may not remain constant, it does not change systematically with density. However, as noted in previous chapters, k often tends to increase with density and can be modelled using a variance–mean relationship. If a variance–mean relationship, such as Taylor's Power Law (TPL) (Equation 3.14) holds for the pest in question, and the negative binomial distribution describes sample counts, the value of k can be regarded as a function of μ:

$$k = \frac{\mu^2}{\sigma^2 - \mu} = \frac{\mu^2}{a\mu^b - \mu} \tag{7.10}$$

This is fine if we can believe that σ^2 is perfectly estimated from μ by TPL. That would be too much to ask, but we should be able to assume that σ^2 can be estimated by TPL with a little variability. We showed in Chapter 5 (Appendix) how we could include variability around TPL in the simulations to estimate OC and ASN functions for full count sampling plans. The extension to binomial count sampling plans is reasonably straightforward: the value of k generated with or without variability around TPL (Equation 5A.4 or 5A.5) is used, along with μ, to estimate the probability of at least T pests on a sample unit.

The OC functions obtained in this way may be regarded as averages of the OC functions that would be obtained for each possible value of k. They are therefore quite different from the OC functions obtained using a constant k, as in Exhibit 7.2. So which approach should be used? Examination of OC functions using constant values of k can provide insight into 'worst case' scenarios. In contrast, OC functions obtained using TPL with variability yield an OC that would be expected on average. Ideally, both types of OC functions should be used to judge the acceptability of a proposed sample plan.

Exhibit 7.3. *Binomial classification sampling with the negative binomial distribution and TPL.*

This example is a continuation of Exhibit 7.2, sampling European red mite. Now we use TPL to estimate k. Nyrop and Binns (1992) estimated a = 4.3, b = 1.4 and mse = 0.286 ($\sqrt{mse}$ = 0.54). We can use 0.54 as an estimate of σ_ε in Equation 5A.4. Three SPRT sample plans (μ_1 = 6, μ_0 = 4, $\alpha = \beta$ = 0.1, *minn* = 5 and *maxn* = 50) were compared: full count sampling, binomial sampling with T = 0 and no variation about TPL, binomial sampling with T = 0 including variation about TPL (σ_ε = 0.54). The OC and ASN functions are shown in Fig. 7.12. Without variation about TPL, the OC functions for the complete count and binomial count models are nearly the same. Adding variation about TPL caused the OC function for the binomial count plan to become significantly flatter.

The effect of variability in TPL on the OC functions for binomial count plans can be lessened by increasing the tally number. The third plan above was compared with two others with different tally numbers, 2 and 5. The results are shown in Fig. 7.13. The OC function became steeper as T was increased. In Figs 7.9 and 7.10, we can see that for T = 5, the OC functions for binomial count sample plans with and without variation about TPL are nearly the same.

Continued

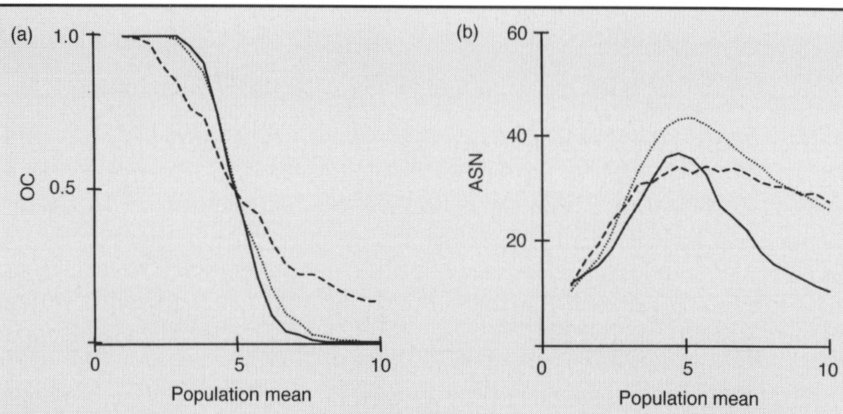

Fig. 7.12. The OC (a) and ASN (b) functions for three SPRT plans used to classify the density with respect to $cd = 5$. Sample counts were described by a negative binomial distribution and the variance modelled using TPL with $a = 4.3$ and $b = 1.4$. The first plan (——) was based on complete counts. The second and third plans were based on binomial counts with $T = 0$. SPRT parameters for all three sampling plans were $\mu_0 = 4$, $\mu_1 = 6$, $\alpha = \beta = 0.1$, $minn = 5$ and $maxn = 50$. The standard deviation of the variance predicted from TPL was: 0 (——), 0 (···), and 0.55 (- – -).

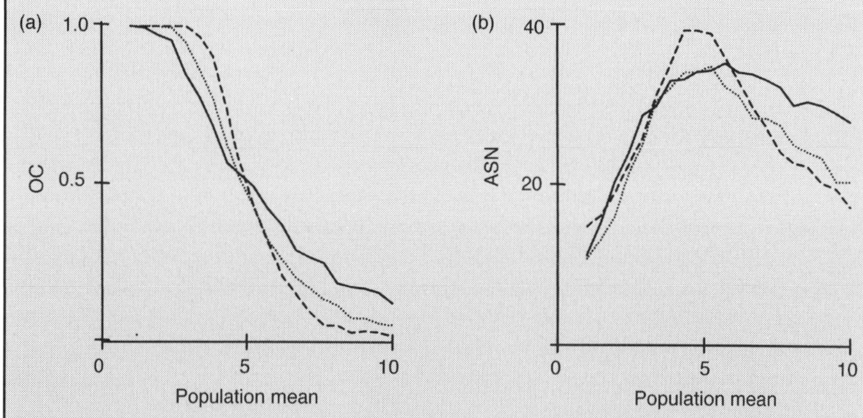

Fig. 7.13. The OC (a) and ASN (b) functions for three binomial count SPRT plans used to classify the density with respect to $cd = 5$. Sample counts were described by a negative binomial distribution and the variances modelled using TPL with $a = 4.3$ and $b = 1.4$. SPRT parameters for all three sampling plans were $\mu_0 = 4$, $\mu_1 = 6$, $\alpha = \beta = 0.1$, $minn = 5$ and $maxn = 50$. The standard deviation of the variance predicted from TPL was 0.55. Tally numbers (T) were 0 (——), 2 (···) 5 (- – -).

7.7 Binomial Sampling Using an Empirical Model

A probability model is not a necessary prerequisite for developing a binomial count sample plan or for estimating their OC and ASN functions. A probability model for the p–μ relationship can be replaced by an empirical model. Several models have been proposed (see, e.g. Ward *et al.*, 1986), but one that has been found most useful and can easily be fitted using linear regression is as follows:

$$\ln(-\ln(1-p)) = c_T + d_T \ln(\mu) \tag{7.11}$$

where μ is the mean and p is the proportion of sample observations with more than T organisms. Note that this model has been shown to fit the relationship for values of T greater than as well as equal to 0 (see, e.g. Gerrard and Chiang, 1970), so the parameters have a subscript T which signifies that they depend on the tally number used. In a strict sense, there are problems with fitting Equation 7.11 using linear regression, because both p and μ are estimated and so d_T will tend to be underestimated (see, e.g. Schaalje and Butts, 1993). However, provided that the data cover a wide range of μ around cd (or, alternatively, if the range 0–1 is nearly covered by p), and the estimates of p and μ are relatively precise, the underestimation should be ignorable. It is more important to check that the data really are linear before fitting the model. Upon fitting the model, the critical proportion, cp_T (note the suffix T) is calculated as

$$cp_T = 1 - e^{-\left(e^{c_T} cd^{d_T}\right)} = 1 - \exp\left(-e^{c_T} cd^{d_T}\right) \tag{7.12}$$

There will always be variability about the relationship expressed by Equation 7.12. This variability is not used to determine the critical proportion, but it is of critical importance for estimating the OC and ASN functions.

Once cp_T has been determined, OC and ASN functions for any sample plan that classifies binomial counts with respect to cp_T can be estimated using a binomial distribution, relating the range of true means (μ_i) to a range of probabilities (p_i) using Equation 7.11. Here, we must consider variability in Equation 7.11 because, while this model may provide a good average description of the data, not all data points fall exactly on the line. Corresponding to any true mean (μ_i) there will be a range of probability values around $p_i = 1 - \exp(-e^{c_T}\mu_i^{d_T})$, and the OC function at μ_i is an average of points, each of which corresponds to one of these probability values.

The procedure is similar to that followed for assessing the consequences of variation about TPL (Chapter 5, Appendix). We assume that actual values of $\ln(-\ln(1-p))$ are approximately normally distributed about the regression line, as determined from the equation

$$\ln(-\ln(1-p)) = c_T + d_T \ln(\mu) + z(0,\sigma_\varepsilon) \tag{7.13}$$

where $z(0,\sigma_\varepsilon)$ is a normally distributed random variable with mean 0 and standard deviation σ_ε. The OC and ASN functions are determined using simulation. Each time a sample plan is used to classify a particular density during the simulation,

Equation 7.13 is used to generate the value of p that will be used to characterize the population of binomial variates being sampled. In other words, if for a particular true mean, 500 simulation runs are used to determine an OC and ASN value, 500 different values for p would be determined using Equation 7.13 and sampling would be simulated from each of these values for p. To generate a value of p, a normally distributed random variable, z, with mean 0 and standard deviation σ_ε, is first generated, a random factor $RF = e^z$ is calculated, and the value of p is $1 - \exp(-e^{c_T}\mu^{d_T}RF)$. If random error is not to be included in the simulations, the random factor is 1: $RF = 1$.

The only question that remains to be answered before calculating OC and ASN functions for empirical binomial models is how to estimate σ_ε. Several writers have addressed this question (Binns and Bostanian, 1990b; Schaalje and Butts, 1992) and it has not yet been fully resolved. However, a reasonable and conservative approximation is

$$\sigma_\varepsilon \doteq \sqrt{\frac{mse}{N} + [\ln(m) - \overline{\ln(m)}]^2 s_d^2 + mse} \qquad (7.14)$$

where mse is the mean square error from the regression, N is the number of data points in the regression, $\overline{\ln(m)}$ is the average of $\ln(m)$, and s_d^2 is the variance of the estimate of the variance of d_T. The astute reader will note that when the same type of variability was considered for TPL, only mse was used to estimate σ_ε. Strictly speaking, an Equation such as Equation 7.14 should be used to estimate variability about TPL. However, we use mse to simplify matters. This simplification is unlikely to have undesirable consequences because the last term in equations such as Equation 7.14, namely mse, is almost certain to account for most of the estimate of σ_ε. If this is so, then why have we included the additional terms in Equation 7.14? We included them because much has been written on estimating the variability about the model expressed by Equation 7.11, whereas the variability about TPL has scarcely been considered, and we wished to maintain consistency with the existing literature. Nevertheless, when we deal with uncertainty in the incidence–mean relationship in order to predict OC and ASN functions for proposed sample plans, mse is the most important component to take into account. The other components are less important, except when the relationship (Equation 7.11) is based on few points and the standard error of the slope parameter d_T is large.

The OC and ASN functions calculated using the procedures just described are averages for the range of p that may occur for a particular true mean. Because these OC and ASN are averages, they say nothing about the extremes that may be encountered when using a sampling plan at a particular place and time. They display the average values that should be expected in the long run, if the sampling plan were used for these mean densities over and over again.

The strategies for estimating OC and ASN functions for the negative binomial and empirical models are similar but not identical, and can be summarized in a table (note that RF is the random factor defined above, and is defined in the same way for both binomial count methods):

	Full count	Binomial count based on negative binomial	Binomial count based on empirical binomial
Criterion	cd	$cp_T = \mathrm{Prob}_{\mathrm{nb}}(x > T \mid cd,k)$	$cp_T = 1 - e^{-e^{c_T} cd^{d_T}}$
Range of means	μ_i	$p_i = \mathrm{Prob}_{\mathrm{nb}}(x \geq T \mid \mu_i,k)$ $k = \dfrac{\mu_i^2}{a\mu_i^b\, RF - \mu_i}$	$p_i = 1 - \exp\left(-e^{c_T}\, \mu_i^{d_T}\, RF\right)$
Distribution for simulation	Negative binomial	Binomial	Binomial

As with all other binomial count models, the sample size and tally number influence the bias and precision of an empirical binomial count sample plan. To consider the effect of T on the OC and ASN, it is necessary to estimate the parameters for Equation 7.11 for several values of T. To minimize bias and maximize classification precision, a useful guide is to use a T for which *mse* is smallest, subject to the condition that the regression remains linear. These concepts are illustrated in Exhibit 7.4.

Exhibit 7.4. *Binomial classification sampling using an empirical model*

This example is based on sampling Colorado potato beetle on stalks of potato plants. Binns *et al.* (1992) described data consisting of 50–200 counts of larvae taken on 74 occasions over 3 years. Critical density for spring larvae can vary depending on other stresses on the plant (poor growing conditions, other pests) and the grower's perception of the threat, but the one suggested by the authors, $cd = 6$, is used here.

The development of a binomial count sample plan using an empirical model follows the steps used in previous examples, when a probability model was used to relate p to μ. The first three plans to be considered were as follows:

1. Fixed sample size binomial count ($n = 50$, $T = 0$).
2. SPRT binomial plan ($\mu_1 = 7$, $\mu_0 = 5$, $\alpha = \beta = 0.1$, $minn = 5$, $maxn = 50$ and $T = 0$) with no variability in the p–μ model.
3. SPRT binomial plan ($\mu_1 = 7$, $\mu_0 = 5$, $\alpha = \beta = 0.1$, $minn = 5$, $maxn = 50$ and $T = 0$) including variability in the p–μ model.

The OC functions for fixed sample size and sequential binomial counts determined without p–μ model variability were nearly identical, but the sequential plan resulted in significantly smaller sample sizes (Fig. 7.14). However, the evaluation of

Continued

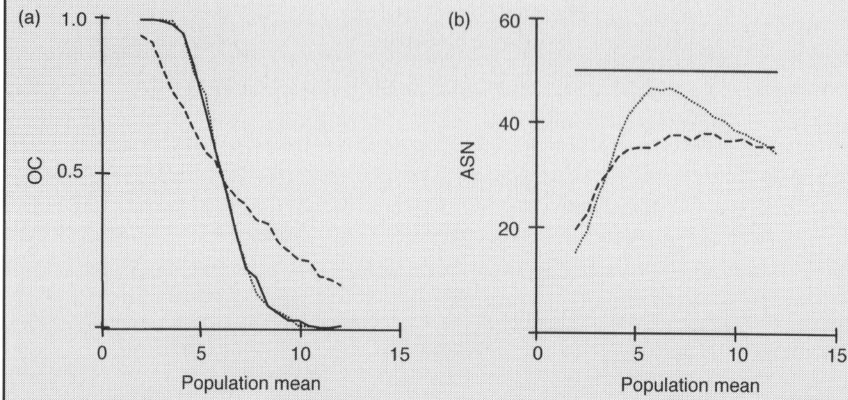

Fig. 7.14. The OC (a) and ASN (b) functions for three binomial count sampling plans used to classify the density with respect to $cd = 6$. The model $\ln(-\ln(1-p)) = c_T + d_T \ln(\mu)$ was used to describe the data with $T = 0$. The first plan (——) used a fixed sample size with $n = 50$. The second (···) and third (- - -) plans were based on the SPRT with parameters $\mu_0 = 5$, $\mu_1 = 7$, $\alpha = \beta = 0.1$, $minn = 5$ and $maxn = 50$. Variation about the p–μ model was included when determining the OC and ASN for the third plan.

the sequential plan gives an overly optimistic assessment of the precision of classification, as it assumes that the relationship between m and p is completely deterministic. When variability in the p–μ model was included, the OC function became much flatter, showing that the sample plan would probably be unacceptable due to the great potential for errors. Further improvement of the sample plan is therefore necessary.

Increasing the tally number can increase the precision of classifications made by the binomial count sample plan. The fitted parameters of the p–m model for $T = 0$, 2 and 4 are shown in Table 7.3. SPRT plans ($\mu_1 = 7$, $\mu_0 = 5$, $\alpha = \beta = 0.1$, $minn = 5$ and $maxn = 50$) comparing these tally numbers were compared (Fig. 7.15). The slope of the p–μ model increases for each increase in T and, for $T = 2$ and 4, mse is less than for $T = 0$. There is a big improvement in the OC when increasing T from 0 to 2. There is also an improvement in the OC when increasing T from 2 to 4, although it is not as great.

Table 7.3. Parameters for the model fitted to Colorado potato beetle data.

Parameter	$T = 0$	$T = 2$	$T = 4$
Model slope (d_T)	0.738	1.092	1.387
Variance of slope, s_d^2	1.295×10^{-3}	1.436×10^{-3}	1.801×10^{-3}
Model mse	0.187	0.121	0.126
Mean of $\ln(m)$	0.773	1.286	1.641

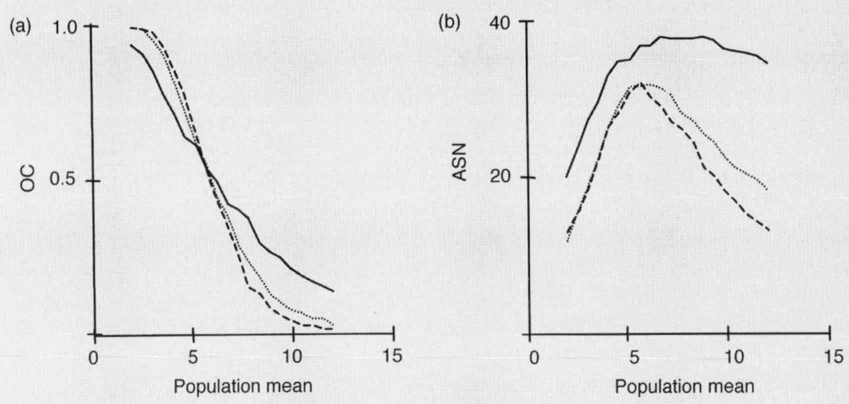

Fig. 7.15. The OC (a) and ASN (b) functions for three binomial count SPRT plans used to classify the density with respect to $cd = 6$. The model $\ln(-\ln(1 - p)) = c_T + d_T \ln(\mu)$ was used to describe the data with $T = 0$ (——), 2 (···) or 4 (- - -). SPRT parameters were $\mu_0 = 5$, $\mu_1 = 7$, $\alpha = \beta = 0.1$, *minn* = 5 and *maxn* = 50. Variation about the p–μ model was included.

7.8 Estimation

Although estimation is not the main emphasis of the book, we need to illustrate the relationship between estimation with full count sampling and estimation with binomial count sampling. Similar methods can be used for binomial sampling as for complete count sampling. In Section 3.2.1, we demonstrated how to calculate the sample size, n, when a given coefficient of variation, CV, is required:

$$n = \frac{\text{variance per sampling unit}}{\mu^2 \text{CV}^2} \tag{7.15}$$

This can be used here, based on the approximate formula in the appendix (7A.4):

$$n = \frac{p(1-p)\left(\mathrm{d}f/\mathrm{d}p\right)^2}{f(p)^2 \text{CV}^2} \tag{7.16}$$

Writing $p = g(\mu)$ and noting the result in Equation 7.8, this can be written in terms of μ as

$$n = \frac{g(\mu)(1-g(\mu))}{\left(\mathrm{d}g/\mathrm{d}\mu\right)^2 \mu^2 \text{CV}^2} \tag{7.17}$$

Using either Equation 7.16 or 7.17, and substituting the incidence–mean relationship,

$f(p)$, for Equation 7.16 and $g(\mu)$ for Equation 7.17, we can calculate the required sample size for a given value of CV. It is usually easier to use $g(\mu)$ and Equation 7.17.

Poisson presence–absence binomial sampling can be used as an example. The equations for $f(p)$ and $g(\mu)$ are

$$f(p) = -\ln(1 - p) \qquad \text{and} \qquad g(\mu) = 1 - e^{-\mu}$$

so

$$\frac{df}{dp} = \frac{1}{1-p} \qquad \text{and} \qquad \frac{dg}{d\mu} = e^{-\mu}$$

For CV = 25% and $\mu = 2$, from Equation 7.17,

$$n = \frac{e^{-2}\left(1-e^{-2}\right)}{\left(e^{-2}\right)^2 2^2\, 0.25^2} = 25.6$$

so 26 sample units are required (exactly the same result is obtained with Equation 7.16). The outcome, 26, can be compared with the number of sample units required for full count sampling, based on the Poisson distribution for which the variance is equal to the mean (Chapter 4). On the basis of Equation 7.15,

$$n = \frac{\mu}{\mu^2 \mathrm{CV}^2} = \frac{2}{2^2 \times 0.25^2} = 8$$

Three times as many sample units are required for the same precision! Using the same formulation, but with $f(p)$ and $g(\mu)$ redefined for T greater than 0, we can calculate the sample sizes required for T = 1, 2, 3 and 4:

T	0	1	2	3	4
n	26	14	12	16	25

Binomial count sampling with T = 2 required only 50% more samples than a full count procedure.

7.9 Summary

1. Binomial counts greatly reduce sampling time and make sampling far less tedious.
2. Care must be taken to ensure that the bias and variability inherent in binomial count sample plans do not seriously compromise the usefulness of the sample information for decision-making.
3. By careful selection of the tally number, a compromise can usually be forged

between (i) low bias and high precision, and (ii) the time required for and difficulty of sampling.

4. Two basic types of p–μ model can be used, based on (i) a probability distribution function and (ii) an empirical relationship.

5. When the negative binomial distribution is used, the effect of imperfect knowledge on k should be checked. If TPL is used to estimate k, the effect of variability around the fitted TPL regression line should be tested.

6. Before an empirical relationship is used, its fit to the regression must be tested. The effect of variability around the fitted regression should be examined.

References and Suggested Reading

Binns, M.R. and Bostanian, N.J. (1990a) Robust binomial decision rules for integrated pest management based on the negative binomial distribution. *American Entomologist* 1, 50–54.

Binns, M.R. and Bostanian, N.J. (1990b) Robustness in empirically-based binomial decision rules for integrated pest management. *Journal of Economic Entomology* 83, 420–427. (Correction note: October 1990, p. vi.)

Binns, M.R., Mailloux, G. and Bostanian, N.J. (1992) Management sampling for larvae of the Colorado potato beetle. *Researches in Population Ecology* 34, 293–307.

Cuperus, G.W. Radcliffe, E.B., Barnes, D.K. and Marten, G.C. (1983) Economic injury levels and economic thresholds for potato leafhopper (Homoptera: Cicadellidae) on alfalfa in Minnesota. *Journal of Economic Entomology* 76, 1341–1349.

Gerrard, D.J. and Chiang, H.C. (1970) Density estimation of corn root – worm egg populations based upon frequency of occurrence. *Ecology* 51, 237–245.

Naranjo, S.E., Flint, H.M. and Hennebury, T.J. (1996) Binomial sampling plans for estimating and classifying population density of adult *Bemisia tabaci* in cotton. *Entomologia experimentalis et applicata* 80, 343–353.

Nyrop, J.N. and Binns, M.R. (1992) Algorithms for computing operating characteristic and average sample number functions for sequential sampling plans based on binomial count models and revised plans for European red mite (Acari: Tetranychidae). *Journal of Economic Entomology* 85, 1253–1273.

Schaalje, G.B. and Butts, R.A. (1992) Binomial sampling for predicting density of Russian wheat aphid (Homoptera: Aphididae) on winter wheat in the fall using a measurement error model. *Journal of Economic Entomology* 85, 1167–1175.

Schaalje, G.B. and Butts, R.A. (1993) Some effects of ignoring correlated measurement errors in straight line regression and prediction. *Biometrics* 49, 1262–1267.

Shields, E.J. and Specker, D.R. (1989) Sampling for potato leafhopper (Homoptera: Cicadellidae) on alfalfa in New York: relative efficiency of three sampling methods and development of a sequential sampling plan. *Journal of Economic Entomology* 82, 1091–1095.

Ward, S.A., Sunderland, K.D., Chambers, R.J. and Dixon, A.F.G. (1986) The use of incidence counts for estimation of cereal aphid populations: 3. Population development and the incidence–density relation. *Netherlands Journal of Plant Pathology* 92, 175–183.

Appendix: Variance and Bias of Binomial Count Estimates of the Mean

To change from using a parameter u to a different parameter t, we need to specify t in terms of u by a function, $f(u)$. Presence–absence sampling with the Poisson distribution provides an example. Instead of counting all pests and obtaining a sample estimate of the mean density, μ, we count the number of sample units, r, with at least one pest, and estimate the probability, p, of a sample unit being infested by r/n. Equation 7.2 defines the function $f(p)$ as $\mu = f(p) = -\ln(1 - p)$, so

$$m_{bin} = -\ln\left(\frac{n-r}{n}\right) \tag{7A.1}$$

It is natural to want to know the variance of m_{bin}, and also its bias, if any. We can obtain approximations to these using the first three terms of what is called the Taylor series expansion of $f(u)$ around the expected value, $u_m = \mathrm{E}(u)$, of u (for 'expected value', see Chapter 2):

$$f(u) \doteq f(u_m) + (u - u_m)\frac{\mathrm{d}f}{\mathrm{d}u} + \frac{(u-u_m)^2}{2}\frac{\mathrm{d}^2 f}{\mathrm{d}u^2} \tag{7A.2}$$

where $\doteq$ shows that this is an approximation, and the symbols $\mathrm{d}f/\mathrm{d}u$ and $\mathrm{d}^2 f/\mathrm{d}u^2$ represent the first and second derivatives of the function, $f(u)$. In Equation 7A.2 they are evaluated at the value $u = u_m$ or, in practice, at the best estimate of u_m.

The variance of $f(u)$ can be approximated by using only the first two terms of Equation 7A.2, moving $f(u_m)$ to the left and squaring:

$$\left(f(u) - f(u_m)\right)^2 \doteq (u - u_m)^2\left(\frac{\mathrm{d}f}{\mathrm{d}u}\right)^2 \tag{7A.3}$$

Taking expectations and noting that $\mathrm{E}(u) = u_m$,

$$\mathrm{var}\left(f(u)\right) \doteq \mathrm{var}(u)\left(\frac{\mathrm{d}f}{\mathrm{d}u}\right)^2 \tag{7A.4}$$

To estimate the bias of $f(u)$, we take expectations of both sides in Equation 7A.2. We find that:

$$\mathrm{E}\left[f(u)\right] \doteq f(u_m) + 0\frac{\mathrm{d}f}{\mathrm{d}u} + \frac{\mathrm{var}(u)}{2}\frac{\mathrm{d}^2 f}{\mathrm{d}u^2}$$

so

$$\mathrm{E}\left[f(u)\right] \doteq f(u_m) + \frac{\mathrm{var}(u)}{2}\frac{\mathrm{d}^2 f}{\mathrm{d}u^2} \tag{7A.5}$$

Hence the bias is approximately equal to

$$\frac{\text{var}(u)}{2}\frac{\text{d}^2 f}{\text{d}u^2} \tag{7A.6}$$

Notice that the bias is ignored in Equation 7A.3. There it is relatively unimportant (usually). Note also that var(u) occurs in both Equation 7A.4 and 7A.5. In the binomial sampling considered in this chapter, u is a sample proportion and var(u) is the variance of a sample proportion based on n sample units. Because of Equation 2.3, var(u) is of the form V/n, where n is the sample size, so the variance of a sample proportion, and its (mathematical) bias, decrease as n increases.

When used to estimate the variance, the method described here is sometimes referred to as the 'delta method'. The Taylor series expansion of a function is discussed in calculus texts for undergraduate mathematics students.

We can exemplify the procedure with presence–absence sampling based on the Poisson distribution, where the estimate of μ is given in Equation 7A.1. The derivatives of $f(p) = -\ln(1 - p)$ are as follows:

$$\frac{\text{d}f}{\text{d}p} = \frac{1}{1-p}, \qquad \frac{\text{d}^2 f}{\text{d}p^2} = \frac{1}{(1-p)^2} \tag{7A.7}$$

Therefore, from Equation 7A.4, the variance of m_{bin} is approximately

$$\frac{p(1-p)}{n}\frac{1}{(1-p)^2} = \frac{p}{n(1-p)} = \frac{1-e^{-\mu}}{ne^{-\mu}} = \frac{e^{\mu}-1}{n}$$

and from Equation 7A.5 the bias of m_{bin} is approximately

$$-\ln\left(e^{-\mu}\right) + \frac{p(1-p)}{2n}\frac{1}{(1-p)^2} - \mu = \frac{p}{2n(1-p)} = \frac{1-e^{-\mu}}{2ne^{-\mu}} = \frac{e^{\mu}-1}{2n}$$

In Chapter 2 we discussed bias and mean square error (*mse*). The *mse* is defined as *mse* = variance + (bias)^2. In this example,

$$mse = \frac{e^{\mu}-1}{n} + \left(\frac{e^{\mu}-1}{2n}\right)^2 \tag{7A.8}$$

Because n, the sample size, is usually large relative to μ, we can see that the variance is relatively more important than the bias term in the *mse*. It is partly for this reason that many people ignore the bias. The bias term in Equation 7A.8 does become relatively large as μ increases, but then binomial sampling with $T = 0$ becomes unattractive for large μ. For $n = 50$:

μ	1	2	3	4	5	6	7
mse	0.04	0.13	0.42	1.4	5.1	24	142
Proportion due to bias	0.01	0.03	0.09	0.21	0.42	0.67	0.85

We have seen throughout this chapter that the use of $T = 0$ for mean values much greater than 0 is next to useless: in the range where $T = 0$ is sensible, the bias term is of minor interest.

8 Multiple Sources of Variation

8.1 Introduction

In this chapter we discuss the situation in which variability in sample counts is associated with the structure of the host on which a pest is found or with the spatial arrangement of observed sample units. Host structure refers to the nested ordering of potential sample units, such as leaves within stems and stems within plants. Spatial arrangement of sample units refers to the spatial proximity or clustering of sample units. In each case, it is possible to identify a primary sample unit, such as a plant or a position in a field, and a secondary sample unit, such as a leaf on a plant or an actual sample observation located at a chosen position in a field. The relation between primary and secondary sample units characterizes a nested hierarchy of sample units: secondary sample units are nested within primary sample units.

In the first part of this chapter, we introduce a procedure called the nested analysis of variance (ANOVA) to characterize and quantify the variability of sample data at different levels of nesting. We derive a guiding principle to determine the number of secondary sample units that should be collected from each primary sample unit, to obtain the most precise pest management decision for a given level of sampling effort.

In the second part of the chapter, we introduce a sampling method called 'variable intensity sampling' (VIS). VIS was devised to address a difficulty concerning adequate coverage of the whole of a management unit when the spatial pattern of pests is aggregated and some sort of sequential sampling plan is used. An early exit through a stop boundary with only part of the management unit covered could lead to incorrect management action if the part covered did not represent the whole. VIS ensures good coverage by imposing a two-level structure on the management unit, with locations (primary sampling units) widely spread out over the management unit. VIS also ensures that sample costs are minimized by analysing the data already collected to adjust sequentially the number of sample units (secondary sample units) collected at each location, so as not to waste sample effort when the management decision appears to be clear.

8.2 Nested or Multistage Sampling

There is a general tendency for counts of biological organisms at neighbouring positions on a host or on neighbouring hosts to be more similar than on widely separated hosts. For example, counts of leafhoppers on two leaves belonging to the same plant may tend to be more alike than on two leaves drawn from different plants. Another way of putting this is that there is a positive correlation among leafhopper counts on leaves within plants. In practical terms, this means that if the count on one leaf from a plant is known, then some information concerning the counts on the other leaves is also implicitly known: selecting another leaf from the same plant adds less information than selecting a leaf from another plant. These considerations influence the choice of sample unit. The choice of a 'high-level' sample unit such as a plant means that pests must be counted on all leaves of the plant. Although this provides very good information on the selected plants, it may be time-consuming and may result in only a few sample units being selected in the field. The choice of a lower-level sample unit such as a leaf means that less information is available on each plant, but it does allow many more sample units to be taken across the field.

Some new terminology needs to be introduced here. We have just referred to a plant and a leaf as sample units, either of which might be useful in a sampling plan. We can distinguish between them by appealing to the hierarchy of the host, namely the plant. The higher-level sample unit (the plant) is called a *primary sample unit*, or *psu*, and the lower-level sample unit (the leaf) is called a *secondary sample unit*, or *ssu*. The concept of a hierarchy allows us a wide choice of sample unit. Not only can we choose either one of *psu* or *ssu* as the sample unit, but we can make combinations. We can define a sample unit as 'two leaves from one plant' if we like, and use it in a sequential procedure: the data from each sample unit is the total count (or mean count) on two leaves from a single plant, and cumulated sums are plotted on a chart as before, taking data from different plants in turn. The advantage of such a procedure (the sample unit is equal to two leaves on one plant) is that any positive correlation among counts on leaves within a plant can be relied on implicitly to augment the information contained in the leaves that are actually selected. We restrict ourselves to two-stage sampling here mainly because sampling for decision-making rarely uses more than two stages, and extension to three or more stages follows precisely the same principles (Cochran, 1977).

If we think that there is correlation among (small) secondary sample units, *ssu*, within (large) primary sample units, *psu*, and we want to develop a sampling plan which exploits the assumed correlation within the *psu*, we need to decide how many *ssu* (n_s), to select in each *psu*. As is shown below, there are formulae for an optimum choice of n_s based on relative costs and variances. Once n_s is determined, the sample unit that is then used in decision-making is defined as 'n_s *ssu* from one *psu*'. If, in the above example, two leaves within each plant was found to be optimal, the sample unit would be 'two leaves randomly selected from a plant'. Once a decision has been made on the sample unit, all of the theory and practice of the previous chapters can be used with no further change.

8.3 Analysis of Variance and Two-stage Sampling

ANOVA is used to obtain estimates of variability among primary and secondary sample units. The ANOVA is based on a model for the variance of counts on secondary sample units in which the variance is assumed to be the sum of

- a component, VC_b, that represents variability *between psu* (hence the subscript b), and
- a component, VC_w, that represents variability among *ssu within psu* (hence the subscript w)

The purpose of the ANOVA is to estimate each variance component and test the above model. Where there is a positive correlation among *ssu* within *psu*, VC_b is greater than zero. When there is no correlation among *ssu* within *psu*, VC_b equals zero, and all of the variability is accounted for by VC_w. Another way of looking at this is that if there is no correlation among *ssu* within *psu*, there is no point in travelling from *psu* to *psu* more than is necessary to collect enough data, because all the variation in the field is between *ssu*, regardless of what *psu* they belong to. This is clearly unrealistic. Therefore, if an estimate of VC_b is zero or negative (as it can be), this result is either because of random error (or, very unlikely, the model for the ANOVA is wrong).

The data required for the ANOVA are counts of individuals on $n_p\ n_s$ sample units. Writing

- x_{ij} as the number of individuals found on the jth *ssu* in the ith *psu*
- $x_{i.}$ as the mean number of individuals per *ssu* found on the ith *psu*, and
- $x_{..}$ as the overall mean number of individuals per *ssu*

the total sum of squares among all the *ssu* is

$$SS_T = \sum_1^{n_p}\sum_1^{n_s}\left(x_{ij} - x_{..}\right)^2 \tag{8.1}$$

This could be used as the basis of an estimate of the variance of the mean count per *ssu*, $V_T = SS_T/(n_p n_s - 1)$ but this would not take into account the correlation among *ssu* within *psu*. SS_T can be subdivided into two parts, SS_b and SS_w, that represent the variability between *psu* and among *ssu* within *psu*, respectively:

$$\begin{aligned} SS_b &= \sum_1^{n_p} n_s\left(x_{i.} - x_{..}\right)^2 \\ SS_w &= \sum_1^{n_p}\sum_1^{n_s}\left(x_{ij} - x_{i.}\right)^2 \end{aligned} \tag{8.2}$$

and some complicated algebra shows that

$$SS_T = SS_w + SS_b$$

In the ANOVA, these sums of squares are laid out along with what are called *degrees of freedom*, or *df*. The *df* are used to divide the sums of squares to obtain estimates of variance, called *mean squares* (MS). The standard layout of results in the ANOVA is shown in Table 8.1. The variance component 'among *psu*' is estimated as

$$VC_b = (MS_b - MS_w)/n_s \tag{8.3}$$

and the variance component among *ssu* within *psu* is estimated as

$$VC_w = MS_w \tag{8.4}$$

The mean square MS_b is the variance estimate per *ssu*, which is appropriate for the two-stage sample (see the Appendix); the variance estimate of the sample mean of all $n_s n_p$ *ssu* is therefore equal to $MS_b/n_s n_p$. Using MS_T instead of MS_b as the variance estimate would ignore the structure of the data, and generally would underestimate the true variance per *ssu* by an amount which depends on the correlation among *ssu* within *psu*. The variance components can be used to predict variances for sampling plans with different values of n_p and n_s. The estimate of the variance of the sample mean based on n'_p *psu* and n'_s *ssu* is

$$V(n'_p, n'_s) = \frac{VC_b}{n'_p} + \frac{VC_w}{n'_p n'_s} \tag{8.5}$$

The above derivations have glossed over some theoretical points which could be important in some instances, but we suggest that they may be safely ignored. We have (i) not mentioned the 'sampling fraction' and we have (ii) assumed that all *psu* contain the same number of *ssu*:

1. The 'sampling fraction', *f*, is defined as the ratio of the sample size to the total number of sample units, and appears in formulae for sampling variances (Cochran, 1977). For all sampling plans in other chapters, *f* is very small and can be ignored, but here there is a sampling fraction for the number of *ssu* sampled in a *psu*, and that ratio may not be small. However, it turns out that this sampling fraction has little effect on the variances that we use, and so it too can be ignored in the context of pest management.

Table 8.1. The layout of nested ANOVA results for partitioning variance components among primary and secondary sample units.

Source of variation	Degrees of freedom, *df*	Sum of squares, *SS*	Mean squares, *MS*
Between primary sample units	$n_p - 1$	SS_b	$MS_b = SS_b/(n_p - 1)$
Among secondary sample units within primary units	$n_p(n_s - 1)$	SS_w	$MS_w = SS_w/[n_p(n_s - 1)]$
Total	$n_p n_s - 1$	SS_T	$MS_T = SS_T/(n_p n_s - 1)$

2. The exact formulae when the number of *ssu* per *psu* is not constant are complicated (Cochran, 1977). Work has been done on the effect of this when the simple formulae are used. We recommend that the simple formulae are quite adequate for sampling in agricultural pest management.

8.3.1 TPL and the application of ANOVA in decision-making

We saw in previous chapters that Taylor's variance–mean relationship can be used for variances when simple random sampling is done. It can also be used for the variance components in the ANOVA, namely VC_b and VC_w. A pair of Taylor's Power Law lines (TPL) that describe the variance components as a function of the mean can be estimated if enough data sets are available. The procedure is as follows:

1. An ANOVA is computed for each data set.
2. The variance components, VC_b and VC_w, are estimated.
3. TPL parameters are estimated by regressing $\ln(VC_b)$ and $\ln(VC_w)$ against ln(sample mean).
4. The two TPLs are

$$VC_b = a_b\mu^{b_b} \quad \text{and} \quad VC_w = a_w\mu^{b_w} \tag{8.6}$$

Note that we use '*VC*' for the true values as well as for the estimates. This is merely to avoid adding more symbols, and we hope that it will create no confusion. Once the TPLs have been estimated, the variance components for any chosen μ can be estimated. The mean squares MS_b and MS_w could be used instead of the variance components to estimate TPL, and the variance components would be derived from them. The two methods are equivalent in principle, although the estimates may be slightly different arithmetically.

8.3.2 Use of the ANOVA to test for correlation within psu

The possible correlation of counts within *psu*, and how much it might affect the variance, can be tested by an *F-test*. The *F*-distribution is another statistical distribution whose properties have been well documented. To use this test here, $F = MS_b/MS_w$ is calculated and compared with tabular *F*-values (as with χ^2 values; Chapter 4). If *F* is larger than the tabular value, then MS_b is significantly greater than MS_w, which implies that the correlation among *ssu* within the same *psu* is significant. In other words, the distribution is aggregated within *psu*, and MS_b rather than MS_T should be used as an estimate of variance of a sample mean. If, for some reason, MS_T continues to be used, the variance of the sample mean will be underestimated. Whether or not this underestimation is important from the standpoint of pest management decision-making can be assessed using the tools that we have presented thus far. Of course, if the variance was never estimated correctly by keeping track of the nested structure of the counts, it would never be known what the correct variance should be, and it would not be possible to assess the influence on decision-making of the use of an underestimate of the variance.

The ratio

$$deff = MS_b/MS_T \tag{8.7}$$

can be used to correct estimates of MS_T to MS_b (Kish, 1965; Cochran, 1977). However, care must be exercised, because estimates of mean squares often are very variable and because variance components, like variances, usually increase with mean density.

8.3.3 Use of the ANOVA to estimate optimum values for n_s

The ANOVA can be used in combination with information on the costs of sample collection to estimate an optimum value for n_s. Costs are usually based on the time it takes to perform certain tasks during sampling. If the cost of finding each *psu* is estimated as C, and (having selected and located the *psu* in the field) the cost of selecting an *ssu* and counting the individuals on it is estimated as *c*, the total cost of sampling can be written as

$$K = Cn_p + cn_pn_s \tag{8.8}$$

The formula for the optimum value for n_s is (Cochran, 1977; Nyrop and Binns, 1991)

$$nopt_s = \sqrt{\frac{VC_w}{VC_b}\frac{C}{c}} = \sqrt{\frac{a_w\,\mu^{b_w}}{a_b\,\mu^{b_b}}\frac{C}{c}} \tag{8.9}$$

For example, if moving from one *psu* to the next *psu* is costly, C will be large relative to *c*, so n_s should be large; in such cases, it is best to take advantage of being at a *psu* by observing more *ssu*. Or, if VC_b is large relative to VC_w, there is a lot of variability among *psu*, and so n_s should be small to allow more *psu* to be looked at; in this situation it would be best not to waste too much time obtaining details at one *psu*, because more information is required at different *psu*. Because the variance components are estimated, Equation 8.9 is an approximation, and so the optimum value for n_s should be close to $nopt_s$, but not necessarily exactly equal to it. Cochran (1977) suggests that values near $nopt_s$ should be nearly equivalent in terms of optimality. Furthermore, because of the variability inherent in variance estimates, the exact values obtained through Equation 8.9 should not be taken as sacrosanct.

Harcourt and Binns (1980) investigated the sampling variability in alfalfa of eggs and mines of the alfalfa blotch leafminer, *Agromyza frontella* (Rond.). They obtained data on several occasions and, on the basis of an equation like Equation 8.9, they estimated optimal numbers of leaves to take per stem. These ranged from 1.1 to 2.1 for eggs, and 1.3 to 2.7 for mines, suggesting that two or three leaves per stem would be best. However, other considerations, such as the expectation that the ratio *c*/C would decrease with sampling experience, moved the authors to recommend four leaves per stem.

Exhibit 8.1. *Estimation of variance components*

Two-stage sampling can be envisaged in terms of the aggregated distributions discussed in Chapter 4. Around any location in a field with an aggregated distribution of individuals, the counts of individuals on sample units close to one another are correlated. There is a small area around each location in the field where the sample units behave as if they were *ssu*'s around a *psu*. Given an aggregated spatial distribution, it is therefore sensible to ask: 'How many sample units should I collect at each location as I go through a field? Should I visit 30 locations, collecting one sample unit at each, or would 10 locations with three sample units at each location be (almost) as good, and less work?' ANOVA on a simulated field can illustrate how to answer this question.

A spatial pattern (Fig. 8.1) was generated using the Poisson cluster process method (Chapter 4, Appendix). A random sample was taken, consisting of 15 locations in the field (*psu*) and five sample units (*ssu*) collected from the square of 3 × 3 sample units surrounding each chosen location, making 75 *ssu* in all. An analysis of variance was done for the two-stage sample, with the results shown in Table 8.2. An F-value equal to 2.51 with these degrees of freedom is significant (at the 1% level). The variance of the sample mean would be estimated by $MS_b/75 = 10.17/75 = 0.14$. If we had used the simple formula using MS_T, we would underestimate the variance by a factor of about 2 (10.17/5.21).

Using the results of the ANOVA, an optimum value for n_s can be estimated. This is best approached by a plot of $nopt_s$ on the ratio of the cost per *ssu* (*c*) to the cost per *psu* (*C*) (Fig. 8.2). Here, if the ratio is about 1 or higher, $nopt_s$ should be 2. But if the ratio is much less, more secondary samples per primary sample should be taken. Plots such as these should be used as guides, rather than as formulae to be followed to the letter.

The results of an ANOVA can have low precision; this is often true for estimated variances. To illustrate this, we took an additional five sets of samples from the hypothetical field shown in Fig. 8.1 and calculated an ANOVA for each. The mean squares for between *psu*'s and for among *ssu*'s within *psu*'s are shown in Table 8.3. These results show that different samples from one field could give a range of values for the variances (and variance components), despite the relatively large sample size (15 *psu*, five *ssu*/*psu*). The take-home message is that ANOVA results should be used, but only as a guide.

Table 8.2. ANOVA for the sample data depicted in Fig. 8.1.

Source of variation	Degrees of freedom, *df*	Sum of squares, *SS*	Mean squares, *MS*
Between primary sample units	14	142.35	10.17
Among secondary sample units within primary units	60	243.2	4.05
Total	74	385.55	5.21

Continued

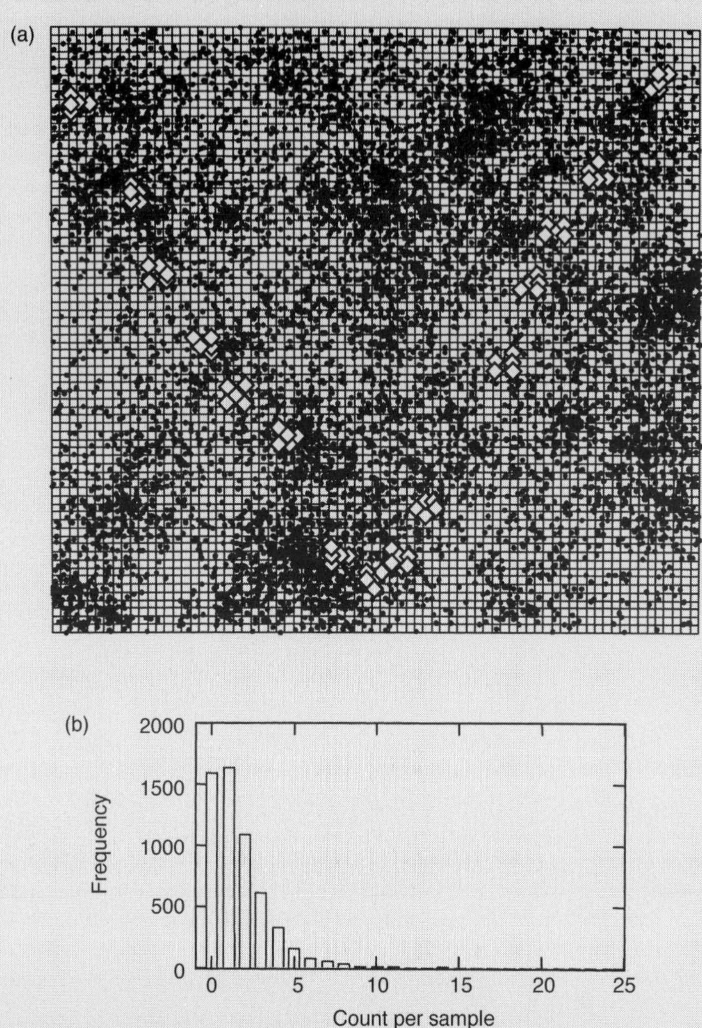

Fig. 8.1. (a) A computer-generated spatial pattern and the location of 75 sample points. The true mean per sample unit is 1.68. Samples were taken using a 'V' path through the simulated field. The location for each set of five samples is identified by a group of diamonds. (b) A histogram of the complete data.

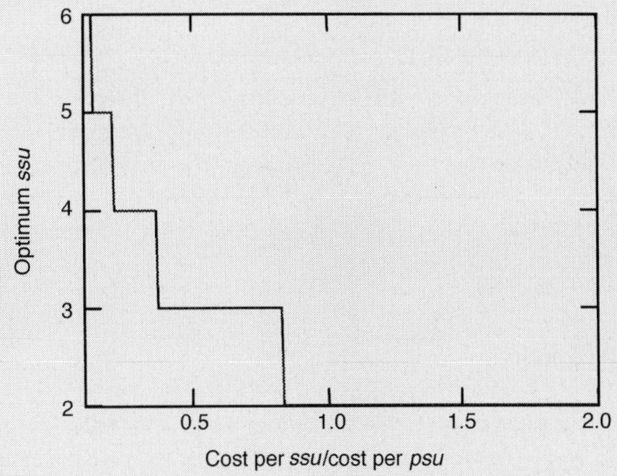

Fig. 8.2. The optimum number of secondary sample units (*ssu*) to take at each primary sample location (*psu*), as a function of the ratio of the cost per *ssu* to the cost per *psu*. The curve was determined using Equation 8.9 based on the mean squares in Table 8.2, rounding up to the next highest whole number.

Table 8.3. Mean squares (*MS*) for five sets of samples taken from the data shown in Fig. 8.1. Ratios of variance components show how variability affects $nopt_s$ (Equation 8.9).

	Sample replicate				
Source of variation	1	2	3	4	5
Between primary sample units	7.94	5.63	3.38	4.46	5.41
Among secondary sample units within primary units	1.77	1.53	1.34	1.43	2.07
Total	2.94	2.3	1.73	2.00	2.70
$\sqrt{VC_w/VC_b}$	0.47	0.52	0.63	0.57	0.62

8.4 Patchy Environments

Pests frequently occur in patches where infestations are noticeably higher than in the rest of the habitat. The possible causes for patchy distributions are many, but can be grouped into three broad categories: (i) quality of resource, (ii) patterns of colonization and subsequent natality, and (iii) patterns of mortality. Patches may occur on a large and predictable scale, or they may occur more or less at random and on a smaller scale throughout a field. For example, onion thrips (*Thrips tabaci*) frequently are more abundant on field margins than in the interior of fields. Also, groups of neighbouring plants usually have more similar numbers of thrips on them than plants a good distance from each other. When patchiness occurs on a large or predictable scale, it may dictate particular management choices. Referring again to the onion example, when sampling for onion thrips it is sometimes recommended that the field margin be managed separately from the field interior, or that counts from the margins of fields be used to schedule management actions for the entire field. When patchiness occurs on a smaller, and more random fashion, it may dictate particular sampling choices.

8.5 Predictable Patches

When patches are somewhat predictable, there are at least four ways in which one might respond from the perspective of sampling for pest management decision-making:

1. Ignore any available knowledge about patchiness and continue as in previous chapters.
2. Treat the portions of the field where pest abundance might differ substantially as separate areas, each to be managed and sampled independently.
3. Sample from the portion of the field where pests are assumed to be most abundant, and use the results to indicate how to treat the whole field.
4. Divide the total number of samples into two, taking some samples from the (assumed) heavily infested area and the rest from the less infested part, and use a weighted average for decision-making (this is called *stratified sampling*).

The choice is obviously dependent on pest dynamics, the cost and effectiveness of control, the cost of pest-inflicted damage, the cost of sampling, and on the practicality and desirability of site-specific management. Bear in mind, though, that the goal is to control the pest, not necessarily to estimate its density. This consideration will often preclude the fourth option above, as being inappropriate. The use of part of the field as a kind of indicator of potential problems (option 3) may be a more efficient alternative than option 2. Option 1 has the advantage of simplicity, but carries with it an increased variability among sample units, and also the danger of classifying the pest density incorrectly and thus applying inappropriate management.

8.6 Unpredictable Patches – VIS

Patchy spatial patterns require that samples be collected from throughout a management unit; otherwise, there is a risk that a sample will be taken that is not representative. For example, if all samples are taken from one part of a field that harbours very few pests, but pests are abundant in the remainder of the field, an error would probably be made to not intervene when in fact some action was warranted. This scenario could easily unfold through the use of a sequential procedure in which the minimum sample size was small. This problem can be circumvented by collecting a fixed number of samples from throughout the field or, if using a sequential procedure, by examining a sufficient minimum number of samples taken from throughout the field to ensure a representative sample. There are shortcomings to these solutions. As has been repeatedly demonstrated, an unnecessarily large number of samples may be taken if a fixed sample size is used when the pest abundance is either much less or much greater than the level that dictates action. Collecting the initial sample for a sequential procedure from throughout the field requires that the field be traversed twice. An alternative is to use VIS (Hoy *et al.*, 1983; Hoy, 1991).

Keeping the idea of sequential sampling in mind, there is no reason to take a lot of sample units if the mean pest density appears to be very much higher or lower than the critical density *cd*, but enough primary sample units must be taken over the whole field to minimize the risk of missing pest clusters. The idea behind variable intensity sampling is to adjust the number of secondary sample units taken from each primary sample unit based on how close the estimated mean is to the critical density. If, after examining the *ssu* from some of the *psu*, the estimated mean is close to the critical density, the maximum number of *ssu* should be taken from the next primary sample. On the other hand, if the estimated mean is far from the critical density, perhaps only one *ssu* need be observed from the next *psu*. The notation required to describe VIS more completely is extensive, and is laid out in Table 8.4.

Suppose that to cover the field $npsu_{tot}$ sample locations (*psu*) are required. With VIS sampling there is a formula to calculate how many second-stage units, *nssu*, need to be selected at the next primary unit during sampling. The formula reflects how far away the current estimate of the mean (m_c) is from *cd*, and comes in two parts:

1. Re-estimate the required total number of second-stage units ($nssu_{tot}$).
2. Re-estimate the average number of second-stage units ($\overline{nssu}$) required in each of the *psu* not yet sampled to achieve $nssu_{tot}$ units in all.

The first involves thinking about the confidence interval (Chapter 3) at the end of sampling:

$$m \pm z\sqrt{\frac{V}{nssu_{tot}}} \tag{8.10}$$

What we should like is that *cd* be outside this interval, because the classification decision could then be made with some degree of confidence. After $nssu_c$ sample units have been examined, we can predict the final confidence interval based on the current estimates of the mean and variance (m_c and V_c):

$$m_c \pm z\sqrt{\frac{V_c}{nssu_{tot}}} \tag{8.11}$$

Note that in Equation 8.11 we use the current estimates of the mean and variance,

Table 8.4. The notation used in variable intensity sampling (VIS).

	Symbol	Description
Prespecified	$nssu_{max}$	Maximum number of *ssu* at any *psu*
	$npsu_{tot}$	Total number of *psu*
Calculated from data	$nssu_c$	The total current number of *ssu* examined
	$npsu_c$	The current number of *psu* from which *ssu* have been examined
	m_c	The current estimate of the mean
	V_c	The current estimate of the variance
Specification for next *psu*	$nssu$	Number of *ssu* to be taken from the next *psu*
	$nssu_{tot}$	Required total number of *ssu* at the end of sampling
	$\overline{nssu}$	The average number of *ssu* to be taken from the remaining *psu*

but the number of sample units at the end of sampling, $nssu_{tot}$. Of course, we do not know $nssu_{tot}$ at this time, but what we should like is that, at the end of sampling, the actual value of $nssu_{tot}$ is large enough for *cd* to be outside the confidence interval Equation 8.11. In other words,

$$cd < m_c - z\sqrt{\frac{V_c}{nssu_{tot}}} \quad \text{or} \quad cd > m_c + z\sqrt{\frac{V_c}{nssu_{tot}}} \tag{8.12}$$

Equation 8.12 provides a way of determining what $nssu_{tot}$ should be. Using either of the inequalities, making it an equality and solving for $nssu_{tot}$ yields

$$nssu_{tot} = \frac{z^2 V_c}{(cd - m_c)^2} \tag{8.13}$$

However, we have already collected $nssu_c$ sample units, and there are only $npsu_{tot} - npsu_c$ locations left to obtain the right value for $nssu_{tot}$. There is good hope that the final confidence interval will exclude *cd*, if $nssu_{tot}$ can be realized by the following formula, which involves the average number of sample units to be taken at each subsequent location ($\overline{nssu}$):

$$nssu_{tot} = nssu_c + npsu_{tot} - (npsu_c)\,\overline{nssu}$$

which can be rearranged as

$$\overline{nssu} = \frac{(nssu_{tot} - nssu_c)}{(npsu_{tot} - npsu_c)} \tag{8.14}$$

Combining Equations 8.13 and 8.14,

$$\overline{nssu} = \frac{\left(\dfrac{z^2 V_c}{\left(cd - m_c\right)^2} - nssu_c\right)}{\left(npsu_{tot} - npsu_c\right)} \tag{8.15}$$

Because partial sample units cannot be examined, $\overline{nssu}$ is rounded up to the nearest integer. VIS can then be implemented, as follows:

1. Decide how many primary samples (locations) to use, $npsu_{tot}$, and the maximum number of sample units to take at each location, $nssu_{max}$. Equation 8.9 may be helpful in determining $nssu_{max}$, provided that estimates of the variance components and sampling costs are available.
2. Go to the first location (*psu*) and take $nssu_{max}$ sample units.
3. Estimate the mean and variance and determine $\overline{nssu}$ using Equation 8.15. Calculate *nssu* as the smaller of $nssu_{max}$ and $\overline{nssu}$.
4. Go to the next location and take *nssu* samples.
5. Repeat steps 3 and 4 until all $npsu_{tot}$ locations have been visited.
6. Make a classification by straight comparison of the final sample estimate of the mean with *cd*.

In the above, we have glossed over the calculation of V_c, the estimate of the variance after samples have been taken from *c* locations. In principle, it should be estimated as a two-stage variance according to formulae like Equation 8.5. In fact, the formulae are even more complicated, because the number of *ssu* collected at each *psu* is not constant. However, as we have stated earlier, the simple formulae should be adequate for sampling in agricultural pest management. An estimate of *deff* (Equation 8.7) may be used to adjust the simple formula for the variance each time $\overline{nssu}$ is estimated. An alternative to estimating the variance is to predict it using variance components modelled as a function of the mean; for example Equation 8.6.

In the following exhibit, we illustrate the application of VIS using a computer-generated aggregated distribution.

Exhibit 8.2. *An illustration of variable intensity sampling*

VIS can be used on the spatial distribution shown in Fig. 8.1. Suppose that $cd = 1.0$, and 15 sample locations are required to cover the field. The maximum number of *ssu* per location, $nssu_{max}$, was set to 4, using Fig. 8.2 and a cost ratio of about 0.25. The confidence interval was defined by $z = 2$. Using a 'V' path through the field, an illustration of sample units examined is shown in Fig. 8.3a, along with the history of the estimated means (Fig. 8.3b) and *ssu* (Fig. 8.3c).

Continued

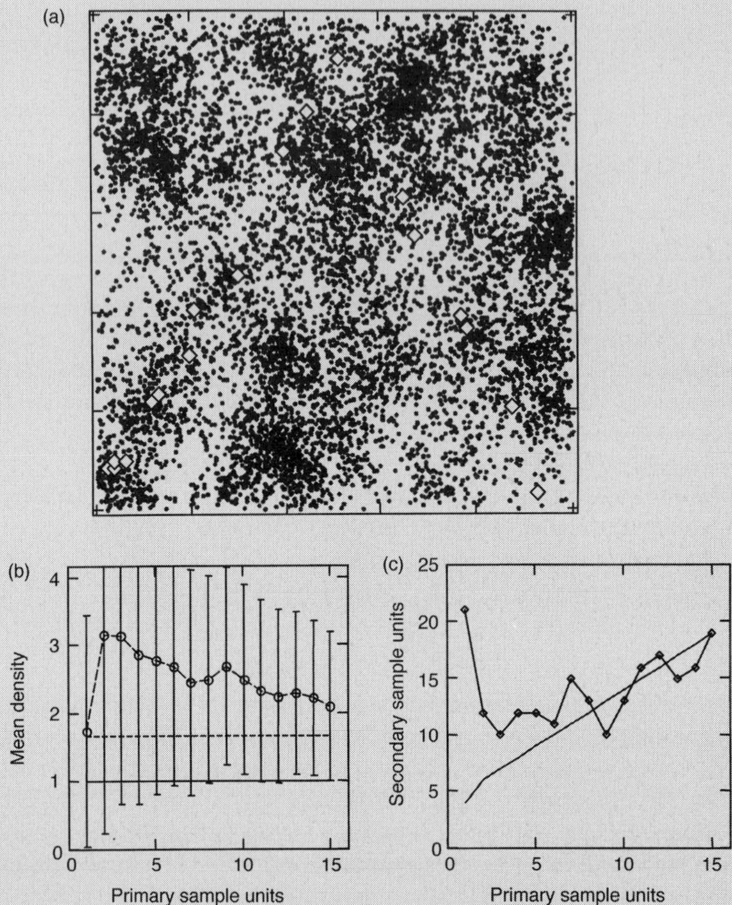

Fig. 8.3. (a) A computer-generated spatial pattern and the location of samples (◇) for variable intensity sampling (VIS) with critical density, *cd* = 1.0 (true mean = 1.68). Samples were taken using a 'V' path. (b) The estimated mean density (m_c, ○) and the confidence interval after samples were collected at each primary sample location, the true mean density (- – -), and the critical density (—). (c) The predicted total number of required secondary sample units ($nssu_{tot}$, ◇) and actual number of secondary sample units examined (cumulated total of *nssu*, —) as a function of the number of primary sample locations examined.

Sampling started at the lower left corner of the field (Fig. 8.3a). Four *ssu* were collected ($nssu_{max}$ = 4), and the sample mean per *ssu* was 1.8. Based on this, the predicted number of sample units needed is 21. Four of these ($nssu_c$) have already been examined (Fig. 8.3c). There are 14 locations still to be visited, so the number of secondary samples to take at the next location is (21 − 4)/14, or about 1.2. But we must examine whole samples, so *nssu* for the next sample location is 2 (which can be seen in Fig. 8.3a, where the second sample location is in the lower left corner).

Overall, 19 samples were taken and the estimated mean was 2.105. This is greater than the critical density, so a decision would be made to intervene. But suppose that $cd = 1.5$. Using the same simulated field, another (randomly selected) sample path is shown in Fig. 8.4a and the sampling history is depicted in Figs 8.4b

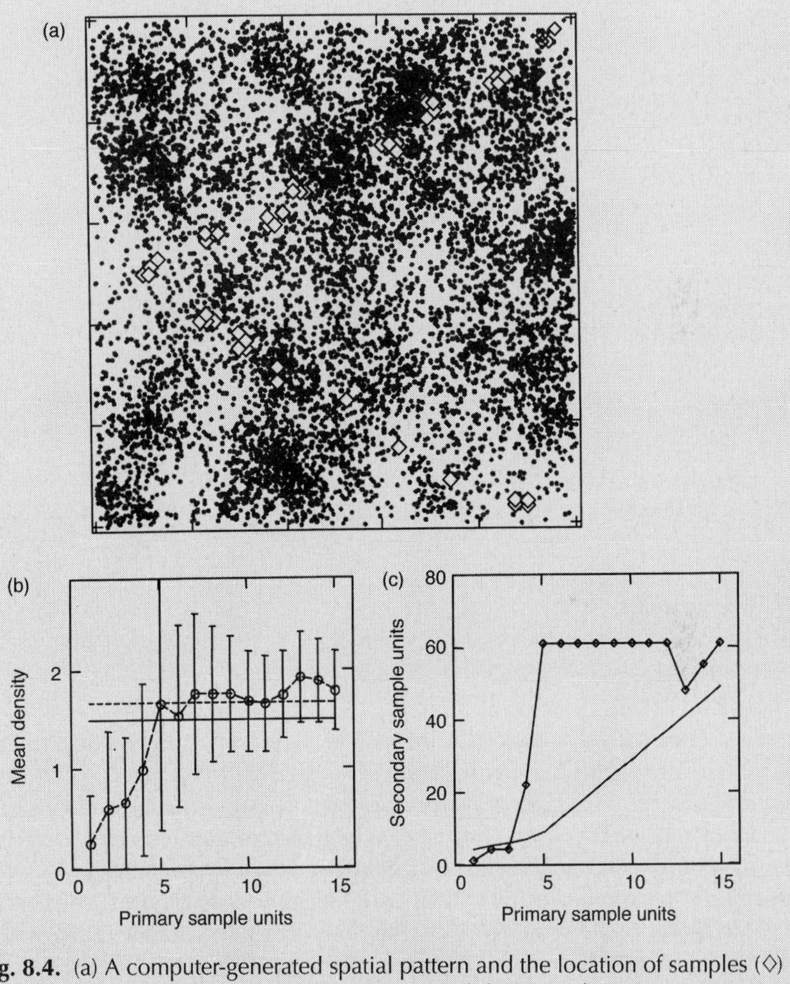

Fig. 8.4. (a) A computer-generated spatial pattern and the location of samples (◇) for variable intensity sampling (VIS) with a critical density, $cd = 1.5$ (true mean = 1.68). Samples were taken using a 'V' path. (b) The estimated mean density (m_c, ○) and the confidence interval after samples were collected at each primary sample location, the true mean density (- – -), and the critical density (——). (c) The predicted total number of required secondary sample units ($nssu_{tot}$, ◇) and the actual number of secondary sample units examined (cumulated total of *nssu*, ——) as a function of the number of primary sample locations examined.

Continued

and c. The final sample size was 49, which reflects the proximity of the population mean to *cd*. Note that after samples were examined from the first three *psu* (lower right in Fig. 8.4a), the estimated mean was quite low. In fact, after processing these samples (six *ssu*), the confidence interval for the estimated mean was less than *cd* (Fig. 8.4b), so a sequential procedure based on Iwao boundaries would have terminated sampling unless a minimum sample size greater than six was specified. Even if a minimum sample size was in effect, if all the samples were taken from the lower right portion of the simulated field, as in Fig. 8.4a, an incorrect decision would likely be made. This illustrates the need for placing samples throughout a field or management unit when a spatially aggregated pattern of pests might occur.

We have illustrated a single application of VIS on the field depicted in Fig. 8.1. What would happen in the long run using the two critical densities? We would expect an operating characteristic (OC) value with $cd = 1.0$ to be lower than for $cd = 1.5$ and for an average sample number (ASN) value also to be lower when the lower *cd* value was used. We can test this by simulating variable intensity sampling using the two critical densities several times, recording and then averaging the results. In the simulations, different sample paths (a random selection of the orientation of the 'V' and the positions of *psu* and *ssu*) are taken, so the same sample units are not repeatedly counted.

After 100 simulations, the OC values were 0.03 and 0.4 for the *cd* values 1.0 and 1.5, respectively. The corresponding ASN values were 34.5 and 43.7. Our expectations are confirmed: The higher critical density for this field yields a higher OC value and greater sampling effort. We will return to the question of OC and ASN functions for VIS in a later exhibit.

8.6.1 A VIS look-up chart

A hand-held calculator or small computer can be programmed to implement VIS. However, using such devices is not always convenient. It is also possible to calculate a chart beforehand that provides the number of *ssu* to take from the next *psu*, given the number of primary and secondary sample units sampled thus far and an estimate of the current mean. The calculations require that estimates of the variance components, VC_b and VC_w, be made with a variance mean model. A mathematical link between mean and variance is a requirement to calculate the chart.

Calculation of the chart proceeds as follows. For each combination of *psu* and cumulative *ssu* examined, we want to know the number of pests counted that indicate a need for $nssu_{max}$, $nssu_{max} - 1, \ldots, 1$ *ssu* from the next *psu*. If the current estimate of the mean is close to *cd*, then the maximum number ($nssu_{max}$) of *ssu* will be needed at the next *psu*. The further the estimated mean is from *cd*, the fewer the number of *ssu* required at the next location in the field. These relationships can be quantified by modifying Equation 8.12. We replace the variance estimate, $V_c/nssu_{tot}$, by the formula based on variance components using the numbers of primary and secondary sample units examined and yet to be examined (see Equations 8.5 and 8.15):

$$\left.\begin{array}{l} cd < m_c - z\sqrt{\dfrac{VC_b}{npsu_{tot}} + \dfrac{VC_w}{nssu_c + \left(npsu_{tot} - npsu_c\right)\overline{nssu}}} \\ \text{or} \\ cd > m_c + z\sqrt{\dfrac{VC_b}{npsu_{tot}} + \dfrac{VC_w}{nssu_c + \left(npsu_{tot} - npsu_c\right)\overline{nssu}}} \end{array}\right\} \tag{8.16}$$

For simplicity, the variance components are not subscripted with c, although they depend on the value of m_c through the TPL (see Equation 8.6). The value of m_c, the current sample mean after c *psu*, varies as more *psu* are examined. Note also that the quantity used to divide VC_w is the total number of *ssu* expected at the end of sampling, but expressed as the number examined thus far plus the number yet to be examined ($nssu_c + (npsu_{tot} - npsu_c)\,\overline{nssu}$).

The inequalities in Equation 8.16 can be made equalities and m_c placed on one side of the inequality so that m_c, rather than cd, is defined in terms of the other parameters. If we now specify $\overline{nssu}$, we can calculate the currently estimated pair of values of the current sample mean (one less than and one greater than cd) that signify the need to examine $\overline{nssu}$ secondary sample units from the remaining *psu*. The equations are as follows:

$$\left.\begin{array}{l} mL_{nssu} = cd - z\sqrt{\dfrac{VC_b}{npsu_{tot}} + \dfrac{VC_w}{nssu_c + \left(npsu_{tot} - npsu_c\right)\overline{nssu}}} \\ \text{and} \\ mU_{nssu} = cd + z\sqrt{\dfrac{VC_b}{npsu_{tot}} + \dfrac{VC_w}{nssu_c + \left(npsu_{tot} - npsu_c\right)\overline{nssu}}} \end{array}\right\} \tag{8.17}$$

If VC_b and VC_w are assumed to be constant and known, Equation 8.17 can be used to calculate mL_{nssu} and mU_{nssu} directly. We can also use TPL models (Equation 8.6) to predict them from the sample mean, but it is the mean that we are trying to solve for! This conundrum can be solved using iteration. Values for VC_b and VC_w are first calculated using the TPL models with μ equal to cd. These variances are then used in Equation 8.17 to estimate starting values for calculating mL_{nssu} and mU_{nssu} by iteration. The iterations continue, one for mL_{nssu} and one for mU_{nssu}. The starting value is used to compute a new pair of variance components; these variance components are again used in Equation 8.17 to obtain a new value for the mean; and the process is repeated until the mean changes by no more than some specified small amount. These means, mL_{nssu} and mU_{nssu}, can then be converted to total counts (counts are easier to use in a look-up table) by multiplying them by the number of *ssu* examined thus far. The calculation of a VIS look-up chart and its use is illustrated in the next exhibit.

Exhibit 8.3. *A VIS Look-up Chart*

The spotted tentiform leafminer (STLM) (*Phylonorycter blancardella* [Fabr.]) is a pest found on the leaves of apple trees. A critical density for the first generation is two eggs per leaf or six eggs in a cluster of three leaves, which is the secondary sample unit. A two-stage structure was set up for sampling: trees as *psu* and clusters of leaves on trees as *ssu*. TPL was estimated for the variance components and the parameters estimated to be $a_b = 0.206$, $b_b = 1.37$, $a_w = 1.872$, $b_w = 1.39$. It was proposed that the number of trees (*psu*) to sample should be seven, and the number of clusters per tree (*ssu*) should be one, two or, at most, three (Nyrop and Binns, 1991). Part of the look-up chart (it is large) is shown in Table 8.5.

The chart is used in the following way. Suppose that three trees (*psu*) and a total of seven leaf clusters (*ssu*) have been sampled. In order for but one leaf cluster to be taken from the next tree, the total number of STLM found must be less than or equal to 29, or greater than or equal to 67. For two leaf clusters to be examined at the next tree, the corresponding range of STLM numbers are: greater than 29 and less than or equal to 30, or greater than or equal to 63 and less than 67. If the number of STLM found thus far is greater than 30 but less than 63, the maximum number of leaf clusters, three, is to be taken from the next tree sampled.

Table 8.5. A VIS chart for sampling spotted tentiform leafminer (STLM). The maximum number of trees (*psu*) is equal to seven, and the maximum number of leaf clusters (*ssu*) per tree is equal to three.

Number of *psu* already sampled, *c*	Number of *ssu* already sampled, $nssu_c$	Maximum number of STLM already counted for taking one *ssu* at next *psu*	Maximum number of STLM already counted for taking two *ssu* at next *psu*	Minimum number of STLM already counted for taking two *ssu* at next *psu*	Minimum number of STLM already counted for taking one *ssu* at next *psu*
1	3	12	13	27	30
2	4	16	17	37	40
2	5	20	22	45	49
2	6	25	26	54	58
3	5	20	21	46	50
3	6	24	36	55	59
3	7	29	30	63	67
3	8	33	35	71	75
3	9	38	30	80	83
4	6	24	25	57	60
4	7	28	30	65	68
4	8	33	34	73	77
4	9	37	39	81	85
4	10	42	43	89	92
4	11	47	48	97	100
4	12	51	53	105	108
...	...	...	...	...	...
6	18	79	79	154	156

The size of VIS look-up charts can be an obstacle to using these charts in practice. Provided that the maximum number of *ssu* or the number of *psu* is not too large, the chart can be manageable. However, one can also see that these charts are more difficult to use than stop boundaries for sequential sampling plans. This aspect of VIS may limit its adoption.

8.6.2 OC and ASN functions for VIS plans

We demonstrated in Exhibit 8.2 how it was possible to estimate OC and ASN values for VIS using a simulated field pattern. The use of spatially explicit data is not practical for evaluating and designing VIS plans, because it would require many simulated field patterns covering a range of mean densities. However, we can generate sample data by simulating the nested structure of sample data and use the generated data to estimate OC and ASN functions. With the exception that the number of *ssu* per *psu* is updated as sample data are collected, the procedure is similar to that described previously for fixed sample size plans: there are no stop boundaries, a fixed number of *psu* are collected, and the final sample mean is compared with *cd* to make a classification.

A method of simulating the nested structure is as follows:

1. Specify a range of true means, μ.
2. For each μ, do *sr* simulations, each time starting with $nssu = nssu_{max}$ (see Table 8.4).
3. Generate a *psu* mean, μ_{psu}, from a negative binomial distribution with mean = m and variance = $a_b\mu^{b_b}$:
- **3a.** Generate *nssu* sample values for the *ssu* at this *psu* (negative binomial distribution with mean = μ_{psu} and variance = $a_w\mu_{psu}^{b_w}$).
- **3b.** Use all of the data so far in this simulation run to calculate the sample mean, m_c, and the sample variance, V_c.
- **3c.** If this is the last *psu* in the sampling plan, go to step 4; otherwise, calculate *nssu* for the next *psu* using Equation 8.15, and go back to step 3.

4. Repeat step 3 for *sr* simulations, always resetting $nssu = nssu_{max}$.
5. Calculate the proportion of the *sr* simulations when m_c was less than or equal to *cd* (OC), and the average number of *ssu* collected (ASN).

The estimation of OC and ASN functions for a variable intensity sampling plan is illustrated in Exhibit 8.4.

Exhibit 8.4. *Simulating OC and ASN Functions for VIS*

We continue to use the STLM example from the last exhibit. Three variable intensity sampling plans were evaluated. Each used seven trees as *psu* and up to three leaf clusters as *ssu*. The critical density was six and the TPL parameters were as given in Exhibit 8.3. The sampling plans differed in the value of *z* used to determine the number of *ssu* required from each primary unit; these values were 1.96, 1.64 and 1.28, corresponding to α values of 0.05, 0.1 and 0.2, respectively. The OC and ASN functions based on 500 simulations are shown in Fig. 8.5. As we have seen in other situations, increasing the sample size by lowering α did not result in an appreciable improvement in classification accuracy.

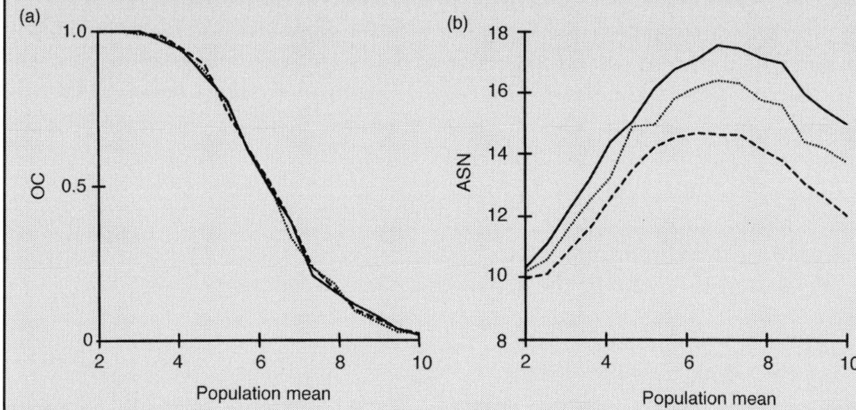

Fig. 8.5. The OC (a) and ASN (b) functions for three VIS procedures used to classify STLM with respect to a critical density of six eggs per leaf cluster. The VIS plans differed in the values of α used; 0.05 (——), 0.1 (···), and 0.2 (- – -).

8.7 Summary

In this chapter we have addressed two further aspects of variability in sample counts that are important to pest management decision-making. The first of these occurs as a result of the nested structure of sample counts. When primary and secondary sample units are used, it is important to estimate the variance correctly using ANOVA. Having done this, the appropriate number of *ssu* to observe from each *psu* can be better determined.

When unpredictable aggregation occurs, it is important that samples be collected from throughout a management unit, so that representative information on pest abundance can be obtained. One way in which this can be achieved is through variable intensity sampling. The main limitation to VIS is the cumbersome look-up chart that must be used to adjust sample sizes, unless a portable computing device is available as samples are collected.

References and Suggested Reading

Cochran, W.G. (1977) *Sampling Techniques*. John Wiley, New York, 413 pp.

Harcourt, D.G. and Binns, M.R. (1980) A sampling system for estimating egg and larval populations of *Agromyza frontella* (Diptera: Agromyzidae) in alfalfa. *Canadian Entomologist* 112, 375–385.

Hoy, C.W. (1991) Variable intensity sampling for proportion of plants infested with pests. *Journal of Economic Entomology* 84, 148–157.

Hoy, C.W., Jennison, C., Shelton, A.M. and Andaloro, J.T. (1983) Variable intensity sampling: a new technique for decision-making in cabbage pest management. *Journal of Economic Entomology* 6, 139–143.

Kish, L. (1965) *Survey Sampling*. John Wiley, New York, 643 pp.

Nyrop, J.P. and Binns, M.R. (1991) Quantitative methods for designing and analyzing sampling programs for use in pest management. In: Pimental, D. (ed.) *Handbook of Pest Management in Agriculture*, 2nd edn, Vol. II. CRC Press, Boca Raton, Florida, pp. 67–132.

Appendix: The Variance of the Sample Mean in Two-stage Sampling

Write X_i as the total count in the ith primary sample unit. Other notation is as in Section 8.3. Using the same formula as in Chapter 2, Equation 2.1, the variance of X_i is estimated by V_X:

$$V_X = \frac{\sum_1^{n_p}\left(X_i - \overline{X}\right)^2}{\left(n_p - 1\right)} \tag{8.A.1}$$

but

$$X_i = \sum_1^{n_s} x_{ij} = n_s x_{i\,.} \tag{8A.2}$$

so

$$V_X = \frac{\sum_1^{n_p}\left(n_s x_{i\,.} - n_s x_{..}\right)^2}{\left(n_p - 1\right)} = \frac{\sum_1^{n_p} n_s^2\left(x_{i\,.} - x_{..}\right)^2}{\left(n_p - 1\right)} = n_s \mathrm{MS}_b \tag{8A.3}$$

There is a result which we now need from statistics, but it is reasonably intuitive. If $y_1, y_2, \ldots, y_n$ are independent and each has the same variance, V, the variance of the sum is equal to nV:

$$\text{variance } (y_1 + y_2 + \ldots + y_n) = n \text{ variance}(y_1) = nV \tag{8A.4}$$

But each X in the above is the sum of n_s secondary sample units, so the variance, V_x, of a secondary sample unit, in the context of two-stage sampling, is given by

$$V_X = n_s V_x \tag{8A.5}$$

so

$$V_x = \mathrm{MS}_b \tag{8A.6}$$

Therefore, using Equation 2.3, the estimated variance of the sample mean of all $n_s n_p$ secondary sample units is $\mathrm{MS}_b/n_s n_p$.

9 Resampling to Evaluate the Properties of Sampling Plans

9.1 Introduction

In previous chapters we have relied on several kinds of models to evaluate the performance of sampling plans (operating characteristics (OC) and average sample number (ASN) functions). In Chapter 2 we used the normal probability distribution (and the assumption of constant variance) to model the distribution of sample means. In Chapter 3 we introduced variance–mean relationships (Iwao and Taylor's Power Law (TPL)) to model the variance of individual sample observations. In Chapter 5, we used the four theoretical distributions introduced in Chapter 4 to model the properties of sample data. In Chapter 7, we used probability models and incidence–mean relationships to model binomial count sample data.

Sometimes, one or more of these models breaks down. In particular, the sample data collected for a given pest may fit none of the four probability models of Chapter 4. This does not reduce us to where we were in Chapter 3 (using the normal distribution model): we can use the actual sample data already collected as a *data-based model* for future sampling. The key element is that real sets of sample data collected at previous times and/or sites and summarized as frequency distributions take the place of a theoretical distribution. OC and ASN functions are estimated by simulation using these data-based distributions, as if they were probability distributions from which subsequent sample unit observations can be derived. This method of estimating the OC and ASN functions is called 'bootstrap resampling' or just 'resampling' (Efron, 1982).[1]

In this chapter we explain how resampling is done to estimate OC and ASN functions and how these functions typically differ from those based on theoretical distribution models. We also look at the effect of the quality of the basic data sets, used for resampling, on the estimated OC and ASN curves.

[1] The word 'bootstrap' is derived from the expression 'to pull oneself up by one's bootstraps'.

9.2 Principles of Resampling

Resampling requires data to sample from. To accumulate data representing the distribution of the pest over the range of density and environmental conditions that are relevant in decision-making, a substantial number of sample data sets must be collected. These are the *basic* data sets. Each of these data sets must be of sufficient size (*nbasic* sample units) to allow it to be representative of the true distribution at that site.

Each basic data set specifies a frequency distribution of sample observations, representing the sampling distribution of the pest at the density or incidence level at the site. In previous chapters, we represented these distributions using mathematical models, but now we use the observed data directly, without first fitting theoretical models to them.

To estimate OC and ASN values for one of the observed sampling distributions, the sampling process is simulated many times, using the observed frequency distribution as if it were a probability distribution. Sampling is simulated with replacement (see Section 2.7). To generate OC and ASN functions, this procedure is repeated for each available data set. For instance, if we have 40 data sets, resampling would be done for each set, resulting in 40 values to characterize the OC function and 40 values to characterize the ASN function. These indicators would be plotted, as before, against some location parameter for the population, such as mean density or incidence calculated from the entire basic data set.

The only thing that is different from evaluation of sampling plans in earlier chapters is that observed data rather than models are invoked to represent the sampling distribution. As a result, the distribution of data points over the density or incidence axis of OC and ASN functions is determined by pest abundance in the basic data sets. It is not controlled during the simulation. We illustrate the process with a simple example (Exhibit 9.1).

Exhibit 9.1. *Resampling from two aphid data sets*

We illustrate the technical principles of resampling using two sets of aphid data collected by Müller (1953). On two occasions during one season the abundance of black bean aphids (*Aphis fabae* Scop.) was estimated on individual field bean plants (*Vicia faba* L.). Abundance was measured on all plants in the field according to a nine-point scale:

Score	Description
0	No attack
1	Initial infestation by small colonies of single-winged immigrant mothers with or without offspring. The colonies are not conspicuous at superficial inspection of the plant
2	Top shoot so densely infested with aphids that it conspicuously stands out as 'black' at superficial observation
3	In addition to the top shoot, other shoots look 'black' with aphids; however, the main stem is not infested

4 In addition to the top shoot, at least three internodes below the top of the main stem are covered with a black mantle of aphids; the plant is starting to suffer
5 The whole main stem is black with aphids from top almost to the bottom; the plant is suffering severely
6 The stage beyond 5 when depopulation results from departure of winged aphids; the black mantle is 'gappy', nevertheless white exuviae indicate the denseness of the former colony; the plant is often severely injured
7 As 6 with the top cracked; the plant is withered, but with some more or less green tissue remaining
8 The plant is completely black or brown, dried up or collapsed

The frequencies of data collected on 27 June and 24 July 1952 are shown in Fig. 9.1. The negative binomial distribution (our versatile workhorse for count data) does not fit either set of frequencies. We therefore resort to resampling from the two observed distributions to assess the performance of sampling. In doing so, we act as if each observed frequency distribution represents a probability distribution for infestation categories.

We do the resampling using the second simulation method described in Section 2.7. There are three preliminary steps:

1. Obtain the sample frequencies; for the first data set (27 June), these are 404, 345, 55, 36, 23, 11, 6, 0, 0 for the nine classes (0, 1, 2 … 8).
2. Calculate the cumulative frequencies 404, 749, 804, 840, 863, 874, 880, 880, 880.
3. Divide by total sample size (880): 0.459, 0.851, 0.914, 0.955, 0.981, 0.993, 1.000, 1.000, 1.000. We call these cumulative relative frequencies.

Sampling is done from these cumulative relative frequencies as follows:

1. A random number, X, between 0 and 1 is generated.
2. The smallest cumulative relative frequency is found which is greater than or equal to X, and its corresponding sample data value is recorded. For example, if $X =$

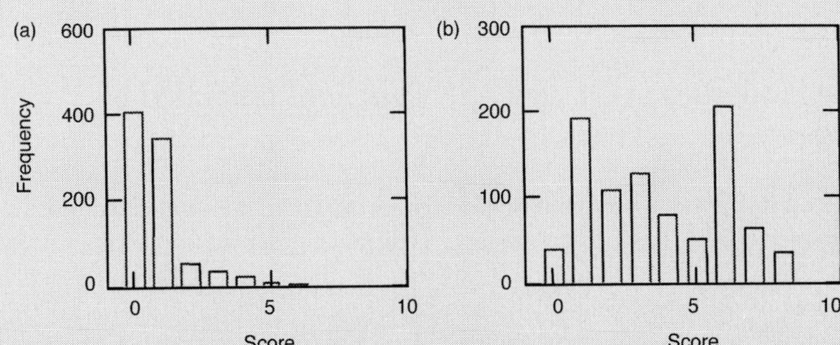

Fig. 9.1. Frequency distributions for Müller's two fields. (a) 27 June, mean score = 0.848; (b) 24 July, mean score = 3.68.

Continued

0.6, the smallest cumulative relative frequency is 0.851, and the corresponding sample data value is 1. The generated random sample value, therefore, is 1.

These two steps are done *sr* times, where *sr*, as before, represents the number of simulation replicates. For example, if $sr = 10$:

X	0.956	0.539	0.462	0.862	0.780	0.997	0.611	0.266	0.840	0.376
Score	5	1	1	2	1	6	1	0	1	0

with sample mean equal to 1.8. The value '1.8' is interpreted as 'close to the severity level for score equal to 2, but not quite so bad'.

Resampling from the two data sets was used to estimate the OC function for fixed sample size full count plans (n = 25, 50 and 100) with a critical infestation category equal to 1. The results (Fig. 9.2) are not satisfactory as a means to evaluate a sampling plan, because with only two basic data sets, we have only two points on the OC curves. Clearly, more basic data sets are required to gain a sound understanding of sampling performance.

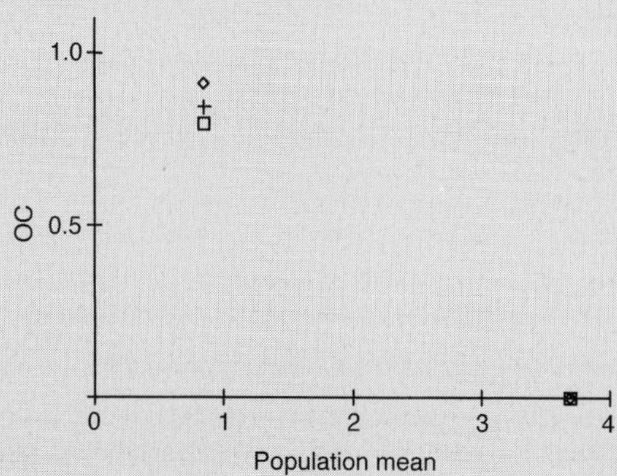

Fig. 9.2. OC functions for Müller's data, with $cd = 1$, 1000 simulations. Fixed sample sizes: $n = 25$ (□), $n = 50$ (+) and $n = 100$ (◇).

9.3 Typical Properties of OC and ASN Functions Estimated by Resampling

Provided that enough simulation replicates are used, OC and ASN functions based on theoretical distributions are smooth, as we have observed in previous chapters. The reason is that shapes and properties of theoretical distributions change smoothly as the mean density, μ, changes. OC and ASN functions estimated by resampling, however, are rarely smooth. The reason is that, even when their means are close together, the shapes of observed frequency distributions can be somewhat different (for example, their variances may differ greatly). These differences can result in considerable variation (or jaggedness) in the OC or ASN functions. Ordinarily, this variation would be due to natural random variation, but there is always the possibility that one or more of the data sets may not properly represent the distribution of the pest. OC and ASN functions which illustrate natural variation are desirable in that they show what could happen in practice. Mistaking inadequate representation for random variation is undesirable in that the OC and ASN functions may give a false impression of what could happen in practice. It is therefore important to analyse the basic data used in resampling.

Analysis includes examining the data sets where variation is large, and deciding whether one or more data sets must be rejected because they do not truly represent the pest distributions in the crops. It also includes assessing whether the coverage of means (or incidences) is adequate for displaying the evaluation properties of a sampling plan. If coverage is inadequate, more data sets should be collected. Coverage and representativeness are key elements. We illustrate some aspects of the analysis in Exhibit 9.2.

Exhibit 9.2. *Resampling to estimate the OC function of a full count fixed sample size plan*

This exhibit uses resampling from 40 data sets of caterpillar density in cabbages to estimate OC functions for fixed size sampling plans. It illustrates that the OC curves obtained by resampling are more variable (jagged) than those obtained in simulations based on variance or probability models, and it explores the reasons why.

The pest The diamondback moth, *Plutella xylostella,* is a pest of cabbages. In New York State, the number of larvae plus pupae is used as an indicator of damage potential. In this exhibit, we use a critical density of 0.5 larvae and pupae per plant. This corresponds to a critical proportion of 30% incidence (tally number, $T = 0$), which has been used in practice to assess incidence levels in cabbage plants grown for processing into sauerkraut.

The basic samples During one season, five fields of cabbages were each sampled on seven, eight or nine occasions. The total number of sampling occasions was 40. Between 236 and 308 sample units (cabbage plants) were examined on each occasion by up to three observers. The numbers of diamondback moth larvae and pupae were counted on each sample unit. For simplicity, possible differences among observers are ignored.

Estimating the OC and ASN functions by resampling Fixed sample size plans were used with sample sizes, n, equal to 25, 50 or 100 sample units. The OC functions obtained by 1000 simulation replicates for each of the 40 data sets, are shown in Fig. 9.3. As we have become used to noticing, OC functions become steeper as the sample size increases. Because there were few data sets with mean values near the critical density, we cannot be very confident about the effect of sample size on the accuracy of classification here. Furthermore, the OC function near and just below 0.4 is jagged, especially for $n = 25$. Nevertheless, the general effect of increasing sample size – that is, increasing the steepness of the OC functions – corresponds to what we have seen in previous chapters.

Jaggedness and variability of an OC function estimated by resampling OC functions estimated by resampling are rarely smooth. Numerical methods can be used to smooth them, and this is often useful to understand the general effects of

Continued

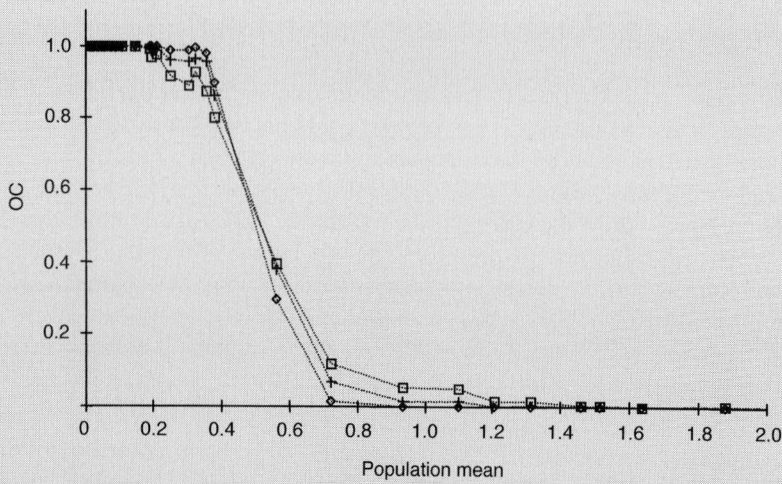

Fig. 9.3. OC functions for the diamondback moth data, with $cd = 0.5$, 1000 simulations. Fixed sample sizes: $n = 25$ (□), $n = 50$ (+) and $n = 100$ (◇).

changing the parameters of sampling plans. In Fig. 9.3, the OC function near and just below 0.4 is rather jagged, especially for $n = 25$. We noted in Chapter 2 (Equation 2.7) that the standard error of a point on the OC function estimated by *sr* simulation replicates was

$$\text{simulation } se = \sqrt{\frac{p(1-p)}{sr}}$$

where p is the OC value. When $sr = 1000$ and $p = 0.9$, the standard error is 0.009, which cannot reasonably account for the jaggedness in Fig. 9.3. As illustrated in previous chapters, OC functions based on theoretical probability distributions are smooth (when enough simulations are done) because the shape of the probability distribution changes smoothly as the population mean, μ, increases. In resampling, however, we use distributions based on real data. The properties of these distributions (e.g. variance, standard deviation) can change abruptly even if the mean densities are very close. The implications for classification sampling can be assessed. This is exemplified here, in particular by the six distributions whose mean values were just less than 0.4, and for a sample size equal to 25:

Mean, μ	0.21	0.25	0.3	0.32	0.35	0.38
Standard deviation, σ	0.37	0.71	0.54	0.39	0.44	0.78
$\sigma / \sqrt{25}$	0.07	0.14	0.11	0.08	0.09	0.16
$P\left(z > \dfrac{0.5-\mu}{\sigma/\sqrt{n}}\right)$	< 0.001	0.04	0.03	0.01	0.05	0.23

Compared with their neighbours, the second and third of these distributions have large standard deviations. The standard deviations of sample means based on 25 sample units $\sigma / \sqrt{25}$ can be used to estimate the probability of a sample mean based on 25 sample units being greater than 0.5 (assuming that $n = 25$ is enough for the Central Limit Theorem; see Sections 2.9 and 2.10). The probability does not decrease smoothly as μ increases, and there is more than a negligible chance that the second and third data sets would be classified as greater than 0.5. As a result, the two distributions with the larger values for standard deviation have the lower OC values. To understand more precisely how large the chance is of the sample mean becoming greater than 0.5, we need to look more precisely at the distributions of these means.

We used the normal approximation in the above argument, but it turns out that 25 sample units is not enough here. For the distributions with mean values equal to 0.25 and 0.30, the simulated distributions of sample means based on $n = 25$ are not close to normal (Fig. 9.4a and b). More importantly, the long tails to the right of these simulated distributions extend well beyond the critical density, $cd = 0.5$, and well beyond the upper classification limits calculated above assuming a normal

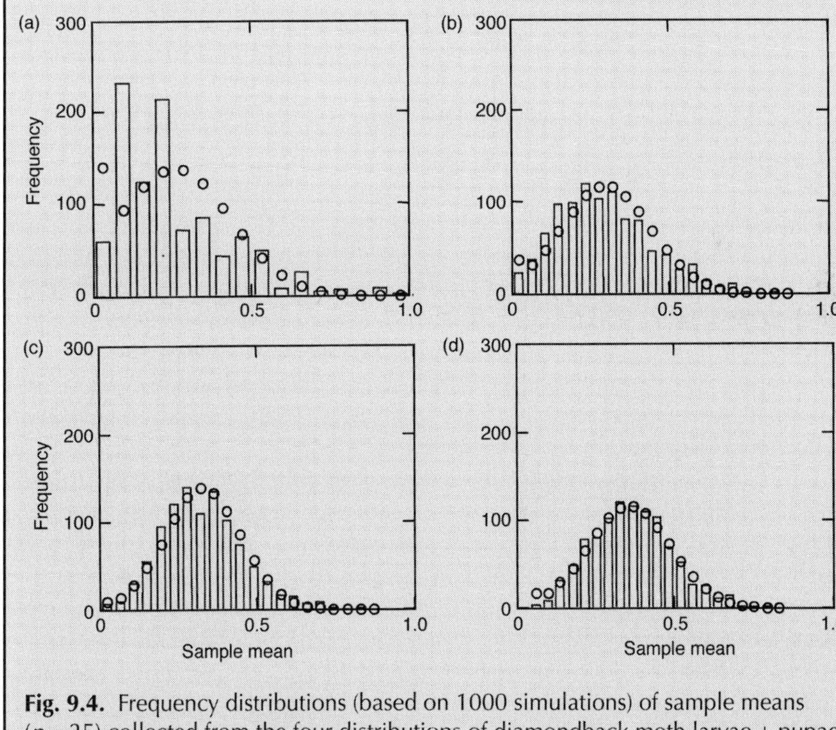

Fig. 9.4. Frequency distributions (based on 1000 simulations) of sample means ($n = 25$) collected from the four distributions of diamondback moth larvae + pupae with means equal to (a) 0.247, (b) 0.304, (c) 0.322 and (d) 0.352. Best fitting normal distributions are shown by circles.

Continued

distribution. Based on these simulations, the probabilities of the sample mean being less than 0.5 are 0.90, 0.91, 0.92 and 0.87 for the distributions with mean values equal to 0.25, 0.30, 0.32 and 0.35, respectively (Fig. 9.4). These probabilities match the corresponding points in the OC shown in Fig. 9.3. When $n = 100$, the effect of the long tail in the original distribution has essentially disappeared, so the OC function is smooth. For $n = 50$, the results are intermediate.

This analysis does not in itself suggest anything amiss with the two data sets with means equal to 0.25 and 0.30. The original data sets and any information on how they were collected may shed more light. Because nothing untoward was found, and because these were two adjacent data sets that showed the effect, the jaggedness can be accepted as due to random variability. As such, it is to be welcomed as illustrating what might happen if this sampling plan were to be used in practice. It is a valuable 'reality check'.

Another noticeable feature of the OC function in Fig. 9.3 is that there are few data sets with means close to *cd* (*cd* = 0.5). There is likely to be as much variability near and above $\mu = 0.5$ as we found near $\mu = 0.3$, but there are no data sets to illustrate it. Potential users might want to know how great the variability is in this region, so that they can better assess the practical properties of the plan. More data sets may be needed. We are thus revisiting the discussion on end-user involvement in designing sample plans (see Section 6.2).

9.4 Collecting Data for Resampling: the Number and Size of the Basic Samples

An important practical question is: How many basic samples are required, and how large should they be? A generally valid and precise numerical answer is impossible, but it is obvious that resampling will give more robust answers if there are more basic data sets and if each of these is based on a larger sample size. The same principles apply – of course – when collecting basic data for the probability models, incidence–mean relationships and variance–mean relationships that are used throughout this book for sampling plan evaluation. There are special guidelines, however, for collecting data sets for resampling.

One of the things that might concern us is that, if the basic sample size (*nbasic*) is small, the sample mean of the data set may not be a very accurate estimate of the true density, and the observed frequency distribution may differ markedly from the 'true' distribution. For instance, if we were sampling from a field where the distribution of pests is long-tailed, we would be ill-advised to collect only a few sample units. If we used *nbasic* = 25, whether or not data in the tail of the true distribution were collected for the basic data set, the sample distribution could not reasonably be expected to represent the true distribution. Hence, the resampling OC and ASN for this particular (sample) mean would not truthfully represent the OC and ASN values at the particular 'true' density of the pest. Fortunately, the situation is not quite as bad as it might seem. We illustrate in Exhibit 9.3 that the means and estimated OC points for basic sample data sets generally move in concord: if the mean is greater than the true mean, then the OC point is less than the true OC point, and vice versa. There is no guarantee of this, but we have found it to occur frequently, provided that *nbasic* is not too small. In Exhibit 9.3, *nbasic* = 25 is shown to be too small. This is not unreasonable when we recall that 25 sample units are certainly not enough to test whether a theoretical distribution fits the data (see Section 4.12).

Another concern, when using resampling from observed distributions to calculate OC and ASN functions, is that we may end up with an uneven distribution of data points on the pest density (or incidence) axis. This would be especially bothersome if a 'gap' were to occur in a range of pest densities which is relevant in our evaluation of the quality of decision-making; for example, close to *cd*. It follows that basic sample data sets should continue to be collected until the range is satisfactorily represented.

As a guideline, Naranjo and Hutchison (1997) suggest that, to use resampling for sampling plan evaluation, at least 10–20 basic samples are needed and the basic sample size should not be less than 50. More samples and greater sample sizes will undoubtedly improve the representativeness of the basic samples. This is certainly true if we need to consider a wide range of pest abundance and if the distributions are long-tailed. A good representation is especially needed for pest densities close to the critical level.

Exhibit 9.3. *The effect of the size of the basic data sets on OC functions of fixed size full count sampling plans*

Introduction using Müller's data set (Exhibit 9.1) Suppose that each of Müller's two data sets contained not 880 units, but fewer. We can investigate what effect this would have on the accuracy of OC functions by taking subsamples from the original basic data sets and using these subsamples as our basic data sets.

Ten random subsamples of *nbasic* = 25 units and ten subsamples of *nbasic* = 100 units were taken without replacement from each of Müller's sets. Using the resulting 42 data sets (two of 880 units, 20 of 100 units and 20 of 25 units), we can evaluate the OC function for a fixed size classification plan with sample size equal to 25 and a critical incidence level equal to 1. Data points are shown in Fig. 9.5.

Twenty one of the points are grouped around the mean class incidence level of 0.8 (27 June) and the other 21 points are grouped around a value of 3.7 (24 July). Not surprisingly, the points (diamonds) for *nbasic* = 25 are more spread out along the horizontal axis than the points for *nbasic* = 100 (pluses). The spread is due to the class mean of a subsample varying according to the random process of selecting *nbasic* units out of 880. As far as the points for the first sampling date (27 June) are concerned, the OC value decreases as the mean incidence class increases. Strikingly, all the 21 points are on a single smooth line, suggesting that, in this example, there are no serious departures in the shape of the distribution in relation to the location parameter, as the basic sample size varies. This illustrates what was suggested above: the means and estimated OC points for basic sample data sets move in concord. The fact that a data set mean is not likely to be equal to the true mean is not a *prima facie* barrier to using resampling.

Continued

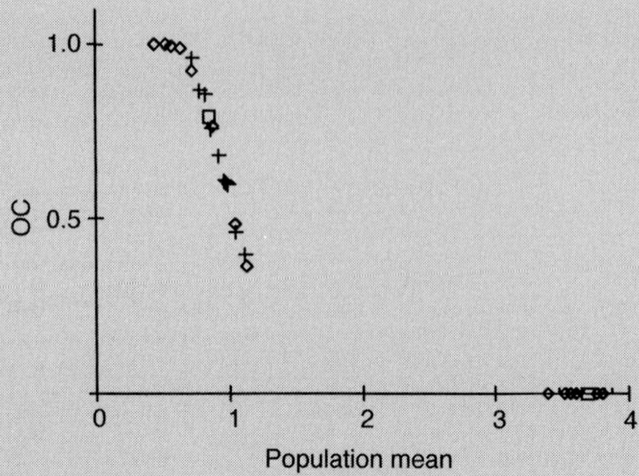

Fig. 9.5. The effect of the size (*nbasic*) of the basic data set on the OC function estimated by resampling for fixed sample size sampling (Müller's aphid data, *cd* = 1, *n* = 25, 1000 simulations). Use of the two original data sets with *nbasic* = 880 sample units each (□); use of 10 random subsets of each original data set as data sets for resampling, each subset containing *nbasic* = 100 sample units (+); use of 10 random subsets of each original data set as data sets for resampling, each subset containing *nbasic* = 25 sample units (◇). Points occur in two clusters that correspond to the two data sets.

The diamondback moth data We now perform a similar exercise using the diamondback moth data, introduced in Exhibit 9.2. Here, the basic sample sizes in our original data varied from 236 to 308. What if we had been more skimpy and instead collected only *nbasic* = 100 or even *nbasic* = 25 sample units for each of the data sets? What would the resampling OC functions look like? As with Müller's data, we simulated these two approaches to get one collection of 40 data sets with *nbasic* = 100 and another collection with *nbasic* = 25. These new series of data sets were then used to estimate the OC function for a fixed sample size full count sampling plan with *n* = 25. The number of simulation replicates was 1000 for each data set.

The OC functions are shown in Fig. 9.6 along with the OC function based on all the data. The results show shifts both in the population mean value, and in the OC value, for each of the data sets. These differences are greatest for the smallest basic sample size. Although the general shape of the OC function is maintained for *nbasic* = 100, it is not maintained for *nbasic* = 25 (e.g. the jaggedness is no longer there). When *nbasic* is small, we can easily obtain unrepresentative results.

To investigate what might happen in general, repeated subsampling was used to obtain 10 different data sets (without replacement) of *nbasic* = 100 sample units from each of the original 40 data sets. All these 400 data sets were then used to estimate the OC function for fixed sample size full count sampling, with *n* = 25 as before. The results are shown in Fig. 9.7, together with the OC function for the original basic data sets.

The OC function for the 400 data sets follows that for the original data set quite well. The jaggedness around $\mu = 0.3$ is sometimes lost or replaced by mild 'variability'. In practice, only one data set is available for one field, so even with *nbasic* = 100 we might have lost the jaggedness/randomness. Nevertheless, ignoring variability, the shape of the OC function would not be altered enough to cause much concern.

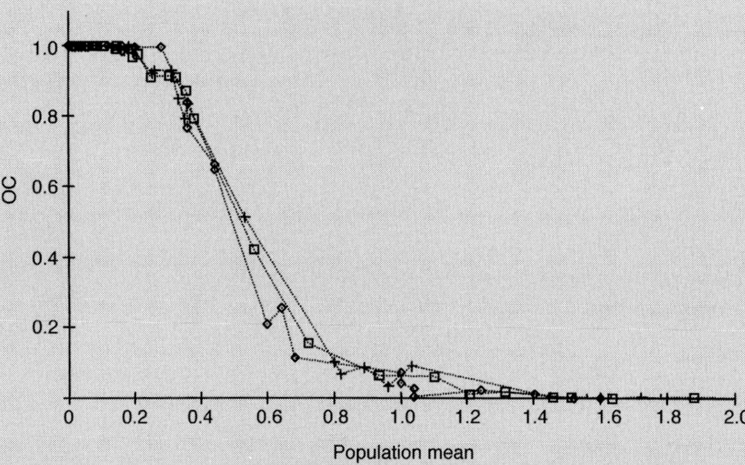

Fig. 9.6. The effect of reducing the size of the original data set on the OC function for fixed sample size sampling with $n = 25$ (diamondback moth data, $cd = 0.5$, 1000 simulations). Resampling from all the data (□); *nbasic* = 100 (+); *nbasic* = 25 (◇).

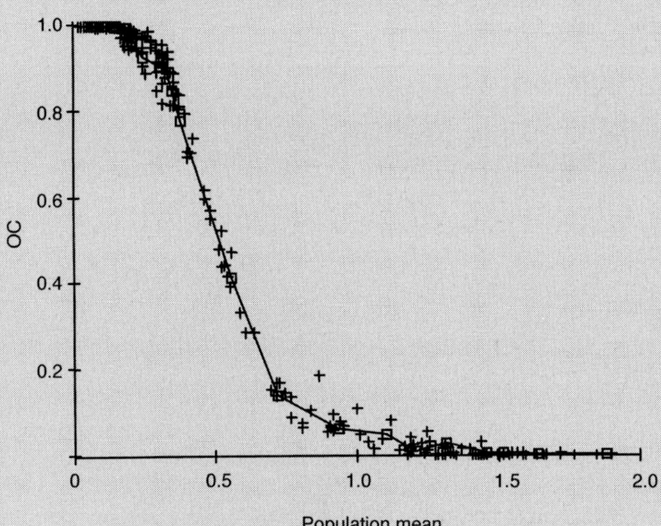

Fig. 9.7. The effect of the size (*nbasic*) of the basic data set on the OC function estimated by resampling for fixed sample size sampling (diamondback moth data, $cd = 0.5$, $n = 25$, 1000 simulations). Use of the 40 original data sets (□); use of 10 random subsets of each original data set as data sets for resampling, each subset containing *nbasic* = 100 sample units (+).

9.5 Sequential Sampling

Resampling can just as well be used to evaluate sequential sampling plans. In addition to the OC function, we also estimate the ASN function. As is clear from the definitions of the Iwao and Converging Line stop boundaries (Equations 5.1 and 5.6), all that is needed are estimates of the critical density or proportion (*cd* or *cp*) and the variance at this critical value. Although, in theory, Wald stop boundaries require a distributional model, they can be used with a normal distribution model and estimates of μ_0 and μ_1.

Plans are set up exactly as in Chapter 5, optionally making use of a variance–mean relationship if desired to estimate the variance at *cd* (see Table 5.1). Simulation is done as before, by resampling from the cumulative probabilities based on the data. Any variability that exists is contained in the data distributions used in the simulations. Therefore, variation around TPL is not considered during the simulation process.

The general effects of changing the parameters of sequential plan stop boundaries (Table 5.2) are unchanged. The major difference is that it is not possible, or even sensible, to test the effect of not knowing σ^2 well (i.e. testing the accuracy of the variance, *V*, used to set up the stop boundaries). However, sample variances encountered in future applications of a sampling plan may be greater or smaller than those found in the basic data sets. A partial test for this would be to replace the variance used to set up the plan with one that is smaller or larger than that estimated from the data, and repeat the simulations.

Exhibit 9.4. *Resampling to estimate OC and ASN functions of sequential sampling plans*

Three full count sampling plans for diamondback moth with $cd = 0.5$ are compared using resampling. The fixed sample size plan with $n = 25$ of Exhibit 9.2 was compared with Iwao and Converging Lines plans with a minimum sample size, *minn*, equal to 5 and a maximum sample size, *maxn*, equal to 50. Additional parameters are $\alpha = 0.2$ for Iwao, and $\alpha U = 0.001$ and $\alpha L = 0.01$ for Converging Lines. The parameters were chosen to make the OC functions similar. TPL was used to estimate the variance, with parameters estimated from the data: $a = 2.93$ and $b = 1.25$. The stop boundaries are shown in Fig. 9.8a.

As the OC functions are similar (Fig. 9.8b), the plans can be compared by their ASN functions (Fig. 9.8c). On average, the Converging Lines plan generally required fewer sample units than the other two plans. We noted a tendency for Converging Lines to need fewer sample units than either Wald or Iwao in Exhibit 5.4. Here, we have another demonstration of the superior efficiency of Converging Lines plans.

Note that the OC functions for the sequential plans display the same jaggedness around $\mu = 0.3$ that we noted when evaluating the fixed sample size plans with $n = 25$ in Fig. 9.3. Again, it is the variability in the shape of the frequency distribution that causes these effects. In Fig. 9.3 the extent of the jaggedness diminishes as the sample size increases; for sequential plans, this would require wider stop boundaries and/or larger maximum sample sizes.

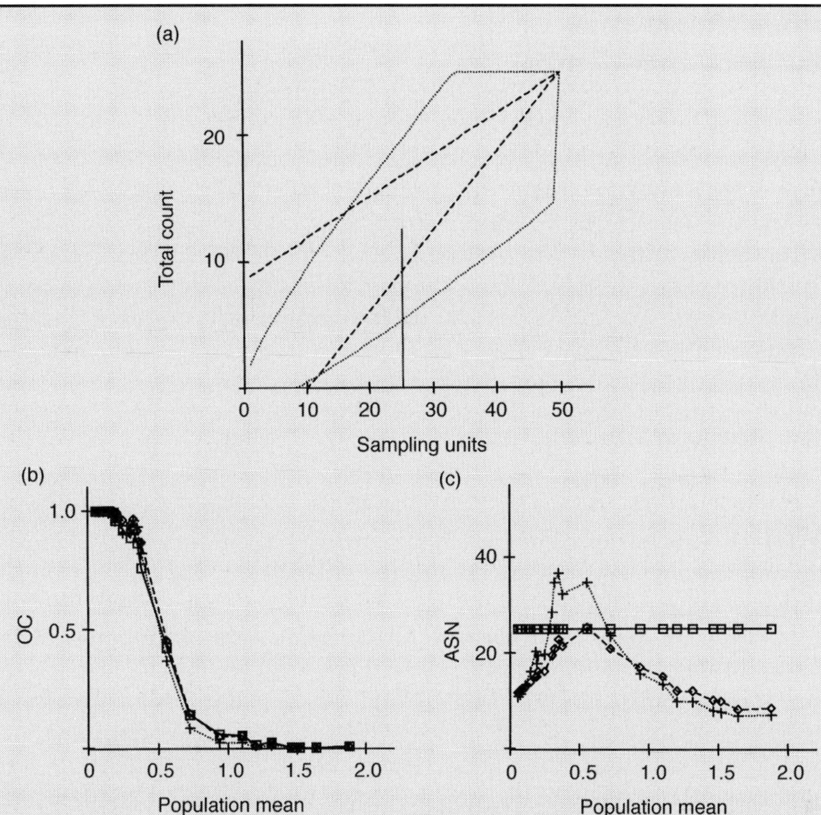

Fig. 9.8. Stop boundaries (a), and OC (b) and ASN (c) functions, for three sampling plans with $cd = 0.5$: fixed sample size, $n = 25$ (□); Iwao, $minn = 5$, $maxn = 50$, $\alpha = 0.2$, V determined from Taylor's Power Law, TPL (+); Converging Lines, $minn = 5$, $maxn = 50$, $\alpha U = 0.001$, $\alpha L = 0.01$, V estimated from TPL (◇). TPL parameters $a = 2.93$, $b = 1.25$.

9.6 Binomial Sampling

We can use resampling methodology to estimate the OC and ASN functions of binomial sampling plans. Three things need to be established first: (i) a tally number T, (ii) a critical proportion, cp_T, and (iii) the proportion, p, of sample units with more than T pests in each of the basic data sets. Determining a tally number proceeds as before by considering the critical density, sampling plan performance for alternative choices of T, and practical considerations.

When not using a probability distribution model, cp_T was determined in Chapter 7 by fitting an incidence–mean model:

$$\ln(-\ln(1-p)) = c_T + d_T \ln(\mu) \tag{9.1}$$

Substituting *cd* for μ, and solving for cp_T:

$$cp_T = 1 - e^{-e^{c_T} cd^{d_T}} \tag{9.2}$$

Equation 9.2 can be used with resampling. However, variability about the model (Equation 9.1) has consequences for resampling estimates of the OC function. These are best described by referring to an example.

The data and fitted empirical binomial model for the diamondback moth data are presented in Fig. 9.9, which shows the critical density ($cd = 0.5$) and the critical proportion with $T = 0$ ($cp_0 = 0.274$). When the model-based methods of Chapter 7 are used to assess sampling performance, probabilities, p, corresponding to a range of mean values, μ, are calculated from the fitted model and the OC function is estimated by simulation sampling from binomial distributions with probabilities equal to the calculated p values (if random variability is requested, the p values are adjusted according to Equations 7.13 and 7.14). When resampling is used to assess sampling performance, the p values are given by the data points shown in Fig. 9.9. The differences to be expected between the empirical binomial method and resampling are related to how well the data points in Fig. 9.9 fit the model, especially near the critical density, *cd*.

If a data point in Fig. 9.9 is *above* the fitted empirical model, the corresponding point for the OC function estimated by resampling lies *below* the OC function estimated by the empirical binomial method (and vice versa if it is below the model). This is because a higher value of p means that the 'intervene' classification is, according to resampling, more likely. How far these OC points differ from the empirical binomial model OC function depend on how far the data points are from the fitted model and how near they are to cp_T. If a data point and its corresponding

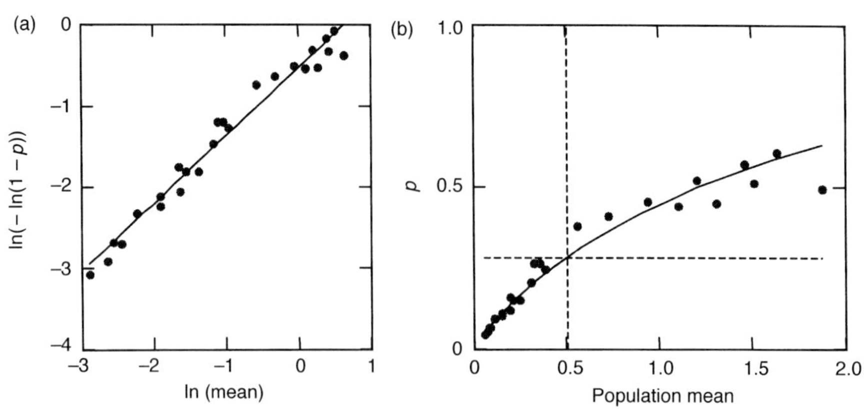

Fig. 9.9. Data and a fitted empirical binomial relationship for the diamondback moth data ($T = 0$). (a) In the linearized form; (b) in the untransformed form, with the critical density (0.5) and critical proportion based on the model (0.274) indicated by dotted lines.

model value are far from cp_T and sample size is high, it does not matter how bad the fit is: the OC will be near 0 or 1 by both methods. But as the data point or model approaches cp_T, then the differences described above will begin to appear. This is illustrated in Exhibit 9.5.

If the data points not only fit the model badly over a range of μ, but the departures from the model are consistently above or below the model or there is a pattern to the departures, the possibility must be addressed that the model (Equation 9.1) may be inappropriate. If the model is inappropriate, then another way of estimating cp_T must be found. Because the data near cd in Fig. 9.9 are above the fitted model, we might consider that the model is incorrect there, and use a different method to estimate cp_T. Using linear interpolation, we would obtain $cp_T = 0.336$. We can also note that the critical proportion used in New York State decision sampling is 30% (Exhibit 9.2), which is between the value estimated by the model (i.e. 0.274) and 0.336. Without discussions with those involved in controlling the pest, along the lines discussed in Chapter 6, all we can do here is recognize that this (choosing a value for cp_T) is an issue that requires attention. We shall proceed on the assumption that the model is appropriate and that 0.274 is a good estimate of cp_T.

Random variation in a binomial count OC function estimated by resampling arises from the above considerations (i.e. the values of p relative to cp_T and/or the empirical binomial model) and not from any other properties (e.g. variances) of the frequency distributions of the data sets, other than through their effect on the relationship between p and μ. This means that the jaggedness or random variation in binomial count OC and ASN functions may look different from that found in the OC and ASN functions for full count plans.

Exhibit 9.5. *Resampling, compared with empirical binomial simulation of sampling performance*

The empirical binomial relationship for the diamondback moth data is shown in Fig. 9.9. Near *cd*, all of the data points are above the fitted relationship. At each of these points, therefore, the probability of incidence is greater than determined by the relationship. This means that the OC values estimated by resampling will be lower than those estimated by empirical binomial simulation. As a result, the OC function estimated by resampling will be left-shifted compared to the function estimated by empirical binomial simulation. This is illustrated by using both methods to estimate an OC function for the same binomial count sampling plan. The sampling plan was fixed sample size with $n = 25$, $cd = 0.5$ and $T = 0$, so $cp_0 = 0.274$ (based on the fitted empirical binomial model). This plan was evaluated by using the empirical binomial method described in Chapter 7 and by resampling. The empirical binomial method starts with a user-provided range for μ, calculates values of p from the incidence–mean model (Equation 9.1) and estimates OC and ASN functions by simulation. The resampling method calculates values of p directly from the data sets (the points in Fig. 9.9) and estimates OC and ASN functions by simulation. The estimated OC functions (1000 simulations) are shown in Fig. 9.10.

Continued

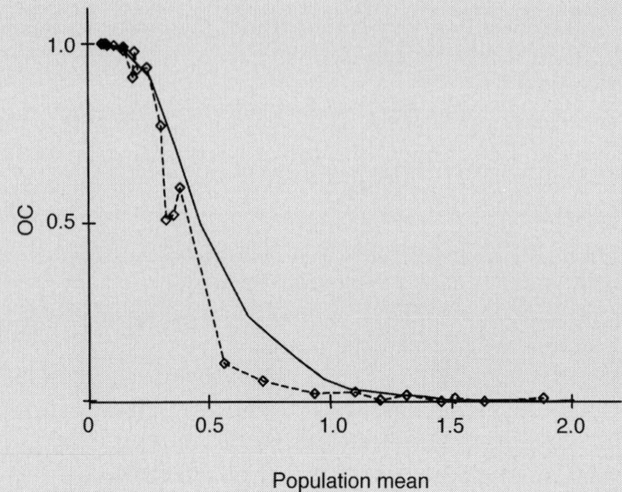

Fig. 9.10. OC functions for a fixed sample size binomial count sampling plan for diamondback moth larvae + pupae ($n = 25$, $T = 0$, $cd = 0.5$, $cp_0 = 0.274$) estimated by two methods (1000 simulations). Use of the empirical binomial relationship with variability (—); use of binomial count resampling (◇).

The OC function for empirical binomial is similar to the full count OC function (e.g. Fig. 9.6) but without the jaggedness. The jaggedness is not there because the empirical binomial model removes the effect of the raw data, and always produces a smooth curve. The OC function for resampling is quite different. Because all of the data points from about $\mu = 0.3$ to $\mu = 1$ are above the fitted model in Fig. 9.9, the OC function estimated by resampling is below the OC estimated by the empirical binomial model; some of the points are well below, because their data points in Fig. 9.9 were well above the fitted relationship.

It is also worth noting that the jaggedness found in OC functions for full count plans is still there, but is relatively minor compared with the effects noted in the previous paragraph.

As in Exhibit 9.3, we must consider the implications of this analysis in terms of the comprehensiveness and accuracy of the data sets. Are the deviations from the empirical binomial OC function due to random variation or are some data sets suspect? As we noted in Chapter 7, estimates based on binomial count sampling are less precise than those based on full count sampling, so we must expect more variability in OC functions based on binomial count resampling. In this light, we have no reason to doubt the representativeness of all the data sets.

9.7 A Comparison Between Resampling and Model-based Methods

We would expect that, if two different methods for estimating OC and ASN functions use the same basic data and account for the same basic principles, they should give similar results. Therefore, if we fitted probability distributions to each individual basic data set and the fits were good, and we then used the fitted distributions to estimate OC values, we would expect similar patterns of scatter and jaggedness as we have seen for OC values estimated by resampling directly from the basic data sets. Using the diamondback moth data, we illustrate some relationships between OC functions estimated by resampling and by models including parameter uncertainty, and demonstrate that uncertainties in OC and ASN values can be adequately captured by using models with parameter uncertainty in the evaluation process.

Exhibit 9.6. *Comparing estimates of OC and ASN functions based on a theoretical distribution and on resampling*

The diamondback moth data used in previous exhibits fit the negative binomial distribution. A summary of χ^2 significance probabilities (P) for all 40 data sets after grouping so that the expected frequency in the tails is at least 1 (see Section 4.5) is as follows:

	Too few frequency classes	< 0.01	$> 0.01–< 0.05$	$> 0.05–< 0.1$	$>0.1–< 0.5$	$> 0.5–< 1$
P	12	0	1	4	11	12

With one P significant at the 5% level out of 28 testable sets of frequencies, it is possible to assert that the negative binomial distribution is an acceptable model for diamondback moth larvae + pupae data (but see Section 4.12). This means that we can use the whole apparatus of Chapters 5 and 7 to estimate the OC and ASN functions, and we can compare the OC and ASN functions estimated by the negative binomial distribution and by resampling.

We now compare three different ways of calculating the OC function for a fixed sample size ($n = 25$) full count plan with a critical density of $cd = 0.5$:

1. Evaluation by resampling, as in Exhibit 9.2.
2. Evaluation by first fitting a negative binomial distribution to each of the basic data sets and using the theoretical distributions based on the fitted values of μ and k.
3. Evaluation by using the negative binomial distribution, letting k be estimated from the fitted TPL (a = 2.93, b = 1.25, mse = 0.138) with random variation (see Section 3.4 and the Appendix to Chapter 5).

The OC functions for evaluation methods 1 and 2 are close to each other (Fig. 9.11). This demonstrates that the negative binomial is a good model for each data set separately. The features of the OC function as estimated by resampling can be captured by simulation based on the negative binomial probability model using the fitted value of k for each data set. The OC for evaluation method 3 is close to the other two below 0.5, but is higher between 0.5 and 1. This is because, for basic data sets with means in the interval 0.5 to 1, the k values were somewhat higher

Continued

than their expectation based on TPL (Appendix and Fig. 9.12). As noted in Chapter 4, a higher *k* means a lower variance. Correct classifications are more easily made at a lower variance (see Section 3.2.2), so for population means greater than *cd*, the OC function should be lower for evaluation methods 1 and 2.

Here, we see an advantage to being able to assume a probability model for counts. If the above model is correct – and it appears to be sustainable for the data to hand – we can be fairly sure that the OC function for plan 3 is a good estimate of what would happen in general if the plan were to be adopted for practical use.

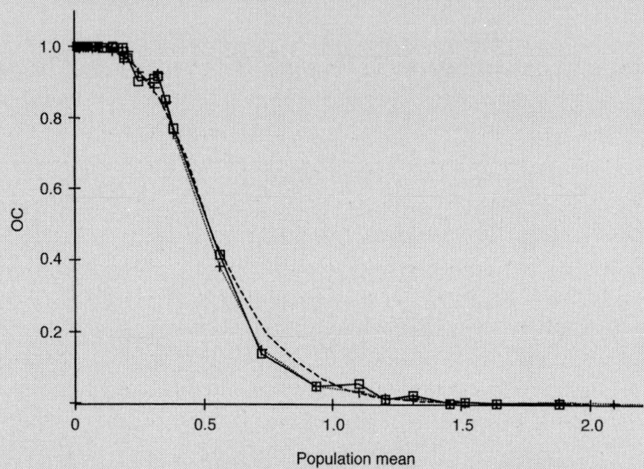

Fig. 9.11. OC functions for a fixed sample size full count sampling plan for diamondback moth larvae + pupae ($n = 25$, $T = 0$, $cd = 0.5$) estimated by three methods (1000 simulations). Resampling (□); negative binomial using *k* values fitted for each data set (+); negative binomial estimating *k* from TPL with random error (----).

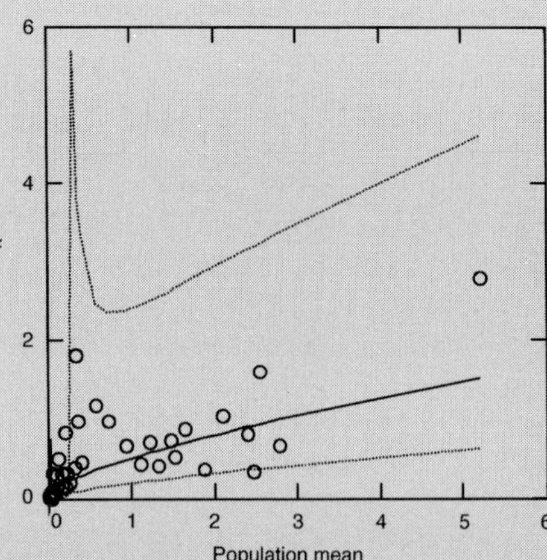

Fig. 9.12. Diamondback moth larvae + pupae data. Model for estimating *k* from TPL (——) and 95% confidence intervals (···); estimated *k* values (○).

Exhibit 9.6 illustrates that, if a theoretical distribution fits the data sets, then including variability about TPL in the simulation can summarize the average OC and the spread around it that would occur in practice. The great advantage of resampling is that it illustrates what can happen with any particular decision-making sample. OC functions based on resampling may be easier to understand, and they can indicate where one must be cautious before recommending a sampling plan. Where a theoretical distribution is an adequate model, the most information is obtained by estimating the OC function in both ways.

9.8 When is Resampling a Better Method to Evaluate Sampling Plans?

Model-based simulation methods and resampling may both be used to evaluate the performance of sampling plans before they are tested and used in the field. Both evaluation methods are useful in the design process towards sampling plans that are effective and efficient. But which approach is better? This is an unanswered, and probably largely unanswerable, question. A more useful approach is to ask under what circumstances is one more informative than the other, and whether they can complement each other.

The main difference between model-based and data-based procedures for estimating the properties of a sampling plan is the way in which the evaluation is done on a computer. Models used in model-based evaluation include probability models (Chapter 4), variance–mean relationships (Chapter 3) and incidence–mean relationships (Chapter 7). During sampling plan evaluation, sample data are simulated using these models. If there is error in the general shape and fit of the models, the OC and ASN functions would be biased. On the other hand, if the models provide good descriptions of the data, the OC and ASN functions should be well estimated.

When observed data are used in the simulations, the results are free of model-based artefacts, which is an advantage. However, if one or more of the frequency distributions among the data sets is not representative of the true distribution in the field (e.g. by not having a long tail when the true distribution has one, or vice versa) parts of the OC and ASN functions may be misleading. If there are few basic data sets, these 'outliers' may obscure the true shape of the OC and ASN functions. This is a weak point of the resampling approach, but we should point out that a model based on only a few data sets is unlikely to be acceptable to other workers anyway. There is no substitute for good comprehensive data.

Sensitivity analysis can be done (Chapters 5 and 6) on the parameters of sampling plans (e.g. *cd* or sample size) for model-based and resampling evaluations. With model-based evaluation, sensitivity analysis can be done on the parameters used in the evaluation (e.g. the variance of sample data). With resampling, this 'sensitivity analysis' is built into the method (e.g. to obtain jagged OC functions).

If many basic data sets are used in resampling, the scatter in OC and ASN functions may be used to assess the field-to-field variability in OC and ASN values, as far as it is related to the variability of the underlying sampling distribution. The visualization of such variability in the OC function can be regarded as an asset rather than a liability, provided that the observed variability of distributions reflects true underlying differences in the shape of distribution, and is not the result of unrepresentative basic data sets. The same variability can be assessed using model-based methods by including parameter variability in the model during the evaluation process (Exhibit 9.6). Note that the variability that we discuss here in OC and ASN functions is not related to sampling error during the evaluation of the plan by simulation; that variation is largely wiped out if enough simulation replicates are done for each population mean or distribution.

What is the bottom line? Both techniques – model-based evaluation and data-based evaluation – have a place, and use of either of the methods may provide important insights during the design phase of a sampling plan. Resampling as a technique is still very much in the course of development (Naranjo and Hutchison, 1997), and its strong and weak points as compared to model-based methods should become clearer as more work is done.

9.9 Summary

1. Resampling can be used to estimate OC and ASN functions of sampling plans, instead of using simulations based on distributional models, variance–mean relationships or incidence–mean relationships. Resampling uses previously collected sample data to generate data distributions which are then used to replace theoretical distributions in simulations of the sampling process.
2. Sample data sets to be used for resampling must satisfy certain criteria; otherwise, the OC and ASN functions may be misleading. Data sets must cover the range of population means that is relevant in decision-making. The more data sets are collected and the greater their sizes, the more reliable are the results of resampling. As a guideline, 20 or more data sets are recommended, collected over more than one season and site. Sample sizes should be large enough – not smaller than 50, but preferably larger – so that the data sets truly reflect the distributions that exist in the fields.
3. Sampling plans and their parameters can be determined following essentially the same strategies as discussed in previous chapters.
4. The jaggedness and variability of OC and ASN estimates obtained by resampling can be insightful, as they may represent reality checks for the sampling plan. Jaggedness in OC and ASN function can also be generated by including parameter variability in models used in the simulation of sampling plan performance. In doing so, model-based methods can mimic resampling.
5. For binomial count plans, care is needed for a proper determination of the critical proportion, in consultations with other workers. The effects of using or not using incidence–mean relationships in the estimation of OC and ASN functions must be considered. Jaggedness and variability are likely to be more pronounced, due to the variability inherent in binomial count data. Minimum requirements for number of basic data sets and their size (2 above) may be more stringent when they are used for the evaluation of binomial plans.
6. Comparing OC and ASN functions determined by resampling and by model-

based models (including parameter uncertainty or not) provides additional insights into possible generalizations or caveats in sampling designs, and is therefore recommended.

References and Suggested Reading

Crowley, P.H. (1992) Resampling methods for computation-intensive data analysis in ecology and evolution. *Annual Review of Ecology and Systems* 23, 405–447.

Efron, B. (1982) *The Jackknife, the Bootstrap and Other Resampling Plans.* Society for Industrial and Applied Mathematics, 92 pp.

Finney, D.J. (1941) On the distribution of a variate whose logarithm is normally distributed. *Journal of the Royal Statistical Society (Supplement)* 7, 155–161.

LePage, R. and Billard, L. (1992) *Exploring the Limits of Bootstrap.* John Wiley, New York, 426 pp.

Müller, H.J. (1953) Der Blattlaus-Befallsflug im Bereich eines Ackerbohnen-und eines Kartoffel-Bestandes. *Beitraege zur Entomologie* Band 3, 229–258.

Naranjo, S.S. and Hutchison, W.D. (1997) Validation of arthropod sampling plans using a resampling approach: software and analysis. *American Entomologist* 43, 48–57.

Appendix: Confidence Intervals for Negative Binomial *k* Estimated by TPL

If TPL is fitted by linear regression, we can generate a value for σ^2 by (see the Appendix to Chapter 5)

$$\left.\begin{aligned} &\ln(\sigma^2) = \ln(a) + b\ln(\mu) + z(0,\sigma_\varepsilon) \\ &\text{or} \\ &\qquad \sigma^2 = a\mu^b e^z \end{aligned}\right\} \tag{9A.1}$$

where z is normally distributed with mean 0 and standard deviation σ_ε. We use the method of moments to estimate k (Equation 5A.4):

$$k = \frac{\mu^2}{\sigma^2 - \mu} = \frac{\mu^2}{a\mu^b e^z - \mu} \tag{9A.2}$$

Equations 9A.1 and 9A.2 give k a distribution based on the distribution of z. In particular, 95% confidence intervals for z provide 95% confidence intervals for k as follows.

For any particular value of μ, a 95% confidence interval for $\ln(\sigma^2)$ can be calculated from Equation 9A.1 as

$$\ln(a) + b\ln(\mu) - 1.96\sigma_\varepsilon < \ln(\sigma^2) < \ln(a) + b\ln(\mu) + 1.96\sigma_\varepsilon \tag{9A.3}$$

If we ignore, for simplicity, the bias due to the back transformation (see, e.g. Finney, 1941), this can be transformed into a confidence interval for σ^2:

$$a\mu^b e^{-1.96\sigma_\varepsilon} < \sigma^2 < a\mu^b e^{-1.96\sigma_\varepsilon} \tag{9A.4}$$

In turn, this can be used with Equation 9A.2 to give an interval for k:

$$\frac{\mu^2}{a\mu^b e^{1.96\sigma_\varepsilon} - \mu} < k < \frac{\mu^2}{a\mu^b e^{1.96\sigma_\varepsilon} - \mu} \tag{9A.5}$$

Provided that these expressions do not require division by zero, or the interval includes negative values, they can be calculated for a range of values of μ to obtain confidence ranges for k, as in Fig. 9.12.

Sampling over Time to Classify or Estimate a Population Growth Curve

10

10.1 Introduction

In previous chapters we have discussed classification sampling at a single point in time – a single sample *bout*. We have implicitly ignored the fact that pest problems develop over time and that resampling on subsequent occasions may provide useful additional data. Although information from one sample bout is often sufficient to guide decision-making, this is not always so. Populations grow and decline, and a population estimate at a single point in time may be useless without additional information that extends that estimate to an assessment of pest pressure into the future. If such information is required, then we must deal with uncertainties over time due to the dynamics of the system. The question that we start to address in this chapter is how the concepts of sampling for decision-making can be extended to include decision-making over time. We refer to taking samples for decision-making over time as *monitoring*.

A main consideration is whether the current density can be used to make a prediction about the future development of the population. For some diseases, the development of epidemics is heavily dependent on weather conditions. When weather conditions are the dominant variable affecting population increase and risk to the crop, a warning system may be based primarily on predicted and measured abiotic conditions, with sampling playing only a minor role. For most arthropod pests, however, current density is often a valuable predictor of the risk of heavy pest pressure later on. Several approaches have been proposed in the literature to combine elements of population sampling and population forecasting into pest management decision-making. In this chapter we discuss two procedures that assume a coherent pattern for the population trajectory of the pest over the growing season. These represent the two sides of sampling that we have met before: sampling for classification and sampling for estimation.

The first procedure is a method, proposed by Pedigo and van Schaick (1984), which extends Wald's Sequential Probability Ratio Test (SPRT) into the time domain. In the SPRT, sample units are collected until the sample information is convincingly (specified by α and β) in favour of one or other of two pest densities,

$\mu = \mu_0$ or $\mu = \mu_1$. In the method of Pedigo and van Schaick, the two population *densities* (μ_0 and μ_1) are replaced by two population *trajectories* ($\mu_0(t)$ and $\mu_1(t)$), and sample data are collected over time until the accumulated sample information gives convincing support (again specified by α and β) to either $\mu_0(t)$ or $\mu_1(t)$. Both trajectories are completely specified as points over time. The one, $\mu_0(t)$, is a low ('endemic') trajectory that does not need intervention, while the other, $\mu_1(t)$, is a high ('outbreak') trajectory that does require control.

The second procedure, proposed by Plant and Wilson (1985), assumes that the increase of incidence levels can be described by an incidence growth curve whose shape is defined by two parameters. Plant and Wilson's method estimates the parameters of this growth curve on the basis of data collected in successive sample bouts over time. As the data from each sample bout become available, the growth curve is re-estimated. Each new estimate is assumed to be closer to the true growth curve, so better predictions can be made after each new sample bout. In particular, a decision can be made whether intervention is required (i.e. incidence is, or soon will be, above a critical level) or whether it is worth collecting more sample data and, if so, when it would be best to resample.

We defer to the next chapter a discussion of monitoring in situations in which we do not wish to assume a coherent pattern for the population growth pattern of the pest over the whole growing season.

10.2 Time-sequential Classification of a Population Trajectory

The classification method described here is a development of Wald's SPRT (see Section 5.4), so it is useful to review some aspects of SPRT. At the heart of Wald's procedure is a function of the data called the *likelihood ratio*. For discrete distributions (e.g. Poisson, negative binomial, positive binomial) this is the ratio of two probabilities, namely:

1. The probability of the data when the true density is μ_1.
2. The probability of the data when the true density is μ_0.

For example, with the Poisson distribution (see Chapter 4), the ratio for a single observation (x_1) is

$$\text{likelihood ratio, } LR(x_1) = \frac{e^{-\mu_1}\left(\mu_1^{x_1}/x_1!\right)}{e^{-\mu_0}\left(\mu_0^{x_1}/x_1!\right)} = e^{-(\mu_1-\mu_0)}\left(\frac{\mu_1}{\mu_0}\right)^{x_1} \tag{10.1}$$

which is easily extended to many observations $x_1, x_2, \ldots, x_n$:

$$LR(x_1,\ldots,x_n) = e^{-n(\mu_1-\mu_0)}\left(\frac{\mu_1}{\mu_0}\right)^{\Sigma x} \tag{10.2}$$

It is convenient to take logarithms:

$$\ln(LR_n) = -n(\mu_1 - \mu_0) + \Sigma x \ln(\mu_1 - \mu_0) \tag{10.3}$$

Referring back to Section 5.4, sampling continues as long as

$$\ln\left(\frac{\beta}{1-\alpha}\right) < -n(\mu_1 - \mu_0) + \Sigma x \ln(\mu_1 / \mu_0) < \ln\left(\frac{\beta}{1-\alpha}\right) \tag{10.4}$$

which is equivalent to

$$\frac{\ln\left(\frac{\beta}{1-\alpha}\right)}{\ln(\mu_1 / \mu_0)} + n\left(\frac{(\mu_1 - \mu_0)}{\ln(\mu_1 / \mu_0)}\right) < \Sigma x < \frac{\ln\left(\frac{1-\beta}{\alpha}\right)}{\ln(\mu_1 / \mu_0)} + n\left(\frac{(\mu_1 - \mu_0)}{\ln(\mu_1 / \mu_0)}\right) \tag{10.5}$$

The formulae shown in Table 5.4 for the slope and intercept can be recognized from Equation 10.5.

The simplicity of the results in Table 5.4 – but not their applicability – depends on the parameters μ_0 and μ_1 remaining the same for all the data. If μ_0 and μ_1 are allowed to vary with time, we can still write formulae that correspond to Equations 10.2–10.4 for sample data collected at different times. It is less confusing if we change the notation slightly: sampling takes place at times t_j, the number of sample units collected at time t_j is n_j, the sample mean at time t_j is m_j, and the current sample bout is referred to as bout number *i*. Note that $n_j m_j$ is the total count at time t_j. The formulae that correspond to Equations 10.2–10.4 are

$$LR_i = \exp\left[\sum_{j=1}^{i} n_j \left(\mu_1(t_j) - \mu_0(t_j)\right)\right] \prod_{j=1}^{i} \left(\frac{\mu_1(t_j)}{\mu_0(t_j)}\right)^{n_j m_j} \tag{10.6}$$

$$ln(LR_i) = \sum_1^i n_j \left(\mu_1(t_j) - \mu_0(t_j)\right) + \sum_1^i \left(n_j m_j \ln\left(\mu_1(t_j) / \mu_0(t_j)\right)\right) \tag{10.7}$$

$$\ln\left(\frac{\beta}{1-\alpha}\right) < -\sum_1^i n_i \left(\mu_1(t_j) - \mu_0(t_j)\right) + \sum_1^i n_j m_j \ln\left(\mu_1(t_j) / \mu_0(t_j)\right) \ln\left(\frac{1-\beta}{\alpha}\right) \tag{10.8}$$

Comparing Equation 10.4 with Equation 10.8, we can see that the running total of sample counts, Σx, has been replaced by a weighted sum of sample counts and the slope is no longer constant (the slope between two times t_j and t_{j+1} depends on $\mu_1(t_j) - \mu_0(t_j)$ and on $\mu_1(t_{j+1}) - \mu_0(t_{j+1})$). However, such an extended SPRT is feasible and can be implemented. For simplicity, a special notation is used for the terms in Equation 10.8, which is rewritten as

$$h_L + \sum_1^i n_j \left(\mu_1(t_j) - \mu_0(t_j)\right) < \sum_1^i \left(\ln\left(\mu_1(t_j)/\mu_0(t_i)\right) n_j m_j\right) < h_U + \sum_1^i n_i \left(\mu_1(t_i) - \mu_0(t_i)\right) \tag{10.9}$$

or even more simply as:

$$h_L + b(t_i) < \sum w_j n_j m_j < h_U + b(t_i) \tag{10.10}$$

Expressions for where h_L equals $\ln\left(\frac{\beta}{1-\alpha}\right)$ and h_U equals $\ln\left(\frac{1-\beta}{\alpha}\right)$, $b(t_j)$ and w_j are given in Table 10.1 for the Poisson, negative binomial and positive binomial distributions. These appear to be formidable to calculate, but all of the calculations can be done before the first bout. The only extra difficulty at sampling time beyond

Table 10.1. Time-sequential sampling stop boundary parameters, $b(t_i)$ and $w(t_i)$, for three distributions. Sample size and sample mean at time t_i are $n(t_i)$ and $m(t_i)$, respectively. Endemic and outbreak trajectories at time t_i are $\mu_0(t_i)$ and $\mu_1(t_i)$, respectively. These parameters are used to define the weighted sums and stop boundaries for the stopping criterion (combining Equations 10.8–10.10):

	$b(t_i)$[a]	$w(t_j)$[a]
Poisson	$\sum_1^i n(t_j)\left(\mu_1(t_j) - \mu_0(t_j)\right)$	$\ln\left(\frac{\mu_1(t_i)}{\mu_0(t_i)}\right)$
Negative binomial	$k\sum_1^i n(t_j) \ln\left(\frac{k+\mu_1(t_j)}{k+\mu_0(t_j)}\right)$	$\ln\left(\frac{\mu_1(t_i)}{\mu_0(t_i)}\right) - \ln\left(\frac{k+\mu_1(t_i)}{k+\mu_0(t_i)}\right)$
Positive binomial	$\sum_1^i n(t_j) \ln\left(\frac{1-p_1(t_j)}{1-p_0(t_j)}\right)$	$\ln\left(\frac{p_1(t_i)}{p_0(t_i)}\right) - \ln\left(\frac{1-p_1(t_i)}{1-p_0(t_i)}\right)$

[a] Note that $b(t_i)$ is a sum but $w(t_j)$ is not.

ordinary SPRT is having to multiply the total count (n_jm_j) by a weighting factor (w_j) before adding it to the previous sum.

The idea of classifying population trajectories using SPRT was originally put forward by Pedigo and van Schaik (1984). They proposed their method to give forewarning of potential trouble caused by green cloverworm (*Plathypena scabra* (F.)) on soybeans, but suggested that it could be used in a more general situation. The idea was to take samples of adults at intervals of 2–3 days over the period when they are active, and use these data to indicate whether the larvae might or might not become a problem. If adult numbers are high, larval sampling is required; if adult numbers are low, larval sampling may be skipped (barring other eventualities which might discount the information from adult sampling). Specifically, they noticed that when the pests (in the larval stage) remained under control, the spring adults had seemed to follow similar trajectories from the start. By averaging such trajectories, a typical 'under control' or *endemic* trajectory could be specified as $\mu_0(t)$. In the same way, a typical *outbreak* trajectory could also be specified, as $\mu_1(t)$ (Fig. 10.1). Pedigo and van Schaik used these two trajectories as extensions of Wald's two critical mean densities, μ_0 and μ_1, allowing them to extend Wald's sequential procedure to compare trajectories, as described above. They called their procedure *time-sequential sampling* (TSS), but we prefer the more specific name *time-sequential probability ratio test* (T-SPRT), because it emphasizes that classification is the goal and that the SPRT is its basis.

There is a difference between the ordinary SPRT and T-SPRT in the construction of the charts with stop boundaries. In SPRT each sample unit, as it is collected, is added into a cumulative sum (Chapter 5). In T-SPRT, we have sample bouts, and sample units per sample. It is advisable (of course!) to collect more than one sample unit at each time to represent the pest population at that time. The

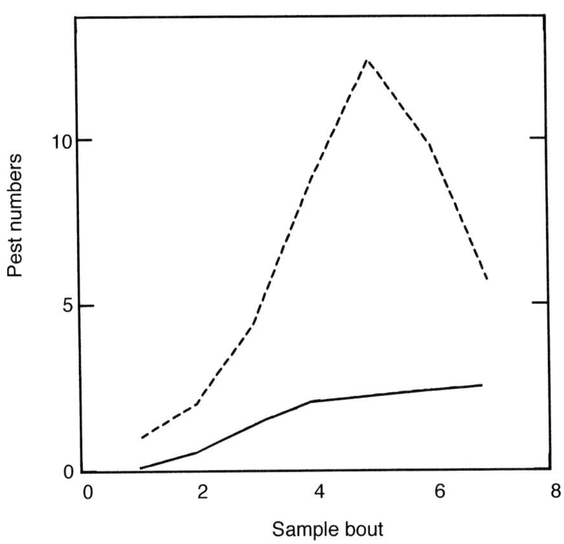

Fig. 10.1. Typical endemic (——) and outbreak (- - - -) trajectories for time-sequential sampling, based on Pedigo (1994).

total of these sample units is what is added, after weighting, to the sum and placed on the chart with the stop boundaries. Fixed sample sizes are used at each bout:

	Sequential sampling	Time-sequential sampling
Number of sample bouts	1	More than 1
Chart: *x*-axis	Number of sample units	Number of sample bouts
Chart: *y*-axis	Cumulative number of individual pests	Accumulation of (weight × bout total)

It would be possible to use a sequential procedure for collecting sample units at each sampling time. For various reasons, including its practicality, this has not been pursued in the literature, and we regard it as relatively unimportant. As in ordinary sequential sampling, if the last scheduled time for sampling is reached without a decision having been made, the final weighted sum is compared to the mid-point between the upper and lower limits.

Implementation of T-SPRT proceeds as follows:

1. Based on the endemic and outbreak trajectories, a time period for sampling is determined. For Fig. 10.1, sample bouts on the first seven indicated times would be acceptable, but there may be scope for skipping some of these, or adding more.
2. Obtain the mean values for the endemic and outbreak trajectories at each time: μ_0 at times 1, 2, ..., and μ_1 at times 1,2, ...
3. Decide on a probability distribution for the sample data. Decide how many sample units to collect at each bout – sample size can change between bouts.
4. Calculate the stop boundaries and weights for the cumulative sums, using Table 10.1, and prepare a simple sampling field sheet for the person who will do the actual sampling.

After the season has started and some sampling has already been done, it is possible to change the next scheduled time (t_{i+1}) for sampling, if there are good practical reasons for doing so. The boundaries and weights beyond the current time (t_i) must be changed, because the values of $\mu_1(t)$ and $\mu_0(t)$ will be different (see Table 10.1). This is not too onerous, but requires some resetting of stop boundary values in the decision aids that are used (e.g. field sheets and computer programs). Eliminating or changing some of the scheduled times alters the properties of the test, but if the changes are small the effects should also be small. For example, advancing or delaying the times of sample bouts by 1 day should make no perceptible difference. Eliminating one sample bout would make a noticeable difference only if the true trajectory at that time was exceptionally high or low, relative to the previous and subsequent bouts.

Although presented initially as an early warning procedure, T-SPRT can be used directly on the damaging stage of a pest. The classification given by the procedure leads in that case directly to intervention or non-intervention. The consequences of 'getting it wrong' are then more critical and it is important to judge the performance of the method, as characterized by the OC function. We do this in the following exhibit.

Exhibit 10.1. *Time-sequential classification of a population growth curve*

This example is based on Pedigo and van Schaick's work on green cloverworm (*Plathypena scabra* (F.)) on soybeans. The endemic (μ_0) and outbreak (μ_1) trajectories, which are the targets of decision-making are shown in Fig. 10.5. As a first exercise, we take a trajectory between the endemic and outbreak trajectories, but which looks more like the latter. It is defined as

$$\text{test trajectory} = endemic^{1-\zeta} \times \text{outbreak}^{\zeta}$$

with $\zeta = 0.7$. As a start, we investigate whether a reasonable classification of this trajectory can be obtained with just one sample unit at each sampling time. The parameters of the plan were $\alpha = 0.01$, $\beta = 0.01$, $n = 1$ and $k = 1.16$. One test run stopped at the sixth bout, classifying the trajectory in the outbreak class (Fig. 10.2a).

It is instructive to note how much variability there was in these data (Fig. 10.2b). The standard errors belong to the individual means, not accumulated means. Clearly, there was so much variability that an incorrect decision was quite likely, although here the decision was 'correct' in that the trajectory was nearer outbreak than endemic. When the sample size was increased to $n = 10$, the decision was the same, but the smaller variability suggests that a 'correct' decision was more likely (Fig. 10.3).

That a larger n gives more reliable results was confirmed by a number of simulation runs with the same parameters as above, but comparing $n = 1$, $n = 10$ and $n = 50$. A smaller n resulted in more sample bouts and many more decisions for non-intervention (overall operating characteristic (OC)):

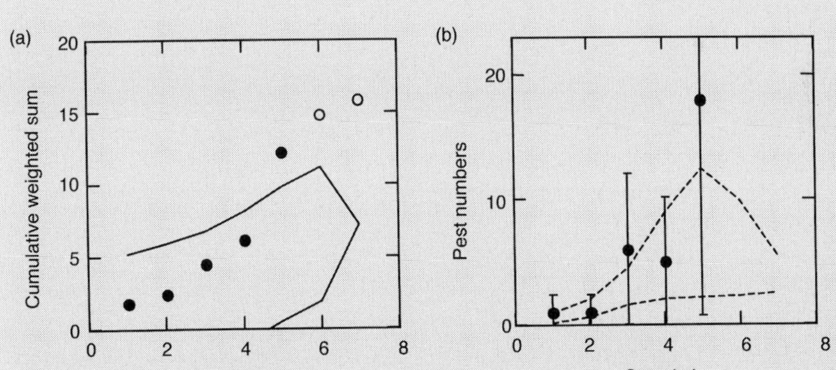

Fig. 10.2. An example of a T-SPRT run with trajectory defined by $\zeta = 0.7$, taking one sampling unit at each bout, and $\alpha = \beta = 0.01$. A negative binomial distribution with $k = 1.16$ is assumed. (a) Weighted cumulative sums in the adapted Wald SPRT (open circles refer to potential sample bouts which were not used); (b) sample data at each bout, with approximate 95% confidence intervals.

Continued

Sample units, n, per bout	Overall OC	Average bouts	Average sample units
1	0.29	6.2	6.2
10	0.12	2.59	25.9
50	0.07	1.21	60.5

Most of the non-intervention decisions with $n = 1$ were taken at the last sample bout, when a decision had to be made (Fig. 10.4). Consideration of Fig. 10.4 suggests that taking more sample units at the first bout and fewer at subsequent bouts might help reduce the number of bouts. Theoretically, this is true, but the usefulness of such a strategy depends heavily on the endemic and outbreak trajectory values at the first bout. If these values are not well established, there will be little or no improvement.

We could estimate an OC function for a range of values of the ζ used to define the test trajectory shown above. Results for $\zeta = 0.7$ have been given here. Letting ζ range over an interval, we can estimate other points on an OC function for trajectories of this type. Some values are shown for $n = 20$ in Fig. 10.5. This illustrates that, if trajectories can be ordered in a sensible way, then the overall OC and ASN functions look like those we encountered in earlier chapters.

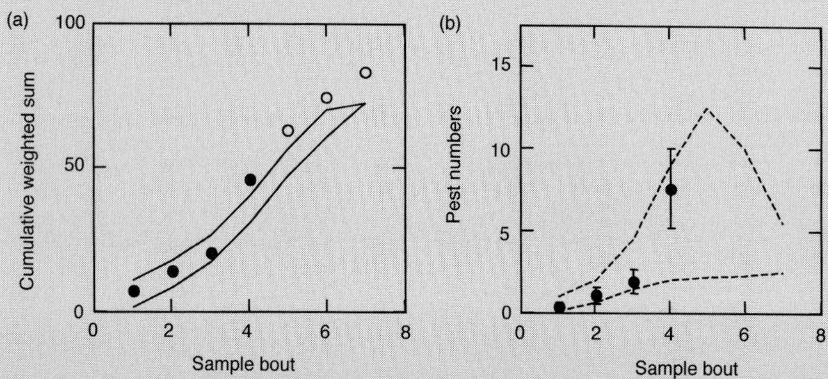

Fig. 10.3. An example of a T-SPRT run with trajectory defined by $\zeta = 0.7$, taking ten sample units at each bout, and $\alpha = \beta = 0.01$. Negative binomial distribution with $k = 1.16$ is assumed. (a) Weighted cumulative sums in the adapted Wald SPRT (open circles refer to potential sample bouts which were not used); (b) sample data at each bout, with approximate 95% confidence intervals.

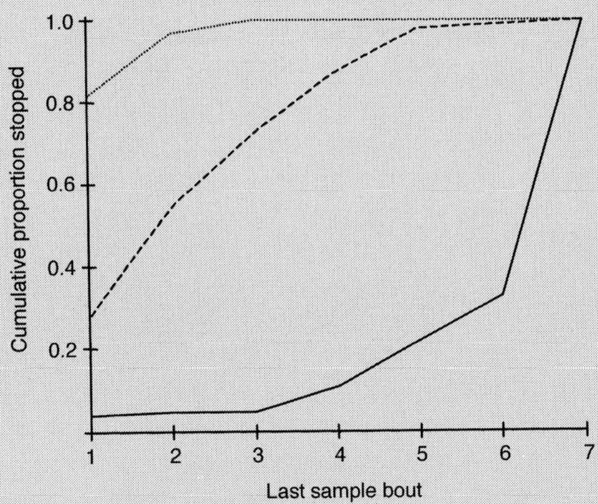

Fig. 10.4. Stopping times for T-SPRT based on trajectories in Fig. 10.1. Cumulative proportions of 100 simulations stopping at each sample bout for sample sizes at each bout equal to 1 (——), 10 (- - -) and 50 (···) . Other parameters: $\alpha = \beta =$ 0.01, $\zeta = 0.7$, $k = 1.16$.

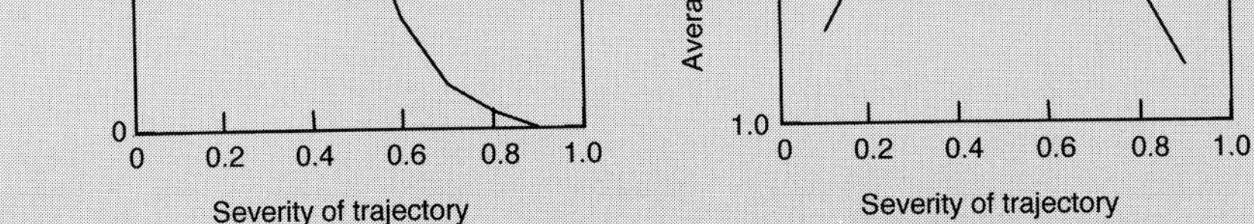

Fig. 10.5. Overall OC (a) and average number of sample bouts (b) for a T-SPRT plan with 20 sample units at each bout, $\alpha = \beta = 0.01$. The horizontal axis refers to values of the parameter ζ in the text. TPL was used with parameters $a = 2.8$ and $b = 1.3$. For $\zeta = 0$, the trajectory corresponds to the endemic trajectory, while for $\zeta = 1$, the trajectory represents the outbreak trajectory.

10.3 Time-sequential Estimation of a Population Growth Curve

The method described here derives from experience with spider mites (*Tetranychus* spp.) in cotton. Plant and Wilson were concerned about the potential for explosive population growth and damage of spider mites in cotton, following early pesticide use which could disrupt spider mite predators. This disruption encouraged further prophylactic acaricide treatments. In extreme cases, up to 70% yield reductions had been noted. Plant and Wilson (1985) proposed an approach to management based on an economic threshold.

They had observed that the relationship between the proportion of leaves with at least one mite and cumulative degree days could often be represented by a logistic curve. The logistic curve that they used has two parameters: initial incidence θ, and a growth parameter, r:

$$\text{pest infestation at time } t = \frac{\theta}{(1-\theta)e^{-rt} + \theta} \tag{10.11}$$

where t represents cumulative degree days.

The parameters r and θ are the growth rate and incidence, respectively, at time 0. We have chosen to replace r by a parameter, S, which is easier to estimate and for which a prior distribution is easier to determine. S is the time at which 50% of the plants are infested. Another interpretation of S is that it is the point of inflexion of the incidence curve; that is, the time at which the *slope* of the (rising) curve begins to decrease. The curve is also symmetric about S:

$$\text{pest infestation at time } t = \frac{1}{\left(\dfrac{\theta}{1-\theta}\right)^{\frac{t}{S}-1} + 1} \tag{10.12}$$

Such a model can rise above any prespecified action threshold, critical proportion (cp), during the season, but on the other hand it might never reach it, depending on the parameters of the model (Fig. 10.6). What if samples were to be taken from time to time during the season, and the parameters re-estimated each time? Could this allow a prediction to be made whether the trajectory will ever rise above cp, and, if so, when? Using Bayesian methods, Plant and Wilson showed how this might be done. They noted that a more involved model for pest incidence over time could replace Equation 10.11: in principle, their method can be used for any parametric model.

10.3.1 Bayesian and regression approaches

The term 'Bayesian'[1] refers to an approach to data analysis and inference that concentrates on a parameter, ϕ, about which we have some information but not enough

[1]Thomas Bayes (1702–1761) was an English clergyman and mathematician.

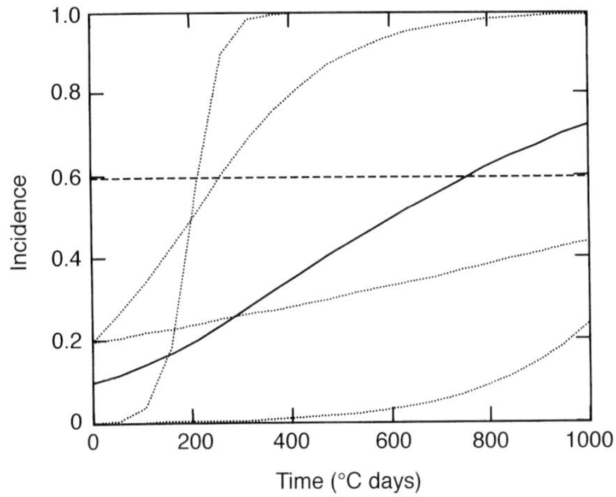

Fig. 10.6. A model of a growth/incidence curve. (——),Prior trajectory; (···), four extreme trajectories using each of $\theta - 0.001$, 0.2 and $S = 200$, 1200; (- – -), indication of *cp*.

for our needs (e.g. to make a classification), and we want to pin down its (ϕ) value more precisely. The information that we have at the start allows us to propose a probability distribution for ϕ. This distribution is called the *prior distribution* of ϕ. As real data are collected, they can be combined with the prior distribution to update *all* of the information that we now possess about ϕ into the *posterior distribution* of ϕ. If we still have insufficient information for our needs, we can repeat the process, using this posterior distribution in its turn as a prior distribution, collect more data and calculate a new posterior distribution. In the example discussed here, ϕ represents the two parameters of Plant and Wilson's model. Once it is calculated, the posterior distribution is used to estimate explicit values of the parameters (and measures of their variability) to put into the model so that predictions can be made. The mathematics required for these calculations are beyond the scope of this book, but can be found in textbooks (see, e.g. Barnett, 1982).

Plant and Wilson's Bayesian procedure can be summarized as follows. Previous experience may suggest upper and lower limits for θ and r (or S), and, further, that within these ranges, some values are more likely than others. This knowledge is used to suggest *prior probability distributions* for the parameters, giving more probability to values that have occurred more often in previous years, or which seem more plausible. It is then possible to display a prior 'expected' incidence curve for the season. More to the point, as sample data are collected over time, they can be incorporated into these prior probability distributions, using Bayesian methodology, to calculate *posterior probability distributions*. These in turn provide updated estimates of the incidence curve. In time, the incidence curve for the current season is well estimated, and useful decisions can be made based on it, such as when next to sample. Decisions on whether or not to intervene can be made using the sample data alone, the predictions, or both.

Fortunately, it is not necessary to use (complicated) Bayesian mathematics to implement the gist of Plant and Wilson's proposal. Instead of Bayesian mathematics, non-linear regression (see, e.g. Ross, 1990) may be used for estimating and updating the model. In this approach, the initial prior distribution is replaced by 'ghost data'. These are data points which lie on or near the long-term average incidence curve. A first estimate of the curve is made using the ghost data. As real sample data are collected over time, new estimates of the parameters are made (by non-linear regression), each time using all the data, including the 'ghost data'. In this approach, the variances and covariances of parameter estimates, resulting from the regression calculations, represent the estimates of uncertainty, which in Bayesian mathematics are contained in the prior and posterior distributions for the parameters.

The advantage of using regression is a simplification of the calculations, making the idea of time-sequential estimation of a time-based model easier to implement and understand. Regression is therefore what we use in the remainder of this section. Provided that the model does not change during the season, the parameters should be estimated with better precision after each bout. On each occasion when the model is fitted, the predicted trajectory can be drawn, with confidence intervals around the predictions. Moreover, the time, *tcp*, at which the trajectory would cross an intervention threshold (and an estimate of its accuracy) can be expressed in terms of the parameters of the model. Therefore *tcp* can be estimated along with approximations to its variance and bias (by methods similar to those described in the Appendix to Chapter 7). All of this information can be used to schedule sampling. Decisions on whether or not to intervene can be made using the sample data alone, the predictions, or both. An illustration is given in Exhibit 10.2.

10.3.2 Implementation of the regression approach

For the regression approach to mimic the Bayesian approach the ghost data must follow certain guidelines. They must be available at two or more time points, to represent a curve and not a single point; they should lie on the prior incidence curve, to represent the prior information exactly; they should be based on relatively small sample sizes, not to outweigh the sample data. The time points chosen for the ghost data influence the weighting of the prior incidence curve relative to the sample data (see, e.g. Ross, 1990), but the theory is beyond the scope of this book; time points near the start and the middle of the time period being studied are sensible. Our implementation of the regression approach is summarized in the following steps.

Preliminary set-up

1. On the basis of all of the knowledge available, estimate prior values for r (or S) and θ. Draw an 'expected' incidence curve based on Equation 10.11 or 10.12.

2. Obtain ghost data.

2a. Choose two values of degree-days for the ghost data, t_{G1} and t_{G2}.

2b. Calculate the incidence proportions p_{G1} at t_{G1} and p_{G2} at t_{G2} based on the prior incidence curve – the ghost data are equal to these proportions.
2c. Choose sample sizes for the ghost data, n_{G1} and n_{G2}.
2d. (Optional) display the prior incidence curve.

Sampling
3. Choose a time (t_1) and a sample size (n_1) for the first sample bout, based on the prior incidence curve, and set $i = 1$.
4. Sample bout *i*:
4a. Obtain a sample estimate, p_i (proportion of infested sample units), from the field.
4b. Fit the model, by non-linear regression, to all the sample proportion data (ghost data and sample data up to and including bout *i*): $\{n_{G1}, p_{G1}\}$, $\{n_{G2}, p_{G2}\}$, $\{n_1, p_1\}, \ldots, \{n_i, p_i\}$.
4c. Calculate predictions into the future, along with confidence intervals.
4d. Calculate *tcp*, the expected time when the incidence curve will reach the critical proportion, *cp*, and its estimated variance.
4e. Display the fitted model, its predictions into the future, along with confidence intervals, and *tcp* (see step 4d).
4f. Make a decision to stop or continue, based on step 4e:

- stop – go to step 5
- continue – increment *i* to i+1, decide on the time, t_i and sample size, n_{ij} for the next sample bout and go to step 4a.

Conclusion
5. Make a management recommendation: intervene or not.

Exhibit 10.2. *Estimating a population growth curve, using time-sequential sampling*

This example follows the gist of the method outlined by Plant and Wilson (1985), but uses non-linear regression instead of Bayesian estimation. Based upon experience, the population growth curves that may be encountered in practice are characterized by ranges of the two parameters for the logistic growth curve (Equation 10.12):

- initial proportion of infested plants, θ: (0.001–0.2)
- inflexion point along time axis, *S*: (200–1200 degree-days)

A prior growth/incidence curve can be characterized by the parameters $\theta = 0.1$ and $S = 600$ degree-days. Extreme and average curves are presented in Fig. 10.6.

Curves with a high initial proportion and high growth parameter would constitute 'worst case' scenarios that require early detection and intervention. Curves with a low initial proportion and a low growth rate would constitute scenarios that do not require control; they should not be treated in order to let biocontrol have its way, and should be sampled as 'lightly' as possible in order to prevent waste of time and resources. Prior knowledge is represented by the prior growth curve (Fig. 10.6). To represent this knowledge, we introduce 'ghost data' at times $t_{G1} = 0$ and $t_{G2} = 600$, lying exactly on the average curve, and collected using a minutely low sample size ($n_{G1} = n_{G2} = 3$) in order to limit the influence of these data points on the regression.

Continued

In order to do the simulation, we have to assume a 'true' trajectory, from which samples are simulated. For this exhibit, we use a growth curve characterized by an initial proportion $\theta = 0.134$, and an inflexion point at $S = 396$ degree-days. The critical proportion, 0.6, is reached at $tcp = 484$ degree-days.

As real sample data are collected, the curve is re-estimated, using *all* of the data (including ghosts). Once the model has been fitted, the predicted trajectory can be drawn, with confidence intervals around it. It is reasonable to schedule each sample bout for a time in the future when we can no longer be confident that the population is below the critical proportion (0.6); that is, when the critical proportion is included inside the confidence intervals. In the example, we use binomial sample plans with a fixed sample size. Plant and Wilson described how to choose a sampling plan based on previous data. We use fixed sample size for simplicity, where choice of n_i is made subjectively after viewing the current estimated trajectory.

We begin our management control of the field with a sample of 40 units at time 200. Early starting seems a good idea, in view of the risk of quickly rising trajectories (Fig. 10.6) and a sample size larger than a modest 40 seems excessive at this early time, when intervention is probably not yet necessary.

Ten infested sample units were found among the 40 collected (based on the true trajectory and binomial sampling with 40 units). The trajectory is re-estimated by non-linear regression (Fig. 10.7a). The predicted trajectory reaches *cp* at around 700 time units, but the confidence intervals are wide. Based on the 95% confidence intervals, we would be safe to take our second sample bout at 400 time units, so we decide to collect 50 sample units then. Nineteen of these are found to be infested. The predicted trajectory, re-estimated from *all* of the data, now suggests that we should probably check again at 500 time units (Fig. 10.7b). This is very soon. Therefore, we take a risk and sample at 600 time units, where we collect 100 sample units, because we are beginning to worry about possible damage to the crop and want a reliable decision. Based on these data (69 infested) and on the predicted trajectory, we decide that action is required right away (Fig. 10.7c).

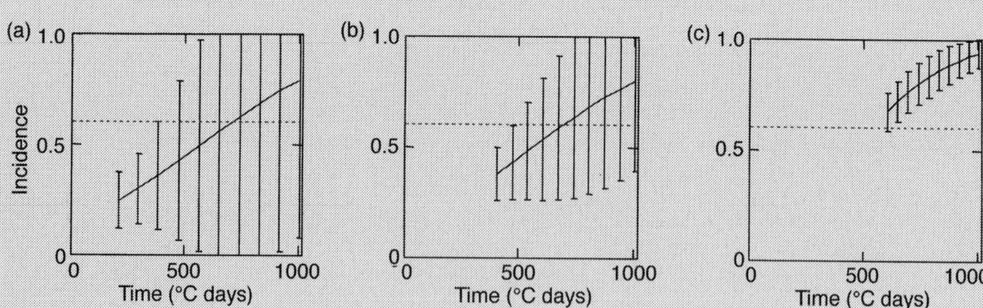

Fig. 10.7. Estimates of growth/incidence curves and predictions (with approximate 95% confidence limits) at times (a) 200, (b) 400 and (c) 600.

We can follow the sequence of fitted trajectories, and compare them with the true trajectory (Fig. 10.8). The sample proportion at 200 raised the fitted trajectory slightly above where the ghost data had placed it (the prior trajectory is not shown, but runs through both ghost data points). Because the sample proportion at 400 happened to be low (by chance), the fitted trajectory had a shallow slope. When a large sample was taken at 600, the fitted trajectory became closer to the true trajectory, although still under it.

Note that the confidence intervals tend to grow larger as the time for prediction increases and decrease as more data are collected (cf. Figs 10.7a–c). The estimates of θ, S and the estimated time, *tcp*, when $cp = 0.6$ is reached at the three sample bouts were as follows:

	t_i	n_i	θ	S	*tcp*
First bout	200	40	0.149	569	701
Second bout	400	50	0.148	557	687
Third bout	600	100	0.094	456	537
True value			0.134	396	482

so S was consistently overestimated, as indicated in Fig. 10.8, mainly because the first two sample proportions were, unluckily, much smaller than the true values (especially the second). This means that our samplers were led to think that the population was growing more slowly than it actually did. Consistently overestimating S meant that *tcp* also was overestimated.

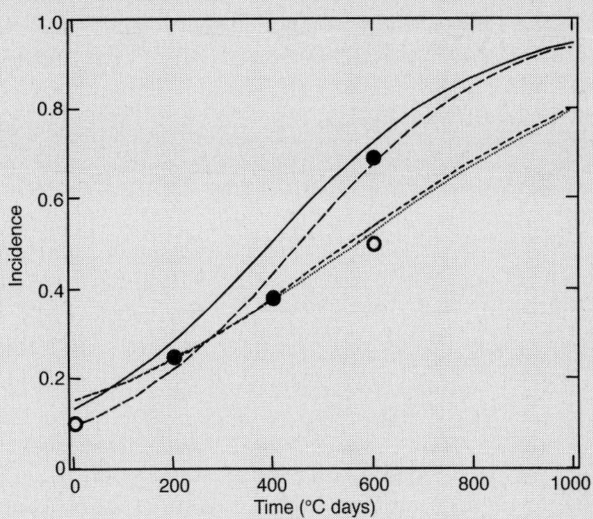

Fig. 10.8. The true model trajectory (——) and the three model estimates made at times 200 (···), 400 (- - -) and 600 (– – –) time units; sample proportions at these times are shown as ●, and ghost proportions as ○; sample sizes were 40, 50 and 100 sample units at 200, 400 and 600 time units respectively.

Continued

The total sample cost includes the cost of collecting three samples, and inspecting 190 sample units altogether. The estimate, *tcp*, of when the trajectory crossed the critical proportion ($cp = 0.6$) was 537 degree-days ($se = 34$). If no effort had been made to control the pest, the losses would have been commensurate with having more than 60% incidence from about 540 degree-days onwards. These costs should be weighed against the sample cost and the cost of implementing some kind of pest control, but if 0.6 was a serious estimate of *cp* relating to unacceptable damage, the decision to control would be correct. However, if *cp* is deemed to be important during only the first part of the season, the predictions can be used to help tell if and when sampling should stop. For example, by time 400 we can see that the trajectory is unlikely to rise above 60% before time 500. If, for biological reasons, we knew that incidence did not matter after 500, a third sample bout would have been unnecessary. We must always look to the practical goal, rather than just trying to obtain very accurate estimates.

In this method, we always have an estimate of the current incidence, as with one-time sampling procedures. In addition, we also have a prediction of when the action threshold will be reached, and a measure of the variability of the prediction. The value of such a prediction will vary with crop and pest. In some instances the value may easily be worth the extra effort of making the extra calculations, but each crop–pest situation must be decided on its own merits. The value will especially depend on the validity of the assumption that population growth over the whole growing season shows regular trends, and can be characterized by growth curves.

10.4 A Comparison Between the Methods and Other Considerations

Both methods presented in this chapter combine information gathered in the past, such as endemic and outbreak trajectories (T-SPRT; Pedigo and van Schaick, 1984), or prior distributions for incidence curve parameters (Plant and Wilson, 1985), with actual sample data on pest population growth, collected through time in the current season. T-SPRT is designed to classify the population trajectory, whereas Plant and Wilson's method tries to estimate it. In T-SPRT, sampling times are predetermined, whereas Plant and Wilson's method allows flexible scheduling of sampling times. In Plant and Wilson's method, samples may be taken more frequently as pest pressure approaches a threshold or if the fitted growth trajectory indicates that the threshold may be reached soon.

Both Plant and Wilson's method and T-SPRT depend on models of pest population dynamics. In Plant and Wilson's method, a functional form is provided, and parameters are determined as data are collected. Practical use of Plant and Wilson's method requires a computer to fit trajectories and make predictions about future pest densities with confidence intervals. In T-SPRT, the models appear in the guise of the basic endemic and outbreak trajectories, and the collected sample data are processed by calculating cumulative weighted sums. Application of T-SPRT in the field requires only simple calculations (multiplication and addition), and a sheet of paper with the weights and stopping boundaries. T-SPRT is therefore easier to implement than the method of Plant and Wilson.

Both methods require that enough previous data are available to specify the following:

- The mathematical form of a typical trajectory and its likely shapes, and a critical damage criterion, or
- Critical upper and lower trajectories to discriminate between damaging and non-damaging populations

Both methods 'remember' the data from previous samples. This means that any basic changes to the trajectory that nature makes during the season could make nonsense of subsequent monitoring. For T-SPRT, the comparison with ordinary sequential sampling is illuminating. In ordinary sequential sampling, remembering earlier data is crucial because all the data are collected at one time from the single field being tested. However, with T-SPRT, because data are collected on different dates, they are much less firmly linked. Although it is not explicitly used in the formulation, predictability in the pest growth process is necessary to ensure confidence in both Plant and Wilson's method and in T-SPRT. For T-SPRT, the link between data collected at different times should fit the endemic/outbreak framework.

In Plant and Wilson's method, any gross change in the model due, for example, to unexpected increases in numbers of natural enemies, can be 'spotted' by examining how well previous and current data fit the model. This would be critical for good decision-making. Spotting a trajectory change in T-SPRT is more difficult. However, it is likely that such changes merely invalidate the theoretical 'promises' offered by the T-SPRT method (i.e. the probabilities of making a wrong decision), but do not result in decision errors.

10.5 Summary

1. Many pests need to be sampled on more than one occasion before their potential for damage can be assessed. A scheme for scheduling sample bouts and specifying sample plans for each is called a monitoring protocol. Every crop–pest complex is unique, and no single monitoring protocol can be recommended across the board.

2. Before embarking on a monitoring adventure, it is important to ask: (i) whether sampling over time is the right procedure for controlling the pest in question, and (ii) what characteristics of the trajectory are related to crop loss.

3. Two potentially useful approaches for time-sequential decision-making with respect to a dynamic pest population are presented. One method is the T-SPRT, first suggested by Pedigo and van Schaick (1984). The principle of the method is that endemic and outbreak trajectories for the pest can be pre-specified, and that data are accumulated over time to choose in a statistically optimal way between these two classifications. One classification leads to a control decision, whereas the other leads to *laissez faire*; that is, to let the system dynamics look after themselves.

4. The second approach, first suggested by Plant and Wilson (1985), is based on the notion that the system dynamics is such that pest incidence follows a specified functional shape, which can be characterized by estimable parameters. By re-estimating the incidence–time function after each sample bout, using all previous sample data and including prior knowledge about likely parameter values, predictions about future pest severity are possible. These predictions are used to schedule sample bouts and to make classification decisions.

5. Methods relying on a model of pest trajectory, whether mathematical or data-based, gain by being able to use all the data. However, the validity of the assumptions underlying these methods should be carefully checked; for example, by plotting the data. Models, especially mathematical models, should be continuously re-examined for their relevance.

References and Suggested Reading

Barnett, V. (1982) *Comparative Statistical Inference*, 2nd edn. John Wiley, New York, 325 pp.

Pedigo, L.P. (1994) Time-sequential sampling for taking tactical action. In: Pedigo, L.P. and Buntin, G.D. (eds) *Handbook of Sampling Methods for Arthropods in Agriculture*. CRC Press, Boca Raton, Florida, pp. 337–353.

Pedigo, L.P. and van Schaik, J.W. (1984) Time-sequential sampling: a new use of the sequential probability ratio test for pest management decisions. *Bulletin of the Entomological Society of America* 30, 32–36.

Plant, R.E. and Wilson, L.T. (1985) A Bayesian method for sequential sampling and forecasting in agricultural pest management. *Biometrics* 41, 203–214.

Ross, G.J.S. (1990) *Nonlinear Estimation*. Springer-Verlag, New York, 189 pp.

Monitoring Pest Populations through Time

11

11.1 Introduction

For some pest management situations it is desirable to know whether pest abundance remains below some critical threshold(s) over a period of time. For example, when biological control is substituted for a chemical pesticide, it may be necessary to monitor the population for some time interval to be sure that control by natural enemies is effective. Pests with multiple generations, such as aphids and mites, often require checking the potential for outbreaks over an extended period of time, and a single sample is not enough.

When a pest population is monitored through time, two questions must be answered: (i) Does the pest density exceed a threshold that dictates management action now? (ii) If the density is below the threshold now, when should the population be sampled again? A simple solution to the second question is to sample very frequently, but this would be unnecessarily costly. A better solution would be to use information about the current density in conjunction with knowledge about population growth to schedule future sampling.

The two procedures described in Chapter 10 are not well suited for answering these two questions. While time-sequential estimation of population growth curves can be used to schedule future sample bouts, the procedure is dependent on specifying a model for the season-long population growth pattern. But if the model does not describe the population trajectory that is being sampled, the management decision may be erroneous. Time-sequential classification of population trajectories depends on specifying an acceptable and unacceptable population trajectory. This may not be a serious limitation provided that all trajectories that require intervention can be identified using the specified 'unacceptable' population trajectory and all trajectories that do not require control can be identified using the specified 'acceptable' trajectory. If this condition is not met, the use of time-sequential classification of population trajectories may lead to incorrect decisions. A greater limitation of this procedure is that it does not use the information collected about pest abundance to schedule future sampling.

Attempts have been made to incorporate information on natural enemies into monitoring strategies (Croft, 1975; Baillod *et al.*, 1989; Nyrop and van der Werf, 1994), but none has found widespread use. The main reasons are the lack of sufficiently precise knowledge about pest and natural enemy dynamics and the influence of biotic and abiotic factors, and (usually) the need to consider more than one natural enemy.

In this chapter, we describe a method for pest monitoring that answers the two questions posed above and is not encumbered by the requirement that pest trajectories follow a specified model, although knowledge of potential pest population growth is needed. The procedure is based on classifying pest density into one of three categories: (i) low density, indicating that damaging pest levels are unlikely to occur in the near future and hence sampling the population again can be delayed; (ii) intermediate density, showing that densities are not currently at a damaging level, but the population should be sampled again soon to make sure this is still the case; and (iii) high density, requiring immediate action. The strategy can be extended to four or more classification alternatives.

We begin our description by introducing some terminology. We then show how sampling plans that classify density into three categories are constructed and we describe the performance criteria that are used to evaluate these plans. We follow this by showing how these sampling plans can be chained together to monitor a population trajectory through time, and we discuss the criteria that can be used to evaluate the effectiveness of the procedure. We conclude the chapter by discussing future perspectives for monitoring.

11.2 Terminology of Monitoring

The specification for collecting sample units on a single occasion and for processing the resulting data to reach a decision is called a *sampling plan*, as noted in previous chapters. A *monitoring protocol* is a strategy for combining one or more sampling plans over a time period to allow for repeated checking of pest abundance. Here, we will only consider use of a single sampling plan in a monitoring protocol. The use of a set of multiple plans (each potentially different) in such a capacity is illustrated by Nyrop *et al.* (1994). We will use sample *bout* to indicate an instance of sampling, potential as well as actual.

11.3 Tripartite Classification Sampling Plans

Pest populations can be monitored by 'chaining' sampling plans over a period of time in such a way that the next sample bout is postponed as long as is deemed prudent in view of observed pest abundance, knowledge of pest population growth and the applicable critical density (Nyrop and van der Werf, 1994). This can be done using a sampling plan to classify density into one of three categories with corresponding management recommendations:

- *Class 0*: pest density greater than the critical value; intervention is required now.
- *Class 1*: pest density less than the critical value; resampling can be delayed for a short period of time (e.g. 1 week) because the density is not expected to reach the critical value within that time.
- *Class 2*: pest density much less than the critical value; resampling can be delayed for a longer period of time (e.g. 2 weeks) because the density is not expected to reach the critical value within that time.

The chaining of such tripartite classification sampling plans through time is illustrated in Fig. 11.1.

A new notation is needed here. When only two classifications are possible, we have been using the terms critical density (*cd*) and critical proportion (*cp*) to denote the population measure below which a sample mean would imply a non-intervention decision. Now that three classifications are possible, we stop using *cd* (to avoid confusion), and use the term *action threshold*, or *at*, for the density above which a sample mean implies intervention, and *waiting threshold*, or *wt*, for the density below which a longer period until resampling is recommended. When the density is between *at* and *wt*, a shorter waiting period is indicated. The relationship between *wt* and *at* is that *at* should be close to what *wt* would become if the pest population was allowed to grow unchecked until the next sample bout. For example, if growth rate per day is expected to be around $r = 0.065$ per day, then in 1 week a pest density, μ, would grow to $\mu e^{0.455}$. A reasonable relationship between *wt* and *at* would then be

$$at = wt\, e^{0.455} = 1.576\, wt \tag{11.1}$$

Therefore, for $at = 10$, $wt = 10/1.576 = 6.3$, so a good initial value for *wt* would be around 6.

Fig. 11.1. Chaining of sampling times when using tripartite classification sampling plans for monitoring a pest population through time. Circles indicate potential sample times (bouts). Arrows originating from a sample time indicate the three possible decisions: (0) intervene because the density is high, (1) check the population at the next sampling time because the density is intermediate or (2) check the population after skipping the next sampling time because the density is low.

Sampling plans that classify density into one of two possible categories have two evaluation criteria: the operating characteristic (OC) and average sample number (ASN) functions. The OC function is the probability of classifying pest abundance as less than some critical value and 1 − OC is the probability of classifying pest abundance as greater than this critical value. With a tripartite classification plan there are three possible classifications and three corresponding probabilities of making these classifications as a function of the true pest abundance (PCi , i = 0, 1 and 2 corresponding to Classes 0, 1 and 2 above). Tripartite classification plans also have ASN functions, and these usually have two peaks instead of one. Probability of classification (PC) and ASN functions must be determined using simulation. The methods are analogous to those we are familiar with for OC and ASN functions. We illustrate constructing a tripartite classification sampling plan and estimating the PC and ASN functions in Exhibit 11.1.

Exhibit 11.1 *Setting up a tripartite classification sampling plan and its PC and ASN functions*

In this exhibit we illustrate the design and evaluation of a sequential tripartite classification sampling plan based on Iwao's stop boundaries. Two Iwao sequential plans were set up based on the negative binomial distribution with Taylor's Power Law (TPL) parameters $a = 3$ and $b = 1.3$. The thresholds were $wt = 6$, $at = 10$, and $\alpha = 0.2$. The stop boundaries are shown in Fig. 11.2a. As for the ordinary Iwao plan, the sum of pests counted is plotted against the number of sample units collected. Sampling stops when:

0) the path crosses the topmost boundary,
1) the path emerges into the space between the two 'arms', beyond where the upper boundary of the lower 'arm' intersects the lower boundary of the upper 'arm', or
2) the path crosses the lowest boundary.

The corresponding decisions are:

0) intervene,
1) resample at the next bout, or
2) resample after skipping the next bout.

The probabilities of making these classifications (PC) and the ASN depend on the true pest density, and their values were estimated using simulation (Figs 11.2b and c).

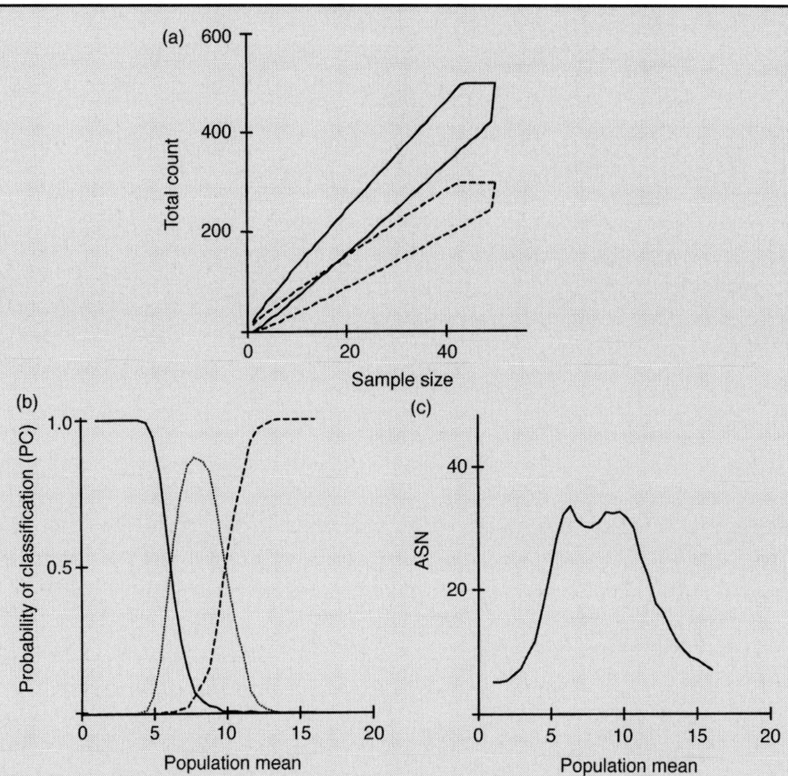

Fig. 11.2. Stop boundaries (a), probability of classification functions (PC) (b) and average sample number function (ASN) (c) for a tripartite sequential classification sampling plan based on Iwao stop boundaries. The parameters for the sampling plan were as follows: action threshold = 10, waiting threshold = 6, $\alpha = 0.2$, minimum sample size = 5 and maximum sample size = 50. The PC and ASN functions were determined using simulation (1000 replicates) by describing counts with a negative binomial distribution and modelling the variance using TPL with $a = 3.0$ and $b = 1.3$. ——, PC2; ···, PC1; - - -, PC0.

11.4 Principles for Quantitative Evaluation of Monitoring Protocols

In previous chapters we have used OC and ASN functions to evaluate the performance of a sampling plan. These performance measures are functions of the pest density or pest incidence being sampled. Each OC or ASN value corresponds to a particular pest density or incidence. However, when we evaluate monitoring protocols, we need to express our evaluation as a function of the population *trajectory*, rather than as a function of one single measure of pest abundance. This is a crucial distinction between sampling once in time and monitoring a population through time. The performance of a sampling plan is judged with respect to the densities being assessed, whereas the performance of a monitoring protocol is judged with respect to densities over time or, in other words, the population path or trajectory being monitored. Thus, each value of a performance measure for a monitoring protocol is indexed to a specific population trajectory.

Including the time factor increases the number of meaningful measures of monitoring protocol performance. These measures include:

- the overall probability of not intervening upon the population trajectory; we call this the monitoring OC (MOC)
- the number of sample bouts (e.g. the number of times the population is sampled)
- the total number of sample units observed
- the pest abundance when intervention is recommended
- the realized cumulative pest abundance (e.g. the area beneath the population trajectory), a measure of the overall stress that the crop has experienced

This list is not exhaustive, but these are the measures that we have found useful when designing monitoring protocols.

These performance indicators can be estimated in two ways. Both methods require specification of a population trajectory to be monitored and a model for the distribution of the sample observations (e.g. negative binomial with TPL).

11.4.1 Direct calculation based on PC and ASN functions and the trajectory

With this method, PC and ASN functions need to have been estimated for the sampling plan proposed for the monitoring protocol. This is done by simulation (Exhibit 11.1). The first step in calculating performance indicators for monitoring is to determine the probability that a sample will be taken at any of the possible sampling occasions. If, for example, the monitoring period is 71 days long and the shortest interval between sample bouts is 7 days, then the possible days on which samples might be taken are days 1, 8, 15, 22, ... and 71, corresponding to bout numbers, $b = 1, 2, 3, 4 \ldots, 11$. Using the population trajectory, the density, μ_b, at any sample bout, b, is known, and the expected outcome from sampling is provided by the PC functions: $\text{PC0}(\mu_b)$ for intervention, $\text{PC1}(\mu_b)$ for resampling at the next bout, and $\text{PC2}(\mu_b)$ for skipping one bout. Using these PC functions, we can calculate the probability of sampling (ps_b) at bout b:

$$ps_b = ps_{b-1}\text{PC1}(\mu_{b-1}) + ps_{b-2}\text{PC2}(\mu_{b-2}) \tag{11.2}$$

for $b = 2, 3, \ldots$.

To make Equation 11.2 work for all possible sample bouts, we start with $ps_1 = 1$ and $ps_0 = 0$. Equation 11.2 reflects the fact that the probability of sampling at any sampling occasion other than the first or second is dependent on the sampling outcome of the previous two bouts.

The monitoring OC (MOC) – that is, the probability of 'accepting' the whole trajectory and never intervening – is estimated as one minus the sum (over bouts) of the probability of sampling at each bout, multiplied by the probability of an intervention at that bout:

$$\mathrm{MOC} = 1 - \sum_{1}^{\mathrm{last}} ps_b \mathrm{PC0}(\mu_b) \tag{11.3}$$

The expected number of sample bouts (*eb*) is estimated as

$$eb = \sum_{1}^{\mathrm{last}} ps_b \tag{11.4}$$

The expected total number of sample units collected in all bouts (*esu*) is estimated as

$$esu = \sum_{1}^{\mathrm{last}} ps_b \mathrm{ASN}(\mu_b) \tag{11.5}$$

The expected pest density at the end of sampling (*ed*) is the sum of the density at each sample bout, multiplied by the product of the probability of sampling and the probability of intervening:

$$ed = \sum_{1}^{\mathrm{last}-1} ps_b \mu_b \mathrm{PC0}\left(\mu_b\right) + \mu_{\mathrm{last}}\left(ps_{\mathrm{last}-1} PC2\left(\mu_{\mathrm{last}-1}\right) + ps_{\mathrm{last}}\right) \tag{11.6}$$

Note that two endpoints are considered in this equation. If the last sample bout is reached, then the contribution to *ed* is the probability of sampling times the density at that time. At the next to last bout, a decision to wait two time periods to sample again is equivalent to not intervening, so the density at the last sample occasion applies here as well. Expected cumulative density (*ecd*) is calculated just as *ed*, except that cumulative density from the start of sampling up to each bout is substituted for density. The implicit assumption is made here that after intervention, the population density is zero for the rest of the season. In instances in which this assumption is unrealistic, the formulae should be modified.

11.4.2 Calculation based on simulation and the trajectory

A shortcoming of the calculations in Section 11.4.1 is that they yield expected values for each of the performance measures, with no indication of the variability that might be encountered. The second way of estimating these performance measures allows for assessment of this variability. This method simulates the actual monitoring process by simulating sampling at each sample bout, by scheduling future sampling using the simulated data and by recording the results of the monitoring. To estimate the performance measures, the monitoring process is simulated many times and average outcomes are calculated. The procedure is as follows:

1. Specify the population trajectory and calculate the mean densities (or infestations) at all possible sample bouts along the trajectory.
2. Determine the density, μ, (or infestation) at the first sample bout.

3. Simulate a single sampling session and record the outcome (i.e. classification 0, 1 or 2) and the number of samples drawn. This simulation makes use of the specified pest abundance, μ, and a model for the distribution of sample observations (e.g. negative binomial with TPL). The next step depends on the classification decision:
   - *Intervene decision*: record the density and cumulative density as the final density and final cumulative density, calculate the total number of sample units collected, record the number of sample bouts realized and proceed to step 4.
   - *Last sample bout reached*: as for 'intervene decision' and proceed to step 4.
   - *Resample at the next bout*: determine the density, μ, at the next sample bout (from the population trajectory) and return to step 3, unless that sample bout falls after the end of the monitoring period, in which case proceed as for 'last sample bout reached'.
   - *Skip a sample bout*: determine the density, μ, that corresponds to the time for the second sample bout in the future and return to step 3, unless that sample bout falls after the end of the monitoring period, in which case proceed as for 'last sample bout reached'.
4. Repeat steps 2 and 3 many (*sr*) times.
5. Calculate the summaries:
   - MOC: the proportion of simulated monitorings in which a decision to intervene was not made.
   - *esu*, *eb*: the expected overall number of sample units (*esu*) and the expected number of bouts (*eb*) are the averages of the total samples drawn and of the number of bouts realized.
   - *ed*, *ecd*: the expected density at the end of sampling (*ed*) and the expected cumulative density (*ecd*) are the averages of these parameters.
   - *Variances, and so on*: because the outcome after each simulated monitoring of the population trajectory is available, the mean, variance and range of these parameters (e.g. the sample size, bouts and density at the end of the monitoring period) can be calculated.

11.4.3 Principles for designing monitoring protocols

The design of monitoring protocols is analogous to the design of sampling plans. Parameters are adjusted until acceptable performance is obtained. Design of monitoring protocols is, however, more difficult than design of sampling plans, because there are more parameters that influence the performance characteristics, and because performance must be judged for the many population trajectories that might be encountered. However, some guiding principles are available.

Some of these principles are similar to those that guide development of sampling plans used once in time (Table 5.2). This is logical because a monitoring protocol is built round its constituent sampling plan(s). Wide stop boundaries in sampling plans result in more accurate decisions, while narrow boundaries give less accurate decisions. Increased accuracy with the sampling plan will result in a greater likelihood that the monitoring protocol will also lead to a correct decision. Of course, increased accuracy carries the burden of increased sample cost. Binomial sampling plans reduce work, but give lower precision (and possibly lower accuracy: see Chapter 7) than full count sampling, especially if a tally number of 0 is used and pest density is high.

One principle is typical for sequential tripartite plans: the two sets of boundaries should allow for an intermediate decision to be reached. This can be accomplished by adjusting *at* and *wt* and by changing the width of the boundaries. Another principle is typical for monitoring: the repeated testing of whether pest abundance exceeds a critical value results in a reduction of the monitoring OC, compared to the OC of constituent sampling plans. This is an important point and is easily visualized. Suppose that obtaining a 'tail' in any flip of a coin is equivalent to deciding to intervene during sampling and the number of possible tosses of the coin is equivalent to the number of times (bouts) a population might be sampled. If only one sample bout is possible, the likelihood of a 'tails' (intervening) is 0.5. With two possible tosses of the coin (two possible sample bouts), the likelihood of obtaining at least one tail is 0.75 (0.5, from the first toss; and $0.5 \times 0.5 = 5^2$, from getting heads first and then getting tails). With three possible tosses of the coin the likelihood of obtaining at least one tail (i.e. intervening) is $0.5 + 0.5^2 + 0.5^3 = 0.875$. Ways to circumvent this reduction in the monitoring OC include raising *at* or reducing the number of sample bouts (by shortening the monitoring period or increasing the time between bouts). However, either of these actions results in a greater likelihood that the monitoring protocol will not recommend timely intervention when intervention is required. A balance between these conflicting objectives must be achieved. This conflict arises for population trajectories that remain close to but below *at* for an extended period of time. Therefore, some of the population trajectories used in evaluations of monitoring protocols should belong to this group. Another way to diminish the MOC-reducing effect of multiple testing in cascaded decision sampling is to use very accurate plans. But this, of course, carries the cost of increased sampling effort.

11.5 Population Trajectories for Use in Evaluating Monitoring Protocols

To see how a sampling plan (or plans) might perform in the context of monitoring, the plan must be evaluated on population trajectories. These trajectories can come from two sources. We can use data on actual trajectories or we can construct artificial trajectories. The benefit of using trajectories based on actual data is that the trajectories are real – or at least as real as the sampling upon which they are based allows. The limitations of these trajectories are that there are relatively few available for use and they may not provide a decisive test of the monitoring protocol. The benefit of artificial trajectories is that they can be formulated to test very specific hypotheses concerning the performance of the monitoring protocol. The pitfall with these trajectories is that they may be overly artificial. We advocate using both types of trajectory. The actual trajectories can provide insight into the overall performance of the monitoring protocol when applied to real-world situations, whereas artificial trajectories can be used to test the monitoring protocol in crucial ways. We now describe desirable criteria for artificial trajectories.

Two types of test need to be made: (i) How quickly does the monitoring protocol detect trajectories which reach densities higher than *at*? (ii) How liable is it to misclassify a trajectory which stays near but below *at*?

First, if densities exceed *at* for more than a brief period of time, any reasonable monitoring protocol will almost certainly lead to an intervene decision (partly due to the repeated testing, as noted above). With such population trajectories, it is the timeliness of intervention which is important. This may be influenced by the relationship between *wt* and *at*, themselves related through the assumed population growth rate (Equation 11.1). Two types of errors may lead to less than timely intervention. First, as a result of sampling uncertainty, a decision may be made to skip the next possible sample bout when in fact sampling would have been beneficial there: not sampling there might allow the population to grow beyond *at*. Second, the growth rate of the population being monitored may be greater than the one used to determine *wt*, thereby also allowing the population to grow to above *at* before sampling is done again. These potential errors indicate that we need to test how the protocol performs when the trajectories have growth rates greater than the one used to determine *wt* (Equation 11.1).

The second type of test that should be made of monitoring protocols is the ability of the protocol not to recommend intervention when trajectories remain below *at*. Incorrect decisions to intervene when the population density is less than *at* depend on how close the density is to *at* and how long it remains there. Densities close to *at* over several sample bouts will probably lead to decisions to intervene, again as a result of the repeated-testing aspects of monitoring protocols based on repeated sampling. This effect can be partially overcome by reducing the potential number of sample bouts either by increasing the time interval between bouts or by increasing *wt*. However, either of these actions will work against timely intervention when intervention is needed.

A single family of population trajectories can be used to make both of the tests described above. These trajectories are not meant to depict reality, but merely to provide patterns that are needed for the tests. The trajectories consist of an exponential increase to a maximum density, maintenance of the maximum density for some period of time, and then exponential decline (Fig. 11.3). The exponential decline is included for convenience as a way of reducing density from the maximum. When using these trajectories, it is also important that they be allowed to start rising over a range of time intervals during the monitoring period. Otherwise, the monitoring protocol will always sample the same densities, resulting in an incomplete picture of performance. For example, if the minimum time between sample bouts is 7 days, then the trajectory should be allowed to start at some random point within the first 14 days. This can be accomplished by allowing the starting time for the trajectory to be a random variable uniformly distributed between 0 and 14.

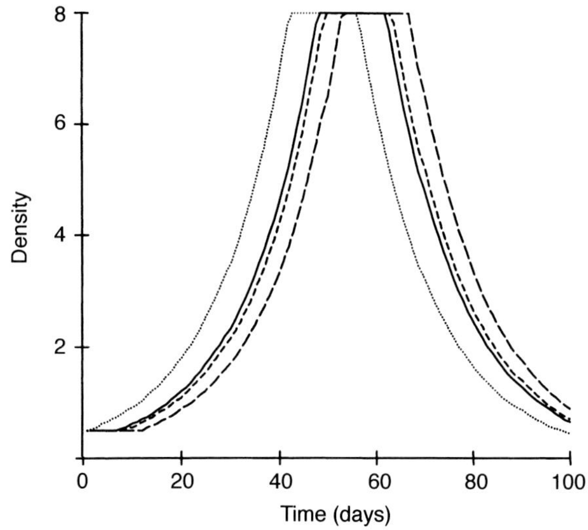

Fig. 11.3. Stimulated population trajectories used to evaluate a monitoring protocol. Three typical trajectories are shown, each with a different dashed line. The solid line is the median trajectory.

11.6 Evaluation of Monitoring Protocols Based on Cascaded Tripartite Classification

In this section, we provide examples of designing and evaluating monitoring protocols based on tripartite classification sampling plans. The material is organized into three exhibits. The analyses presented are loosely based on our work with phytophagous mites, but are not meant to depict any real system. Because the concepts we are working with in this chapter are relatively new, there are few real systems to which they have been applied and therefore there are few choices of systems to use in the exhibits. We chose not to present our work on mites again because that is readily available elsewhere (Nyrop *et al.*, 1994; van der Werf *et al.*, 1997). Nonetheless, the exhibits do illustrate the most salient aspects of design and evaluation of monitoring protocols.

The first exhibit (Exhibit 11.2) illustrates how changing *at* and *wt* affects the performance of monitoring protocols. The sampling plans used are all full count sequential plans based on Iwao boundaries. Two estimated population trajectories are used in the evaluation. In one, the cumulative density is less than the level requiring control, while in the other, the cumulative density exceeds this level. The second exhibit (Exhibit 11.3) compares the performance of three Iwao procedures, all with the same thresholds, where one is a full count procedure, the second is a binomial procedure with a tally number of 2, and the third plan is a binomial procedure with a tally number of 6. The same trajectories as in Exhibit 11.2 are used. The third exhibit (Exhibit 11.4) replaces the single trajectories with trajectories generated to test specific aspects of monitoring protocol performance.

Exhibit 11.2. *The effect of thresholds on performance of monitoring with tripartite full count Iwao plans*

In this exhibit we set up three Iwao plans, each based on a different pair of thresholds, *at* and *wt*, and use the plans to monitor two population trajectories. The objective in the monitoring is to intervene if cumulative density exceeds 550. The monitoring period is 100 days and a single sampling plan is used for the entire period. Three action thresholds (*at*) were tested; 8, 10 and 12. The minimum waiting time between sample bouts was set to 7 days. Assuming a population growth factor of about 1.67 per week, corresponding waiting thresholds (*wt*) were set at 5, 6 and 7 (for simplicity, we used integers for all the thresholds). Remaining parameters for the each of the sampling plans were $\alpha = 0.2$, minimum sample size 5 and maximum sample size 50. Sample counts were described using a negative binomial distribution with TPL parameters $a = 3.0$ and $b = 1.3$. Stop boundaries and PC and ASN functions for the plan with $at = 10$ and $wt = 6$ are shown in Fig. 11.2. We would expect the plan with the lowest thresholds (plan A) to indicate intervention most frequently and earliest, and the plan with the highest thresholds (plan C) to indicate intervention least frequently and latest. The remaining plan (plan B) should provide intermediate results.

Two population trajectories were used to evaluate the resulting monitoring protocol. In one, density had a maximum of 9.5 and cumulative density was 525, while in the second, densities grew to 15 and cumulative density was 725. These population trajectories, while hypothetical, are meant to reflect estimates of actual population changes. Application of the three monitoring protocols to the two population trajectories is illustrated in Figs 11.4 and 11.5. These figures show the two population trajectories and the sample results through time. A dot indicates the

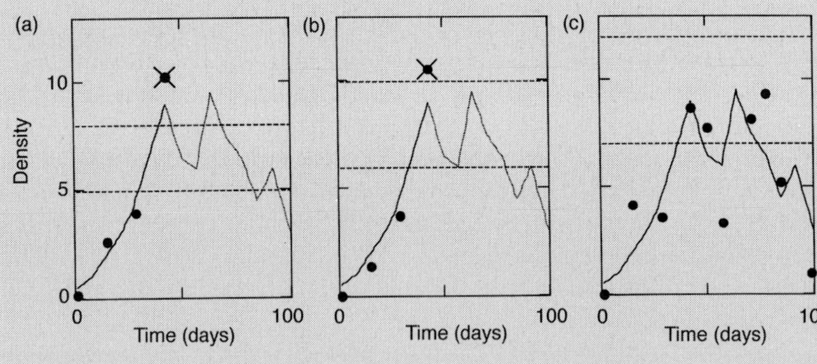

Fig. 11.4. Illustrative monitoring of a population trajectory using three tripartite sequential classification sampling plans A–C (a, b and c, respectively). The dashed horizontal lines show the action and waiting time thresholds for each of the sampling plans. The trajectory is represented by the solid and dotted line, the dotted portion indicating that part of the trajectory which was not monitored because a decision was made to intervene (indicated by the cross). The circles represent the sample information collected at each monitoring bout.

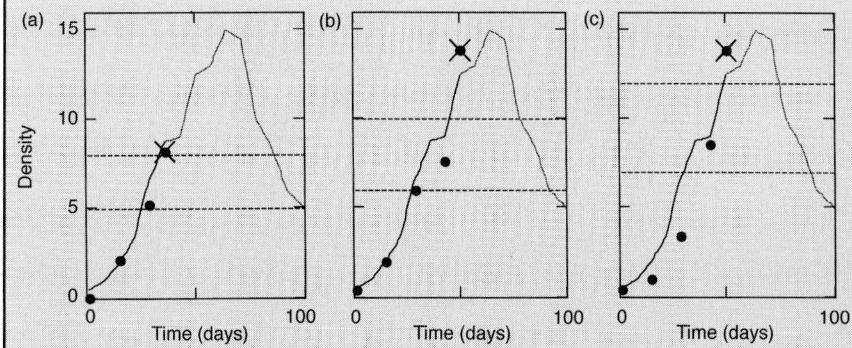

Fig. 11.5. Illustrative monitoring of a population trajectory using three tripartite sequential classification sampling plans A–C (a, b and c, respectively). The dashed horizontal lines show the action and waiting time thresholds for each of the sampling plans. The trajectory is represented by the solid and dotted line, the dotted portion indicating that part of the trajectory which was not monitored because a decision was made to intervene (indicated by the cross). The circles represent the sample information collected at each monitoring bout.

sample information that was collected at each realized sample bout. The density estimate depicted by the dot was calculated by dividing the total count at the end of the sequential sample by the sample size. These estimates are not used in the decision-making (which is based on classification); the points are only shown as an indication of when samples were collected, and what kind of sample information was collected. A cross indicates that a decision was made to intervene. Note that the sample estimates of density (dots) for each of the two trajectories are not the same for each plan. This is because samples are randomly drawn each time a sampling plan is applied to one of the population trajectories. For the first trajectory (maximum density = 9.5), protocols A and B recommended intervention while protocol C did not (Fig. 11.4). All three plans led to fortnightly sampling when the population was lower than *wt*, whereas above *wt*, weekly samples were taken when using plan C. For the second trajectory (maximum density = 15), all three plans recommended intervention, with plan A doing so earliest.

The graphs in Figs 11.4 and 11.5 are useful for visualizing the monitoring process when applied to a population trajectory. However, these figures do not provide an insight into how the monitoring protocol will perform on average when assessing these population patterns. By simulating monitoring of these trajectories many times, average performance can be calculated. The results of 100 simulations using the population trajectory with a maximum density of 9.5 are shown in Table 11.1. Plan A nearly always recommended intervention, plan B recommended intervention about 40% of the time and plan C did so about 10% of the time. As this population trajectory had a cumulative density of 525, the MOC of plan C should be regarded as much superior to those of plans A and B. The number of sample bouts was highest with plan C, reflecting that the population was usually monitored for the entire period.

Continued

Table 11.1. Results of simulating monitoring of a population trajectory 100 times using three full count tripartite sequential sampling plans. The sampling plans differed only in their action and waiting thresholds (*at* and *wt*): for plan A *at* = 8 and *wt* = 5; for plan B *at* = 10 and *wt* = 6; and for plan C *at* = 12, *wt* = 7. The density was less than or equal to 9.5 and the trajectory is portrayed in Fig. 11.4.

	Sampling plan		
Performance measure	A	B	C
Monitoring OC	0.01 ± 0.01[a]	0.58 ± 0.05	0.92 ± 0.03
Sample bouts	4.6 ± 0.12	8.2 ± 0.24	8.8 ± 0.12
Total sample units	85.9 ± 5.4	180.0 ± 8.0	161.0 ± 4.9
Cumulative density	176.1 ± 6.6	411.7 ± 14.7	506.5 ± 7.8

[a]Mean and standard error.

Because *at* for plan A (8) was less than the maximum density for the first trajectory (9.5), it is not surprising that the monitoring protocol based on this plan had an MOC close to zero. More surprising perhaps is the large proportion of interventions scheduled by plan B, whose *at* (10) was slightly greater than the maximum density (9.5). This occurred because classification uncertainty was greatly amplified by the effect of repeated classification.

The results of 100 monitoring simulations using the second population trajectory (some densities > 10) are shown in Table 11.2. Monitoring protocols based on all three sampling plans always recommended intervention. Plan A did so the earliest with resulting lowest density when intervention was recommended and lowest cumulative density. Plan C resulted in the highest density when intervention was recommended and the highest cumulative density. Plan B was intermediate in these regards.

Table 11.2. Results of simulating monitoring of a population trajectory 100 times using three full count tripartite sequential sampling plans. The sampling plans differed only in their action and waiting thresholds (*at* and *wt*): for plan A *at* = 8 and *wt* = 5; for plan B *at* = 10 and *wt* = 6; and for plan C *at* = 12, *wt* = 7. The maximum density was 15 and the trajectory is shown in Fig. 11.5.

	Sampling plan		
Performance measure	A	B	C
Monitoring OC	0	0	0
Sample bouts	4.2 ± 0.1[a]	5.3 ± 0.1	5.6 ± 0.1
Total sample units	78.5 ± 3.4	106.3 ± 3.8	112.2 ± 3.9
Cumulative density	146.7 ± 4.7	236.0 ± 5.2	302.9 ± 6.7
Density at intervention	8.8 ± 0.11	11.2 ± 0.18	12.7 ± 0.11

[a]Mean and standard error.

Provided that a density at intervention of about 13 and a cumulative density of approximately 300 is acceptable, plan C is better than plans A or B, because it recommended intervention many fewer times when intervention was not needed. If these density and cumulative density values are not acceptable, then one must resort to plan B. At the beginning of this exhibit we stated that our objective in monitoring the hypothetical pest was to keep cumulative density below 550. Of course, before accepting any of these three proposed sampling plans – or any other plan – further tests should be made using other population trajectories.

Exhibit 11.3. *Evaluation of monitoring with full count and binomial tripartite Iwao plans*

In this exhibit we continue the investigation into the properties of tripartite plans in the context of monitoring, by introducing binomial count plans. We set up three plans, each with $at = 12$ and $wt = 7$. The first plan (A) used complete counts and was identical to plan C in the previous exhibit. The second plan (B) used binomial counts with a tally, T, equal to 2 and $\alpha = 0.4$. The parameter α was increased to allow for room between the two arms of the stop boundaries, and hence for more classifications of density between wt and at. The third plan (C) was identical to the second except that $T = 6$, which is expected to improve classification accuracy because the tally number 6 is closer to at and wt (Chapter 7). The negative binomial distribution with TPL was again used to describe sample observations.

Stop boundaries for these plans are shown in Fig. 11.6 and probability of classification (PC) and ASN functions are shown in Fig. 11.7. The precision of the classifications is greatest with the complete count plan, with the tally 6 binomial plan offering a slight improvement over the tally 2 binomial plan. The results of 100 simulations with the lower population trajectory used in Exhibit 11.2 (maximum density = 9.5) are shown in Table 11.3. The full count plan classified the density as greater than the action threshold about 5% of the time, the tally 2 plan did so about 60% of the time, and the tally 6 plan did so approximately 30% of the time. The binomial count plans required fewer sample observations, partly because they resulted in fewer sample bouts, but also because the ASN functions for these plans were lower than for the complete count plan (Fig. 11.7).

For the higher population trajectory used in Exhibit 11.2 (maximum = 15), all three sampling plans did equally well at scheduling intervention when needed (Table 11.4). In fact, the binomial count plan with a tally point of 2 did slightly better, in that the density when intervention was recommended was the lowest. This is a result of the increased variability and bias that occurs when using a tally number of 2 and the mean is much greater than 2 (Chapter 7).

This exhibit begins to illustrate a general principle of tripartite classification sampling plans used for monitoring. Sampling plans with low accuracy are often adequate for detecting growing populations that will exceed the action threshold. This property stems from the repeated-testing effect of cascading sampling plans. However, sampling plans with low accuracy are not as effective as sampling plans with higher accuracy at avoiding incorrect decisions to intervene.

Continued

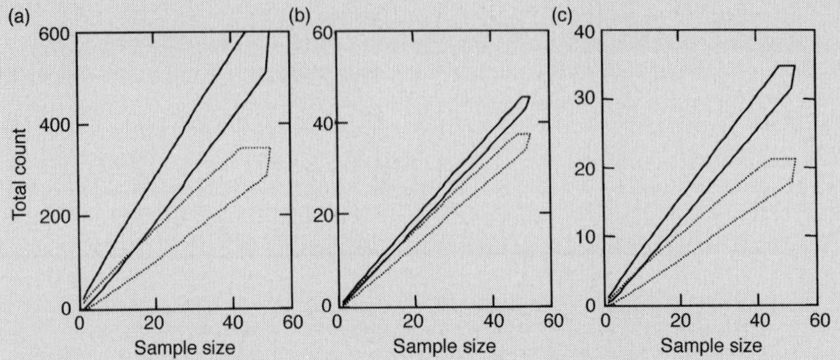

Fig. 11.6. Stop boundaries for three tripartite sequential classification sampling plans. All three plans use an action threshold of 12 and a waiting threshold of 7. Plan A (a) uses complete counts and $\alpha = 0.2$. Plans B (b) and C (c) use binomial counts with $T = 2$ and 6 respectively and $\alpha = 0.4$. All three plans have minimum sample sizes of 5 and maximum sample sizes of 50.

Table 11.3. Results of simulating monitoring of a population trajectory 100 times using three tripartite sequential sampling plans. The sampling plans had identical action and waiting thresholds (*at* and *wt*) of 12 and 7. Plan A used complete counts and $\alpha = 0.2$, plan B used binomial counts with $T = 2$ and $\alpha = 0.4$, and plan C used binomial counts with $T = 6$ and $\alpha = 0.4$. The maximum density was 9.5 and the trajectory is portrayed in Fig. 11.4.

	Sampling plan		
Performance measure	A	B	C
Monitoring OC	0.93 ± 0.03[a]	0.38 ± 0.05	0.68 ± 0.05
Sample bouts	8.8 ± 0.1	6.6 ± 0.2	7.8 ± 0.2
Total sample units	168.4 ± 5.0	84.5 ± 5.2	86.9 ± 4.0
Cumulative density	509.3 ± 7.4	350.6 ± 16.2	438.6 ± 14.1

[a]Mean and standard error.

Table 11.4. Results of simulating monitoring of a population trajectory 100 times using three tripartite sequential sampling plans. The sampling plans had identical action and waiting thresholds (*at* and *wt*) of 12 and 7. Plan A used complete counts and $\alpha = 0.2$, plan B used binomial counts with $T = 2$ and $\alpha = 0.4$ and plan C used binomial counts with $T = 6$ and $\alpha = 0.4$. The maximum density was 15 and the trajectory is portrayed in Fig. 11.5.

	Sampling plan		
Performance measure	A	B	C
Monitoring OC	0	0	0
Sample bouts	5.6 ± 0.07[a]	4.8 ± 0.1	5.4 ± 0.08
Total sample units	113.3 ± 3.9	63.7 ± 3.7	64.0 ± 2.7
Cumulative density	308.8 ± 6.03	251.7 ± 10.7	308.9 ± 8.03
Density at intervention	12.7 ± 0.09	10.9 ± 0.25	12.579 ± 0.15

[a]Mean and standard error.

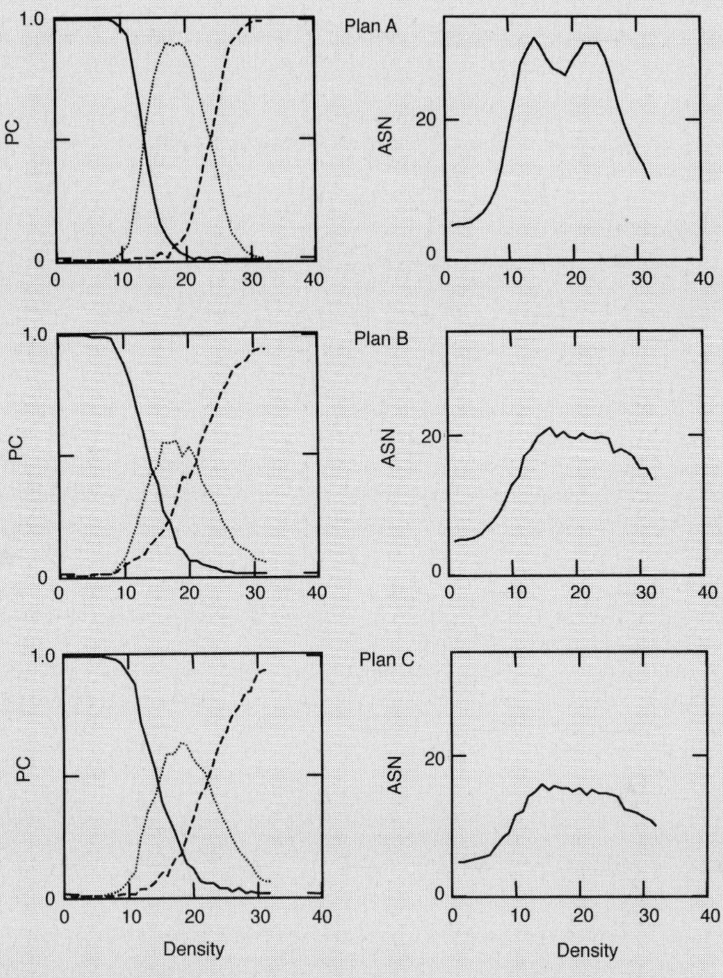

Fig. 11.7. Probability of classification (PC) and average sample size (ASN) functions for the three sampling plans described in Fig. 11.6. The functions were estimated using simulation (500 replicates) by describing counts with a negative binomial distribution and modelling the variance using TPL with $a = 3.0$ and $b = 1.3$.

Exhibit 11.4. *Testing binomial count plans using families of trajectories*

In this exhibit we use five families of population trajectories to further examine the performance of the three sampling plans described in Exhibit 11.3. In each family the trajectory was described by an exponential increase in density, attainment and maintenance of a maximum density for a period of time, and then an exponential decline. Families of trajectories were constructed by allowing the starting point for the trajectories to be uniformly distributed between 1 and 14 days with an initial density equal to 0.5.

Two families of trajectories were used to test the effectiveness of the monitoring protocols at avoiding unnecessary decisions to intervene. Each of these used a growth rate of 0.067 per day and a maximum density of 8. In the first family, maximum density persisted for 14 days, whereas in the second, the maximum persisted for 28 days. The results from 100 simulated monitorings are shown in Table 11.5. For each simulation replicate of monitoring, a new population trajectory was constructed (using a random starting time). The sampling plan that used complete counts (A) was most effective at avoiding unnecessary interventions and the sampling plan that used binomial counts with a tally point of 2 (B) was the least effective. Increasing the length of time during which density was close to the action threshold from 14 to 24 days resulted in little change in the monitoring OC for plan A, but caused a noticeable reduction for both binomial count plans.

The other three families of trajectories were used to test the effectiveness of the monitoring protocols at scheduling intervention when required. Each family had a maximum density of 25 that persisted for 7 days. The growth rates for the three families were 0.067, 0.076 and 0.084 per day. On the basis of 100 simulated

Table 11.5. Results of simulating monitoring families of population trajectories 100 times using three tripartite sequential sampling plans. The sampling plans had identical action and waiting thresholds (*at* and *wt*) of 12 and 7. Plan A used complete counts and $\alpha = 0.2$, plan B used binomial counts with $T = 2$ and $\alpha = 0.4$ and plan C used binomial counts with $T = 6$ and $\alpha = 0.4$. Two families of trajectories with maximum densities of 8 were used; one where the maximum persisted for 14 days, and the other where the maximum persisted for 28 days.

	Sampling plan		
Performance measure	A	B	C
Trajectory with maximum density of 8 for 14 days			
Monitoring OC	0.99 ± 0.01[a]	0.71 ± 0.05	0.91 ± 0.03
Sample bouts	8.4 ± 0.06	7.5 ± 0.15	8.1 ± 0.12
Total sample units	109.3 ± 3.3	75.9 ± 3.3	70 ± 2.7
Trajectory with maximum density of 8 for 28 days			
Monitoring OC	0.96 ± 0.02	0.6 ± 0.05	0.71 ± 0.05
Sample bouts	9.1 ± 0.1	7.5 ± 0.19	8.0 ± 0.17
Total sample units	153.0 ± 4.4	89.7 ± 4.6	79.5 ± 3.7

[a]Mean and standard error.

monitorings, there was little variation in the performance measures among the three sampling plans, and the plan with the least accuracy did about as well as the plan with the highest accuracy (the results are not shown). As expected, all plans always classified density as greater than the action threshold. Perhaps not as obviously, the densities at which a decision was reached to intervene were similar for each sampling plan and did not increase greatly with increasing growth rate of the population being monitored. For the two lower growth rates these densities were about 15 and for the highest growth the densities were about 17. These are all greater than the action threshold (12), and if unacceptable, would require both *wt* and *at* to be reduced. In all cases the cumulative density was less than the target of 550. These results reinforce the principle put forth at the end of the previous exhibit: monitoring protocols using sampling plans with low accuracy are often adequate for detecting growing populations that will exceed the action threshold.

11.7 Further Considerations

There are three additional topics related to monitoring with cascaded sampling plans that warrant introduction: the use of sampling plans with more than three classification choices, the incorporation of time-varying thresholds, and strategies for including information about pest dynamics into the monitoring protocol. None of these topics is developed in detail, because they have not been well studied and have as yet, not shown much practical application. Nonetheless, we think that they may provide further ideas for developing monitoring protocols in the future.

Monitoring can be done with any sampling plan that provides a choice between intervening and resampling at a later moment. Sampling plans that classify pest abundance into one of two categories result in sampling at each successive potential sample bout if pest density is low. This is feasible, but it incurs greater sampling effort than necessary (Nyrop *et al.*, 1994; Binns *et al.*, 1996). Monitoring plans that make use of constituent sampling plans that skip one or more sampling times, as density allows, can save sample bouts. Tripartite classification sampling plans allow for skipping the next possible sample bout. Sampling plans can be designed that allow for skipping more than one future sample bout.

Monitoring with sampling plans that allowed for up to three sample bouts to be skipped before resampling was proposed and evaluated by van der Werf *et al.* (1997). Constituent sampling plans were constructed by combining an upper stop boundary from an SPRT (for intervention) with a fixed sample size plan (to estimate when to resample). Density estimates based on the fixed sample size were grouped into categories that indicated the allowable time to the next sample. These categories were identified by determining the upper confidence limit for the estimated density that, assuming a specific growth rate, would lead to the action threshold one, two, three or four sample bouts in the future. Monitoring with such sampling plans was termed 'adaptive frequency monitoring' to emphasize the adaptiveness of sampling frequency. The procedure was evaluated favourably in computer simulations and via field tests with mites infesting apples in New York State, but has not yet been used in practice. The lack of adoption is probably because the sampling plans are somewhat complicated to use. Action thresholds that increased during the growing season were used in the monitoring protocol, and this resulted in a set of stop boundaries that are indexed to a specific time during the growing season.

In the design of monitoring programs, the notion of time-varying thresholds is highly relevant. For growing pest populations, an early occurrence of a low population density may give forewarning of later trouble. Later in the season, less further growth of the pest population is expected and plants are often less vulnerable to attack. Hence, there is reason to use lower thresholds initially than later on. On the other hand, higher thresholds for the early season may be advocated to minimize the chance that intervention will be recommended when it is not needed. Scenario studies with sampling plans with different thresholds and varying pest trajectories are a great help in determining which thresholds give the best monitoring performance, in view of the objectives of monitoring. These scenario studies can be carried out by independently studying a portion of the monitoring period when a particular threshold applies, or by cascading sampling plans with different thresholds over the entire monitoring period. The advantage of the first approach is that potential problems with specific sampling plans can be more readily identified. The advantage of the second approach is that a picture of overall performance is obtained.

The final topic to be presented here concerns incorporation of knowledge about pest dynamics into the monitoring protocol. As presented thus far, the only information about pest dynamics used in monitoring protocols based on cascading sampling plans is the growth rate used to determine the waiting threshold. There may be instances when the actual growth rate is considerably less than the value used to determine the waiting threshold, as would be the case when natural enemies were abundant or weather conditions did not favour pest population growth. If this knowledge of population growth could be incorporated into the monitoring protocol, the protocol might be less likely to recommend intervention unnecessarily and sampling resources might be conserved. This could be accomplished two ways. First, a higher waiting threshold could be used if it was suspected that the actual growth rate was less than the one originally used to construct the sampling plan. Second, a longer waiting time between sample bouts could be adopted.

This concept also offers an alternative to predicting the outcome of pest – natural enemy ratios for predicting future pest abundance (Nyrop and van der Werf, 1994). Decision guides based on this concept have not been widely adopted, probably because it is often difficult to estimate both natural enemy and pest numbers and because use of a ratio of pests to natural enemies for predicting the outcome of an interaction is only useful when there is one key natural enemy or group of similar natural enemies that drive pest dynamics.

11.8 Summary

Many important crop pests have multiple generations, and pose a risk of outbreak over an extended period of time. In order to design and evaluate sampling methods for the surveillance and management of these pests, we must take the time aspect into account. When sampling plans are used in the context of monitoring, design and evaluation of the plans is more complicated. We not only have to deal with sampling uncertainty, but also with uncertainty about pest dynamics. The notion of designing monitoring protocols and evaluating their performance is fairly new and more work is needed.

There are principles that can be used to guide the development of sampling plans that are cascaded for the purpose of monitoring population trajectories. For tripartite classification plans, the action threshold is related to the waiting threshold via the expected growth rate of the target population. The performance of the monitoring protocol will be influenced by the growth rate used to determine the waiting threshold. A higher growth rate will lead to a lower waiting threshold, more intermediate density classifications and hence more sample bouts. Having more sample bouts guards against a rapidly growing population, but causes more incorrect decisions to intervene when densities are low.

Another principle is that monitoring results in repeated sampling of the population which, even with changing density, leads to an increased likelihood that intervention will be called for when it is not needed. To circumvent this, action thresholds must be raised.

A final principle is that the accuracy of constituent sampling plans influences the performance of the monitoring protocol. This influence is more pronounced when population trajectories remain below the action threshold. With such population patterns, sampling plans with higher accuracy help guard against incorrect decisions to intervene. However, when populations exceed the action threshold, increased sampling plan accuracy may add little to the effectiveness of the monitoring protocol.

References and Suggested Reading

Agnello, A.M., Kovach, J., Nyrop, J.P., Reissig, W.H., Breth, D.I. and Wilcox, W.F. (1994) Extension and evaluation of a simplified monitoring program in New York apples. *American Entomologist* 40, 37–49.

Baillod, M., Antonin, Ph., Guignard, E. and Jermini, M. (1989) Vers une généralisation de la lutte biologique contre les acariens phytophages en verger de pommiers. *Revue Suisse de Viticulture, Arboriculture et Horticulture* 21, 279–284.

Binns, M.R., Nyrop, J.P. and van der Werf, W. (1996) Monitoring pest abundance by cascading density classification. *American Entomologist* 42, 113–121.

Croft, B.A. (1975) Tree fruit pest management. In: Metcalf, R.L. and Luckman, W.H. (eds) *Introduction to Insect Pest Management*. John Wiley, New York, pp. 471–507.

Entwistle, J.C. and Dixon, A.F.G. (1987) Short-term forecasting of wheat yield loss caused by the grain aphid (*Sitobion avenae*) in summer. *Annals of Applied Biology* 111, 489–508.

Nyrop, J.P. (1988) Sequential classification of prey/predator ratios with application to European red mite (Acari: Tetranychidae) and *Typhlodromus pyri* (Acari: Phtoseiidae) in New York apple orchards. *Journal of Economic Entomology* 81, 14–21.

Nyrop, J.P. and van der Werf, W. (1994) Sampling to predict or monitor biological control. In: Pedigo, L.P. and Buntin, G.D. (eds) *Handbook of Sampling Methods for Arthropods in Agriculture*. CRC Press, Boca Raton, Florida, pp. 245–336.

Nyrop, J.P., Binns, M.R., van der Werf, W. and Kovach, J. (1994) Cascading tripartite binomial classification plans to monitor European red mite (Acari, Tetranichidae) through a season; development and evaluation of a new methodology for pest monitoring. *Experimental and Applied Acarology* 18, 123–153.

Schmaedick, M.A. and Nyrop, J.P. (1995) A method for sampling arthropod pests with uncertain phenology with application to spotted tentiform leafminer *Phyllonorycter blancardella* (Lepidoptera: Gracillariidae). *Journal of Economic Entomology* 88, 875–889.

Waters, W.E. (1955) Sequential sampling in forest insect surveys. *Forest Science* 1, 68–79.

van der Werf, W., Nyrop, J.P., Binns, M.R. and Kovach, J. (1997) Adaptive frequency classification: a new methodology for pest monitoring and its application to European red mite (*Panonychus ulmi*, Acari: Tetranychidae). *Experimental and Applied Acarology* 21, 431–462.

12 Epilogue

The intention of this book is to provide an introduction and basic overview of sampling approaches for use in pest management. We have not tried to cover all possible approaches and available literature. Other books provide more comprehensive outlooks; for example, the voluminous multi-author *Handbook of Sampling Methods for Arthropods in Agriculture* (Pedigo and Buntin, 1994). Instead, we have concentrated on the basic design ingredients and performance indicators that are important variables to be considered in any sampling approach. We believe that, armoured with the ideas and methods that are covered in this volume, the reader can confidently approach practical problems and consult the literature to find additional details and helpful suggestions where that seems fit.

Throughout the book, simulation of sampling processes is used as a tool to design and evaluate proposed sampling plans with respect to their practical purpose. Theory is important for proposing methods; simulation is an indispensable tool for predicting the practical performance of proposed methods, to compare methods and alternative sampling designs, and so save a lot of fieldwork. Software tools are placed on the World Wide Web. They allow a hands-on live experience that should complement and enhance the experience of reading the book – hopefully for the better.

While composing this book, we (re)discovered some approaches that hold promise for pest management, but are not much used as yet: sequential sampling plans based on Converging Lines stop boundaries and the calculations for the economic value of sampling (VSI) are examples.

Monitoring is a subject where we have gone deeper than might be necessary for an introductory book. We believe that sampling is often conducted in the framework of a pest surveillance exercise that extends over a lengthy period of time. It can be misleading to evaluate sampling performance as if sampling were done only once. Viewing sampling as a component of time-sequential quality control of cropping systems provides a different outlook, and suggests other questions that should be asked relative to sampling than have been asked for the past 10–20 years. For instance, more thought and work is needed on the question of how often and at what time samples might be taken optimally, and what is an optimal series of thresholds to use through time.

There is a wealth of literature on fitting parameters to sampling distributions of pests and on variance–mean relationships, but there is a lack of underpinning for economic thresholds, and there is not quite enough information on pest trajectories over time. These two aspects have tremendous influence on the design parameters of sampling plans, and they deserve more attention in future studies. It would also be helpful if more data were made available on frequency distributions for pest counts to allow for bootstrapping. The Internet might be a useful place to store and make available such data.

There is also a need to develop tools for management (including sampling) at higher levels of scale, because it is not always appropriate to assume that a pest occurs or develops in a single field, independent of the situation in neighbouring fields. For instance, foci of potato late blight (*Phytophthora infestans*) spread their spores to neighbouring crops (Zwankhuizen *et al.*, 1998), and calculations of infection pressure due to dispersal processes at the between-fields scale are relevant to farmers. This requires observations on the density and source strength of disease foci at the regional scale. Sampling methodology for such sources is non-existent; nevertheless, practice is going ahead in this area (see www.dacom.nl).

Related to the matter of scale is whether agriculture will emphasize curative or preventative methods to combat pests. Sampling is often framed in a reactive mode of pest management. If that were the only application of sampling, pest management should be renamed 'pest mismanagement'. Proactive methods, based on plant resistance, crop rotation, sanitation and so on, are required, and are often most effective when implemented at larger scales. The most recent policy plans for agriculture not only demand a reduction of pesticide usage, but they also call for a shift of emphasis, away from end-of-pipe solutions towards the purposeful design of pest-suppressive agro-ecosystems. There is no reason why observations and sampling should not play a role in designing proactive and potentially area-wide approaches for pest management. The same principles as are described in this book may be useful.

The fact that this book is largely about theory and calculations does not mean that we believe that sampling approaches can be designed behind the computer screen only. There is a mix needed between demand and supply and between practice and theory. Design tools create value if they are matched with practical questions and when practitioners are involved in the design process. But if tools are developed without taking into account user feedback, they are not likely to be adopted.

In the simulations, this book emphasizes descriptive models; that is, models that characterize a process or phenomenon without bothering about the underlying causal relationships. We use such models for frequency distributions (although some of these models represent stochastic processes based on biological parameters), for spatial patterns, for variance–mean relationships, for pest dynamics and for damage. Mechanistic models, if available, could replace the descriptive models. Unfortunately, in many situations process-based models are not available, or they may be available but are not valid beyond a restricted range of conditions, whereas it may be possible to construct meaningful descriptive models without these restrictions. Descriptive models with few parameters have the advantage of ease of use. Therefore, the bias towards descriptive models in this book should be regarded more a matter of practicality than a matter of preference. Especially in the area of crop damage, the use of crop physiological models may give substantial help in establishing time-variable action thresholds (Rossing *et al.*, 1994).

Sampling natural enemies was treated as a possibility. Its practicality, however, seems low because of the increased sampling effort. Yet the social desirability is high, given decreasing consumer and governmental acceptance of pesticide use. We cannot predict the way in which things will develop. It will certainly be attractive to bring more ecology into sampling for pest management, but who will pay the bill? Or will legislation push for adoption of such approaches? Researchers help the implementation of enemy-conserving approaches by developing decision tools, but the adoption of such tools depends on factors that are outside the control of researchers. A worthwhile aspect of including enemies in sampling programs may be the learning component. Some growers seem to appreciate the learning component of decision tools more than direct decision support (Zadoks, 1989; Leeuwis, 1993; Wagenmakers *et al.*, 1999). This aspect may be used to encourage adoption, especially in an era that is seeing expansion of nature-friendly production methods.

A complicating factor is the topic of multiple pests and multiple pest thresholds. Most of the approaches outlined in this book are based on single pests, yet practical decision-making proceeds in the presence of multiple pests, with the possibility of using a single management action to combat multiple pests. It is inevitable that this will tend to lower action thresholds. It is difficult to include these aspects in a general and formal way. Much depends on the particularities of the production system, and methods should be developed in view of actual problems.

Information and communication technology may have profound impacts on the way in which sampling and monitoring in cropping systems are implemented. Many of the current sampling approaches work with charts on which pre-calculated and integrated information are summarized into simple decision criteria that are compared to collected sample information. Hand-held or tractor-mounted computing systems, in conjunction with communication technologies, may enable real-time and very site-specific calculations to be made. The new tools provide ways to put more biological realism and site specificity into decision tools.

As computer hardware becomes faster, smaller and cheaper, and global positioning equipment becomes affordable for large-scale farming operations, we are likely to see more spatially explicit approaches in pest management. Another factor driving this development is the progress in spatial statistics. More fine tuning becomes technically, computationally and economically feasible. This may lead to the disappearance of sampling approaches that take the whole field as a management unit. The time seems ripe to take space into account in crop protection and to start asking questions such as: What are the spatial patterns? How can they be observed efficiently? What techniques can be developed for site specific management? What are the relevant damage relationships when we take spatial variability into account? What benefits for crop protection can be expected when using spatially explicit information at small scales? We have chosen not to deal with the matter in this book, because the area is too new and bare. There are texts that deal with spatial statistics in an ecological context (Bailey and Gatrell, 1995) and papers on site-specific management of pests and diseases in the primary literature (see, e.g. Weisz *et al.*, 1996).

Can we do with less variance–mean modelling? All of the sampling methods presented in Chapter 5 use some kind of variance model, be it a probability distribution, a variance–mean relationship (Taylor's Power Law (TPL)) or a correlation coefficient, to set up stop boundaries. When sample data are collected for classification, any information about variability contained therein is not used. Why is this? There are two reasons: practicality and the generally poor precision of variance estimates, as discussed in Chapter 3 (Section 3.3.1). Therefore, although the use of hand-held devices for data logging and processing brings variance calculations close (and variable intensity sampling (VIS) in Chapter 8 uses it), we are often well off using a variance model.

An important question is: When a sampling plan has been developed, can it be used in another agro-eco-economical setting? In other words, is the plan portable? We touched on this in Chapter 6, but no general answer can be given. From situation to situation, there may be variability in the damage potential of the pest, the price of the product, the pest growth rate, the sampling distribution and so on, which would preclude portability. Nevertheless, asking whether plans could be transported is a worthwhile research question. And most certainly within organism groups variance–mean models and damage relationships show similarities, which would suggest that management approaches should to a certain extent also be portable. For instance, variance–mean models for mites vary within a limited range. Hence, estimating parameters for such models in every new instance seems less than worthwhile, especially because simulation studies such as those in Chapter 5 show that the operating characteristic function of sample plans is not overly sensitive to the choice of parameter value within the biologically plausible range, provided that an appropriate tally number is used if the sample plan is binomial (Nyrop *et al.*, 1999). The two factors that carry the largest risks of causing sample plan failure, and which therefore deserve the most attention during transport from one situation to another are the critical density of the pest, and the acceptability of a plan to growers and advisers in view of their experience, objectives and constraints with the cropping system and its pests.

As to evaluation, Chapter 6 has dwelled on it at length. We think that the hard criteria that may be derived from simulation are important, and an indicator such as average sample size will weigh heavily in any evaluation by practitioners. However, practitioners also include heavily perceived or real risks to their crop, and practicalities such as multiple pests, scheduling farm operations and so on when they think about adopting sample plans for pest management decision-making. Therefore, evaluation cannot be only a laboratory undertaking. This is only one of the stages in the process at which interaction with the users is a key element of evaluation.

You have now reached the end of this book (or you cheated). That is good (not the cheating!) because it shows you started with the end in mind. All of the tools in this book are discussed towards an end – practically, economically and ecologically sound crop protection. We wish you much success in applying these tools.

References

Bailey, T.C. and Gatrell, A.C. (1995) *Interactive Spatial Data Analysis*. Longman, Harlow, UK, 413 pp.

Leeuwis, C. (1993) *Of Computers, Myths and Modelling; the Social Construction of Diversity, Knowledge, Information and Communication Technologies in Dutch Horticulture and Agricultural Extension*. Wageningen Studies in Sociology no. 36, Pudoc, Wageningen, 468 pp.

Nyrop, J.P., Binns, M.R. and van der Werf, W. (1999) Sampling for IPM decision-making: where should we invest time and resources? *Phytopathology* 89, 1104–1111.

Pedigo, L.P. and Buntin, G.D. (1994) *Handbook of Sampling Methods for Arthropods in Agriculture*. CRC Press, Boca Raton, Florida, 714 pp.

Rossing, W.A.H., Daamen, R.A. and Hendrix, E.M.T. (1994) Framework to support decisions on chemical pest control under uncertainty, applied to aphids and brown rust in winter wheat. *Crop Protection* 13, 25–34.

Wagenmakers, P.S., van der Werf, W. and Blaise, Ph. (eds) (1999) *Proceedings of the Fifth International Symposium on Computer Modelling in Fruit Research and Orchard Management*. IOBC/WPRS Bulletin 22(6) and Acta Horticulturae 499. International Society of Horticultural Science, Leuven, Belgium, 303 pp.

Weisz, R., Fleischer, S. and Smilowitz, Z. (1996) Site-specific integrated pest management for high-value crops: impact on potato pest management. *Journal of Economic Entomology* 89, 501–509.

Zadoks, J.C. (1989) EPIPRE, a computer-based decision support system for pest and disease control in wheat: its development and implementation in Europe. In: Leonard, K.J and Fry, W.E. (eds): *Plant Disease Epidemiology*, Vol. II. Macmillan, New York, pp. 3–20.

Zwankhuizen, M.J., Govers, F. and Zadoks, J.C. (1998) Development of potato late blight epidemics: disease foci, disease gradients, and infection sources. *Phytopathology* 88, 754–763.

Glossary

Some of the terms used in this book may be unfamiliar to readers, especially those who have no background in statistics. We give here short definitions of some of these terms and where each one is described first (or where the description is extended).

Term	Chapter	Brief definition
accuracy	2	A measure of how close an estimate is to what it is estimating. It combines bias and precision, and can be estimated by the mean square error, *mse*. The greater the *mse*, the lower is the accuracy, and vice versa.
average sample number (ASN)	1, 5	For a sequential sampling plan, the average number of required sample units. The ASN function is the relationship between the population mean, μ, and the expected number of sample units required before sampling stops.
bias	2, 7	The difference between the expectation of an estimate and the true value.
binomial count sampling plan	7	A sampling plan where the information collected from a sample unit is only whether or not the number of individuals on it is greater than a prespecified tally number, T.
cascaded sampling	11	The use of sampling plans sequentially in time to monitor a population.
Central Limit Theorem	2	A mathematical theorem which states that, under very broad conditions, the mean of n sample units becomes close to a normal distribution as n increases.

Term	Chapter	Brief definition
classification interval	3	A probability interval around *cd* (or *cp*) which can be used to classify a sample mean. For example, if $m < cd - z_{\alpha/2}\sqrt{V/n}$, it can be classified as less than *cd* with probability of being correct approximately equal to $\alpha/2$, and vice versa if $m > cd + z_{\alpha/2}\sqrt{V/n}$ (assuming that *n* is large enough for the Central Limit Theorem). We call $cd \pm z_{\alpha/2}\sqrt{V/n}$ a classification interval. See also 'confidence interval' (*q.v.*).
coefficient of variation (*CV*)	2, 3	The standard deviation divided by the mean: σ/μ. The *CV* of a sample mean is, $\{\sigma/\sqrt{n}\}/\mu$, which can be estimated by $\{\sqrt{V/n}\}/m$.
confidence interval	3	An interval that is used to delimit the value of an unknown population characteristic. For example, if *n* is large enough for the normal distribution to be assumed, a $(1-\alpha)\%$ confidence interval for μ is $(m - z_{\alpha/2}\ sem,\ m + z_{\alpha/2}\ sem)$ where the probability, under the standardized normal distribution, of getting a value greater than $z_{\alpha/2}$ is $\alpha/2$, and *sem* is the standard error of the mean.
Converging Lines sampling plan	5	A sequential sampling plan characterized by straight converging stop boundaries.
critical density (*cd*)	1	A density of individual pests per sample unit used for comparison purposes to decide on a management action. Intervention is regarded as appropriate when $\mu > cd$, and as inappropriate when $\mu \leq cd$.
critical proportion (*cp*)	4, 7	When the sample mean is a proportion (*p*), *cp* plays a similar role to the critical density, *cd*: intervention is regarded as appropriate when $p > cp$, and as inappropriate when $p \leq cp$. When binomial count sampling is used, *cp* may depend on *T*, the tally number, and *cp* is then often written as cp_T.
cumulative probability distribution function	2, 4	The theoretical probability of getting a given data value or any smaller value. For example, the Poisson probability of getting 1 or fewer individuals when the mean is equal to $\mu = e^{-\mu} + \mu e^{-\mu}$.
expected value of an estimate	2	The expected value of an estimate is the long-term average of sample estimates.
frequency distribution	2, 4	A summary of the sample data. The sample data (X_j) are classified into classes: the number of sample units for which X_j is equal to *i* is defined as the frequency, f_i. The collection of all f_i, $i = 0, 1, 2 \ldots$ is the frequency distribution.

Term	Chapter	Brief definition
goodness-of-fit test (χ^2)	4	An objective method for determining if sample data conform to a specific distributional model.
incidence–mean relationship	7	A descriptive mathematical model for the relationship between a population mean and the proportion of sample units with more than *T* pests per unit.
Iwao sampling plan	5	Sampling plan proposed by Iwao, based on classification intervals around the critical density or proportion.
likelihood	4	Given a collection of sample data, X_i, and a probability distribution model $p(X\|\theta)$ with parameter θ, the likelihood is the probability of getting the data as a function of θ. The value of θ for which the likelihood is maximized, written as $\hat{\theta}$, can be used to obtain an optimal probability model for the data: $p(X\|\hat{\theta})$.
mean square error of an estimate (*mse*)	2, 7	The variance plus the bias (squared) of an estimate. If the estimate is unbiased, *mse* = variance.
median	5	The value that divides a frequency or probability distribution into two equal parts. For a frequency distribution, 50% of the data values lie above the median, and 50% below it. For a probability distribution, $p(X \leq \text{median}) = 0.5$.
monitoring protocol	11	A strategy for determining (i) when sample bouts should be made, based on information and data already obtained, and (ii) as data are collected and analysed, how to use the results for decision-making.
normal distribution	2, 3	When enough sample units are collected, the sample mean is distributed approximately according to the normal distribution see 'Central Limit Theorem'. The parameters of the normal distribution are the mean, μ, and the variance, σ^2.
observer bias	2	Sample units are collected by human beings (observers). Depending on the sample unit or other properties of a sampling plan, an observer may consistently under-estimate (or over-estimate) the true mean. This constitutes observer bias.
operating characteristic (OC) function	1, 2	A type of probability of classification function. The OC function is the probability, for any true mean pest density, μ, of recommending non-intervention based on sample data collected according to the sampling plan.
pest density per sample unit	2	The average number of pests per sample unit in the population being studied.

Term	Chapter	Brief definition
pest incidence	4, 7	When the information collected on a sample unit is restricted to 0 or 1, such as whether or not the sample unit is infected by a pathogen ($X = 1$ if infected, $X = 0$ if not), pest incidence is defined as the probability $p(X = 1)$. In binomial count sampling, pest incidence is defined as the probability of more than T pests in a sample unit: $p(X > T)$.
population mean (μ)	2	The expectation, or long-term average, of the number of organisms on a sample unit.
population (of sample units)	2	Total number of sample units (in the area to be sampled), each of which has a chance of being selected.
precision	2, 3	A measure of how close a sample estimate is to its expectation. The precision of an estimate is inversely proportional to its variance.
presence–absence sampling plan	7	A binomial count sampling plan with tally number $T = 0$.
primary sample unit	8	See 'two-stage sampling'.
probability distribution	4	A theoretical model for a frequency distribution. A probability distribution has one or more parameters which can be adjusted to make it fit more closely as a model for a given frequency distribution (see 'likelihood').
probability of classification (PC) function; also called probability of decision function (PD)	1, 11	Once a decision-making sampling plan has been properly specified, the probability of any one of the allowable classification decisions can be estimated for all values of the true mean density, if the distribution of sample data is known. The functions that relate the probabilities of these classifications to the true mean are defined as probability of classification functions. In particular, the probability of a classification recommending a decision not to intervene is called the OC function.
probability density function	2, 4	The theoretical probability of obtaining a given data value. For example, the Poisson probability of getting one individual when the mean is equal to μ is $\mu e^{-\mu}$.
quartile	5	The quartiles and median divide a frequency or probability distribution into four equal parts. For a frequency distribution, 25% of the data values lie below the lower quartile, 50% below the median, and 75% below the upper quartile. For a probablity distribution, $p(X \leq \text{lower quartile}) = 0.25$, $p(X \leq \text{median}) = 0.5$, $p(X \leq \text{upper quartile}) = 0.75$.

Term	Chapter	Brief definition
relative frequencies	4, 9	The frequencies (f_i) divided by the sample size (n).
resampling	9	A method for estimating properties of a sampling plan by resampling from data which have already been collected. It is especially useful when no suitable theoretical distribution can be found to represent the sample data. The resampling is done with replacement.
sample	1, 2	All the sample units collected according to a sampling plan.
sample bout	10, 11	The complete process of implementing a sampling plan, including going to the site. It may also include choosing an appropriate sampling plan.
sample data	2, 4	The data collected on sample units; in this book, usually the numbers of organisms counted on sample units (X_i = data in sample unit i).
sample mean (m)	2	The average of the data collected on sample units: $m = \Sigma X_i/n$.
sample size (n)	2	The number of sample units collected in a sample.
sample unit	2, 6	The smallest physical entity which is collected during sampling, and from which data are recorded.
sample variance (V)	2	An estimate of the true variance, σ^2. V is the sum of the squared differences between the data on each sample unit, X_i, and the sample mean, m, divided by the sample size minus one: $V = \Sigma (X_i - m)^2/(n - 1)$.
sampling with replacement	2, 9	Allowing repeated selection of the same sample units. If n sample units are collected by sampling with replacement, the total number of distinct sample units may be less than n.
sampling without replacement	2	Not allowing repeated selection of any sample unit. Sample units selected in this way are all different.
secondary sample unit	8	See 'two-stage sampling'.
sequential sampling plan	5	A sampling plan in which sample units are collected one by one (or in groups of more than one). As each sample unit is collected, the data are accumulated and compared with preset boundary conditions to decide whether to stop and make a classification decision, or continue sampling.

Term	Chapter	Brief definition
spatial pattern	4	The way in which individual organisms are situated in an area. The organisms may be spread out at random, independently of each other; or their positions may depend on each other, so that the pattern becomes either more aggregated or more uniform.
Sequential Probability Ratio Test (SPRT) sampling plan	5	A sequential sampling plan based on a Sequential Probability Ratio Test: proposed by Wald. The stop boundaries are parallel straight lines for most distribution types used in decision sampling.
standard deviation	2	The square root of a population variance. For example, the standard deviation of the mean is $\sqrt{\sigma^2/n} = \sigma/\sqrt{n}$.
standard error (*se*)	2, 3	The square root of a sample variance, calculated from sample data $s = \sqrt{V}$.
standard error of a mean (*sem*)	2, 3	The square root of the sample variance of a mean. $sem = \sqrt{V/n}$
standard normal distribution	2	The normal distribution with mean, μ, equal to 0 and variance, σ^2, equal to 1. A common notation for a random variable following the standard normal distribution is z. A useful notation for defining values of z is $p(Z \le z_\alpha) = 1 - \alpha$.
tally number (T)	7	In binomial count sampling, the number that defines a classification of sample units. If the number of individuals on a sample unit is greater than T that sample unit is classified as 'greater than T', and as 'less than or equal to T' otherwise. See pest incidence.
Taylor's Power Law (TPL)	3	A descriptive mathematical model for the relationship between the mean density of a population (pests per sample unit) and the variance: $V = a\,m^b$.
theoretical variance (σ^2)	2	The expectation, or long-term average, of the squared difference between the data on each sample unit and the population mean: $\sigma^2 = E(X - \mu)^2$ In general, σ^2 denotes the true variance of any distribution.
time-sequential probability ratio test (T-SPRT)	10	The time-sequential version of the SPRT, proposed for sampling through time by Pedigo and van Schaick.
trajectory (of a population)	10, 11	The density (or infestation) of an organism, as it varies over time.
tripartite sampling plan	11	A sampling plan with three classification alternatives. Tripartite sampling plans are constructed by combining two sets of stop boundaries (either Iwao, SPRT or Converging Lines).

Term	Chapter	Brief definition
two-stage sampling	8	A procedure used when there is a hierarchical arrangement of sample units in a crop, such as leaves on plants, and the pest is most conveniently counted on the smaller units. A number of primary sample units (e.g. plants) are selected, and from these a number of secondary sample units (e.g. leaves) are examined for the pest.
value of sample information (VSI)	6	VSI is a summary statistic, representing an overall numerical evaluation of the economic costs and benefits of sampling for decision-making.
variable intensity sampling (VIS)	8	A method of ensuring good coverage of a management unit while collecting sample data. Locations for sampling are chosen over the whole management unit, and at least one sample unit is taken at each location. As data are collected from each location in turn, they are all analysed together to determine how many sample units to take at the next location.

Index

Accuracy 20–21
Adaptive frequency monitoring *see* Monitoring
Aggregated spatial pattern *see* Spatial aggregation
Analysis of variance *see* Nested sampling
ASN function *see* Average sample number function
Average sample number function (ASN)
  definition of 9
  distribution, in sequential procedures 107–108
  estimation of, *see* Iwao's sequential procedure
  for double sampling 48–52
  see also Batch sequential sampling; Variable intensity sampling

Batch sequential sampling 46–59
  average sample number (ASN) function 53
  double sampling 46–52
  operating characteristic (OC) function 53
Bayesian estimation 236–238
  posterior distribution 237–238
  prior distribution 237–238
Beta-binomial distribution 83–89
  cluster size in 83
  intracluster correlation, ρ 84–89, 128–129
  relationship to other distributions 90
  see also Simulation; Variance–incidence relationship 89, 128–129
Bias 19–23
  dealing with 22–23
  enumeration bias 19, 22
  selection bias 19–22
Binomial count sampling
  bias 156–157, 164, 180–182
  estimation 177–178
  tally number 158–165
  variance 156–157, 164, 180–182
  with empirical model 173–177
  with negative binomial distribution 165–172
  with Poisson distribution 156–165
  with variance–mean relationship 170–171
Binomial distribution 82–83
  relationship to other distributions 89–90
Binomial sampling *see* Binomial count sampling; Resampling
Bootstrap *see* Resampling

cd *see* Critical density
Central Limit Theorem 29–38, 42, 53
χ^2 distribution *see* Goodness of fit
Chi square distribution *see* χ^2 distribution
Classification 45–46
Classification interval 45, 46, 98, 117
Cluster *see* Beta-binomial distribution, Variable intensity sampling
Coefficient of variation (CV) 28, 30, 44–45, 177
Confidence interval 43–44
Converging Lines sequential procedure 116–121, 122–124
  effect of parameters on 102, 117–118
  stop boundary 116–117
Critical density (cd) 5–8, 45, 140

Decision guide 132
- adoption by users 132
- design of 132–134
- evaluation of 132–134, 138–141
- expected value of sample information *see* Expected value of sample information
- field evaluation of 139–140
- robustness of 140–141
- sampling plan in, *see* Sampling plan
- *see also* Management unit

Degrees of freedom *see* Goodness of fit

Economic injury level (EIL) 3–5
Economic threshold (ET) 5
EIL *see* Economic injury level
Error
- random 20
- systematic 19–20

Estimation 42–45
- Green's stop boundary 120, 122
- *see also* Bayesian estimation; Binomial count sampling; Time sequential estimation

ET *see* Economic threshold
Evaluation *see* Decision guide
Expected value of sample information 19–20, 42

Fitting distributions
- effect of sample size 78–80
- goodness-of-fit test *see* Goodness of fit
- maximum likelihood estimates *see* Maximum likelihood
- mechanics of 80–81

Frequency distribution 27–29, 62–63
Full count sampling 156

Gibbs process 94–95
Goodness of fit 67–71
- degrees of freedom 68
- grouping in 69
- significance probability 69
- χ^2 test for 68–69

Green's stop boundary *see* Estimation

Incidence 82–89
Incidence–mean relationship 89, 128
Integrated pest management 2–3
Iwao's variance–mean relationship (IVM) *see* Variance–mean relationships
Iwao's sequential procedure 98–108, 122–124
- effect of parameters on 102, 103
- estimating operating characteristic (OC) and average sample number (ASN) functions 100–101
- stop boundaries 98, 100

Likelihood ratio 108–110, 228–229

Management unit 3, 134–135
Maximum likelihood 66–69, 80
Monitoring 11–12
- action threshold 247
- adaptive frequency monitoring 263–264
- expected bouts 251
- expected density 251
- expected total sample units 251
- monitoring operating characteristic (MOC) 250, 251
- monitoring protocol 11, 246, 249–263
- probability of classification 248, 250–251
- simulation of 251–252
- tripartite classification sampling 246–249, 255–263
- waiting threshold 247

Monitoring protocol *see* Monitoring
Multistage sampling *see* Nested sampling

Negative binomial distribution 71–77
- exponent *k* 72–77, 127–128, 226
- Iwao's variance–mean relationship in 77
- relationship to other distributions 89–90
- Taylor's power law in 77
- variance formula 72
- *see also* Simulation

Nested sampling
- analysis of variance, use in 185–191
- definition of 184
- optimal sample size 188
- primary sample unit 184
- secondary sample unit 184
- two-stage sampling 184, 185–191
- variance components 185–188
- variance of sample mean 186, 187–188, 204
- variance–mean relationships in 187

Normal distribution 29–38, 42–44, 46, 211
  standard normal deviate 33–34
  standard normal distribution 43–44
  see also Central Limit Theorem

OC function *see* Operating characteristic function
Operating characteristic function (OC)
  calculation using Normal distribution 34–38
  definition of 9–10
  estimating, *see* Iwao's sequential procedure
  for double sampling 48–52
  simulating 35
  steepness of 10
  see also Batch sequential sampling; Resampling; Variable intensity sampling

Patches *see* Spatial pattern
Pattern *see* Spatial pattern
Poisson cluster process 94
Poisson distribution, relationship to other distributions 89–90
Poisson probability distribution 63–66
Population 18
Population trajectory 231, 249, 253–255
Precision 20–21, 44–45
Probability distributions 62–63, 91
Probability of classification *see* Monitoring
Probability of decision curves 7–9
Probability of decision functions 11–12

Random pattern 63–64
Random sampling 17–19
Relative frequencies 63
Resampling
  basic samples for 212–215
  comparison with distribution models 221
  evaluation of binomial sampling 217–220
  evaluation of sequential sampling 216–217
  OC functions, estimate of 208–212
  principles of 206–208

Sample path 137
Sample size
  choosing for estimation 44–45, 177–178
  minimum with sequential procedures 101, 141
Sample unit 18–29, 63, 135–136
  see also Nested sampling
Sample variance 27–32
Sampling
  definition of 3
  with replacement 26–27
  without replacement 26–27
Sampling effort 8
Sampling plan 24, 134–138
  criteria 24
  design ingredients 134–138
  practical feasibility 24
  precision 24, 44–45
  relevance 24
  reliability 24, 42, 101
  representativeness 23–24, 42, 101
  see also Sample unit
Sequential sampling 97–98
Significance probability *see* Goodness of fit
Simulation
  of beta-binomial ρ 128–129
  of negative binomial k 127–128
  of sampling process 25–27
  see also Decision guide, evaluation of; Monitoring
Spatial aggregation 71–72
Spatial pattern 61–63
  patches 71, 191–192
  simulation of 94–95
SPRT *see* Wald's sequential probability ratio test
Standard deviation 30–36
Standard error 30
Standard normal distribution *see* Normal distribution
Stop boundary
  for batch sequential sampling 52
  for double sampling 47
  see also Converging Lines sequential procedure; Iwao's sequential procedure; Wald's sequential probability ratio test

Tally number *see* Binomial count sampling
Taylor's power law (TPL) *see* Variance–mean relationships
Time sequential classification 228–235, 242–243
Time sequential estimation 236–243
Time sequential sampling (TSS) *see* Time sequential classification

Tripartite classification sampling *see* Monitoring
T-SPRT *see* Time sequential classification
Two-stage sampling *see* Nested sampling

Uncertainty
  effect on decision making 6–8
  in sample estimates 43

Value of sample information (VSI) 141–153
  loss functions 141
  prior distributions 141
Variable intensity sampling (VIS) 192–202
  look-up charts 198–201
  operating characteristic (OC) and average sample number (ASN) functions 201–202
  use of variance components 198–199
  variance–mean relationship in 201–202
Variance 27–32
  see also Nested sampling
Variance components *see* Nested sampling; Variable intensity sampling
Variance–incidence relationship *see* Beta-binomial distribution
Variance–mean relationships 53–55
  Iwao's variance–mean relationship (IVM) 54, 77
  portability of 54–55
  Taylor's power law (TPL) 54–55, 77, 99, 127
  variability about 55, 127
  see also Nested sampling; Variable intensity sampling

Wald's sequential probability ratio test (SPRT) 108–115, 122–124, 228
  effect of parameters on 102, 111
  stop boundary 110